Studies in Advanced Mathematics

T0187490

Partial Differential Equations
and Complex Analysis

Studies in Advanced Mathematics

Series Editor

STEVEN G. KRANTZ
Washington University in St. Louis

Editorial Board

R. Michael Beals
Rutgers University

Dennis de Turck
University of Pennsylvania

Ronald DeVore
University of South Carolina

L. Craig Evans
University of California at Berkeley

Gerald B. Folland
University of Washington

William Helton
University of California at San Diego

Norberto Salinas
University of Kansas

Michael E. Taylor
University of North Carolina

Titles Included in the Series

Real Analysis and Foundations, *Steven G. Krantz*

CR Manifolds and the Tangential Cauchy–Riemann Complex, *Albert Boggess*

Elementary Introduction to the Theory of Pseudodifferential Operators,
   Xavier Saint Raymond

Fast Fourier Transforms, *James S. Walker*

Measure Theory and Fine Properties of Functions, *L. Craig Evans and
   Ronald Gariepy*

Partial Differential Equations and Complex Analysis, *Steven G. Krantz*

The Cauchy Transform, Potential Theory, and Conformal Mapping,
   Steven R. Bell

STEVEN G. KRANTZ
Washington University in St. Louis, Department of Mathematics

Partial Differential Equations and Complex Analysis

Based on notes by Estela A. Gavosto and Marco M. Peloso

CRC Press
Taylor & Francis Group
Boca Raton London New York

CRC Press is an imprint of the
Taylor & Francis Group, an **informa** business

CRC Press
Taylor & Francis Group
6000 Broken Sound Parkway NW, Suite 300
Boca Raton, FL 33487-2742

First issued in paperback 2019

ISBN-13: 978-0-8493-7155-4 (hbk)
ISBN-13: 978-0-367-40275-4 (pbk)
Library of Congress Card Number 92-11422

Library of Congress Cataloging-in-Publication Data

Krantz, Steven G., 1951–
     Partial differential equations and complex analysis / Steven G. Krantz.
      p.   cm.
     Includes bibliographical references (p.   ) and index.
     ISBN 0-8493-7155-4
     1. Differential equations. Partial. 2. Functions of a complex variable.
3. Mathematical analysis. 4. Functions of several complex variables. I. Title.
     QA374.K9   1992
     515'.35–dc20   92-11422
                                                      CIP

Visit the Taylor & Francis Web site at
http://www.taylorandfrancis.com

and the CRC Press Web site at
http://www.crcpress.com

To the memory of my grandmother,
Eda Crisafulli.

Contents

Preface

The subject of partial differential equations is perhaps the broadest and deepest in all of mathematics. It is difficult for the novice to gain a foothold in the subject at any level beyond the most basic. At the same time partial differential equations are playing an ever more vital role in other branches of mathematics. This assertion is particularly true in the subject of complex analysis.

It is my experience that a new subject is most readily learned when presented *in vitro*. Thus this book proposes to present many of the basic elements of linear partial differential equations in the context of how they are applied to the study of complex analysis. We shall treat the Dirichlet and Neumann problems for elliptic equations and the related Schauder regularity theory. Both the classical point of view and the pseudodifferential operators approach will be covered. Then we shall see what these results say about the boundary regularity of biholomorphic mappings. We shall study the $\bar{\partial}$-Neumann problem, then consider applications to the complex function theory of several variables and to the Bergman projection. The book culminates with applications of the $\bar{\partial}$-Neumann problem, including a proof of Fefferman's theorem on the boundary behavior of biholomorphic mappings. There is also a treatment of the Lewy unsolvable equation from several different points of view.

We shall explore some partial differential equations that are of current interest and that exhibit some surprises. These include the Laplace–Beltrami operator for the Bergman metric on the ball. Along the way, we shall give a detailed treatment of the Bergman kernel and associated metric, the Szegö kernel, and the Poisson–Szegö kernel. Some of this material, particularly that in Chapter 6, may be considered ancillary and may be skipped on a first reading of this book.

Complete and self-contained proofs of all results are provided. Some of these appear in book form for the first time. Our treatment of the $\bar{\partial}$-Neumann problem parallels some classic treatments, but since we present the problem in a concrete setting we are able to provide more detail and a more leisurely pace.

Background required to read this book is a basic grounding in real and complex analysis. The book *Function Theory of Several Complex Variables* by this author will also provide useful background for many of the ideas seen here. Acquaintance with measure theory will prove helpful. For motivation, exposure to the basic ideas of differential equations (such as one would encounter in a sophomore differential equations course) is useful. All other needed ideas are developed here.

A word of warning to the reader unversed in reading tracts on partial differential equations: the *métier* of this subject is estimates. To keep track of the constants in these estimates would be both wasteful and confusing. (Although in certain aspects of stability and control theory it is essential to name and catalog all constants, that is not the case here.) Thus we denote most constants by C or C'; *the values of these constants may change from line to line, even though they are denoted with the same letter.* In some contexts we shall use the now popular symbol $\lesssim$ to mean "is less than or equal to a constant times ...."

This book is based on a year-long course given at Washington University in the academic year 1987–88. Some of the ideas have been presented in earlier courses at UCLA and Penn State. It is a pleasure to thank Estela Gavosto and Marco Peloso who wrote up the notes from my lectures. They put in a lot of extra effort to correct my omissions and clean up my proofs and presentations. I also thank the other students who listened to my thoughts and provided useful remarks.

— S.G.K.

1

The Dirichlet Problem in the Complex Plane

1.1 A Little Notation

Let $\mathbb{R}$ denote the real number line and $\mathbb{C}$ the complex plane. The complex and real coordinates are related in the usual fashion by

$$z = x + iy.$$

We will spend some time studying the unit disc $\{z \in \mathbb{C} : |z| < 1\}$, and we denote it by the symbol D. The *Laplace operator* (or *Laplacian*) is the partial differential operator

$$\Delta = \frac{\partial^2}{\partial x^2} + \frac{\partial^2}{\partial y^2} .$$

When the Euclidean plane is studied as a *real* analytic object, it is convenient to study differential equations using the partial differential operators

$$\frac{\partial}{\partial x} \quad \text{and} \quad \frac{\partial}{\partial y} .$$

This is so at least in part because each of these operators has a null space (namely the functions depending only on y and the functions depending only on x, respectively) that plays a significant role in our analysis (think of the method of guessing solutions to a linear differential equation having the form $u(x)v(y)$).

In complex analysis it is more convenient to express matters in terms of the partial differential operators

$$\frac{\partial}{\partial z} \equiv \frac{1}{2}\left(\frac{\partial}{\partial x} - i\frac{\partial}{\partial y}\right) \quad \text{and} \quad \frac{\partial}{\partial \bar{z}} \equiv \frac{1}{2}\left(\frac{\partial}{\partial x} + i\frac{\partial}{\partial y}\right).$$

Check that a continuously differentiable function $f(z) = u(z) + iv(z)$ that satisfies $\partial f / \partial \bar{z} \equiv 0$ on a planar open set U is in fact holomorphic (use the

Cauchy–Riemann equations). In other words, a function satisfying $\partial f/\partial \bar{z} \equiv 0$ may depend on z but *not* on $\bar{z}$. Likewise, a function that satisfies $\partial f/\partial z \equiv 0$ on a planar open set may depend on $\bar{z}$ but cannot depend on z.

Observe that

$$\frac{\partial}{\partial z} z = 1 \qquad \frac{\partial}{\partial z} \bar{z} = 0$$

$$\frac{\partial}{\partial \bar{z}} z = 0 \qquad \frac{\partial}{\partial \bar{z}} \bar{z} = 1.$$

Finally, the Laplacian is written in complex notation as

$$\Delta = 4 \frac{\partial}{\partial z} \frac{\partial}{\partial \bar{z}} = 4 \frac{\partial}{\partial \bar{z}} \frac{\partial}{\partial z} .$$

1.2   The Dirichlet Problem

Introductory Remarks

Throughout this book we use the notation $C^k(X)$ to denote the space of functions that are *k-times continuously differentiable* on X—that is, functions that possess all derivatives up to and including order k and such that all those derivatives are continuous on X. When X is an open set, this notion is self-explanatory. When X is an arbitrary set, it is rather complicated, but possible, to obtain a complete understanding (see [STSI]).

For the purposes of this book, we need to understand the case when X is a closed set in Euclidean space. In this circumstance we say that f is C^k on X if there is an open neighborhood U of X and a C^k function $\tilde{f}$ on U such that the restriction of $\tilde{f}$ to X equals f. We write $f \in C^k(X)$. In case $k = 0$, we write either $C^0(X)$ or $C(X)$. This definition is equivalent to all other reasonable definitions of C^k for a non-open set. We shall present a more detailed discussion of this matter in Section 3.

Now let us formulate the Dirichlet problem on the disc D. Let $\phi \in C(\partial D)$. The *Dirichlet problem* is to find a function $u \in C(\bar{D}) \cap C^2(D)$ such that

$$\Delta u(z) = 0 \ \text{ if } z \in D$$

$$u(z) = \phi(z) \ \text{ if } z \in \partial D.$$

REMARK   Contrast the Dirichlet problem with the classical *Cauchy problem* for the Laplacian: Let $S \subseteq \mathbb{R}^2 \equiv \mathbb{C}$ be a smooth, non–self-intersecting curve

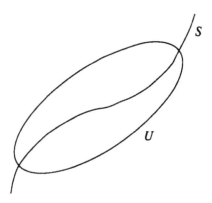

FIGURE 1.1

(part of the boundary of a smoothly bounded domain, for instance). Let U be an open set with nontrivial intersection with S (see Figure 1.1). Finally, let ϕ_0 and ϕ_1 be given continuous functions on S. The Cauchy problem is then

$$\Delta u(z) = 0 \ \text{ if } z \in U$$
$$u(z) = \phi_1(z) \ \text{ if } z \in S \cap U$$
$$\frac{\partial u}{\partial \nu}(z) = \phi_1 \ \text{ if } z \in S \cap U.$$

Here ν denotes the unit normal direction at $z \in S$.

Notice that the solution to the Dirichlet problem posed above is unique: if u_1 and u_2 both solve the problem, then $u_1 - u_2$ is a harmonic function having zero boundary values on D. The maximum principle then implies that $u_1 \equiv u_2$. In particular, in the Dirichlet problem the specifying of boundary values also uniquely determines the normal derivative of the solution function u.

However, in order to obtain uniqueness in the Cauchy problem, we must specify both the value of u on S and the normal derivative of u on S. How can this be? The reason is that the Dirichlet problem is posed with a simple *closed* boundary curve; the Cauchy problem is instead a local one. Questions of when function theory reflects (algebraic) topology are treated, for instance, by the de Rham theorem and the Atiyah–Singer index theorem. We shall not treat them in this book, but refer the reader to [GIL], [KR1], [DER]. ∎

The Solution of the Dirichlet Problem in L^2

Define functions ϕ_n on ∂D by

$$\phi_n(e^{i\theta}) = e^{in\theta}, \qquad n \in \mathbf{Z}.$$

Notice that the solution of the Dirichlet problem with data ϕ_n is $u_n(re^{i\theta}) = r^{|n|}e^{in\theta}$. That is,

$$u_n(z) = \begin{cases} z^n & \text{if } n \geq 0 \\ \bar{z}^n & \text{if } n < 0. \end{cases}$$

The functions $\{\phi_n\}_{n=-\infty}^{\infty}$ form a basis for $L^2(\partial D)$ That is, if $f \in L^2(\partial D)$ then we define

$$a_n = \frac{1}{2\pi} \int_0^{2\pi} f(t)e^{-int}\, dt.$$

It follows from elementary Riesz–Fischer theory that the partial sums

$$S_N = \sum_{n=-N}^{N} a_n e^{in\theta} \to f \tag{1.2.1}$$

in the L^2 topology.

If $0 \leq r < 1$ then observe that

$$\sum_{n=-\infty}^{\infty} a_n r^{|n|} e^{in\theta}$$

is an Abel sum for the Fourier series $\sum_{-\infty}^{\infty} a_n e^{in\theta}$ of f. It follows from (1.2.1) that

$$S(r,\theta) \equiv \sum_{n=-\infty}^{\infty} a_n r^{|n|} e^{in\theta} \to f(e^{i\theta}) \tag{1.2.2}$$

in the L^2 topology as $r \to 1^-$.

In fact, the sum in (1.2.2) converges uniformly on compact subsets of the disc. The computation that we are now about to do will prove this statement: We have

$$S(r,\theta) = \sum_{n=-\infty}^{\infty} a_n r^{|n|} e^{in\theta}$$

$$= \sum_{n=-\infty}^{\infty} r^{|n|} e^{in\theta} \frac{1}{2\pi} \int_0^{2\pi} f(e^{it}) e^{-int}\, dt$$

$$= \frac{1}{2\pi} \int_0^{2\pi} \left(\left\{ \sum_{n=0}^{\infty} \left[re^{i(\theta-t)} \right]^n \right\} + \left\{ \sum_{n=1}^{\infty} \left[re^{-i(\theta-t)} \right]^n \right\} \right) \cdot f(e^{it})\, dt.$$

If we sum the two geometric series and do the necessary algebra then we find that

$$S(r, \theta) = \frac{1}{2\pi} \int_0^{2\pi} f(e^{it}) \frac{1 - r^2}{1 - 2r \cos(\theta - t) + r^2} \, dt.$$

This last formula allows one to do the estimates to check for uniform convergence, and thus to justify the change of order of the sum and the integral.

We set

$$P_r(\psi) = \frac{1}{2\pi} \frac{1 - r^2}{1 - 2r \cos(\psi) + r^2},$$

and we call this function the *Poisson kernel*. Since the function

$$u(re^{i\theta}) = S(r, \theta)$$

is the limit of the partial sums $T_N(r, \theta) \equiv \sum_{n=-N}^N a_n r^{|n|} e^{in\theta}$, and since each of the partial sums is harmonic, u is harmonic. Moreover, the partial sum T_N is the harmonic function that solves the Dirichlet problem for the data $f_N(e^{i\theta}) = \sum_{n=-N}^N a_n e^{in\theta}$. We might hope that u is then the solution of the Dirichlet problem with data f. This is in fact true:

THEOREM 1.2.3
Let $f(e^{it})$ be a continuous function on ∂D. Then the function

$$u(re^{i\theta}) \equiv \begin{cases} \int_0^{2\pi} f(e^{i(\theta-t)}) P_r(t) \, dt & \text{if } 0 \le r < 1 \\ f(e^{i\theta}) & \text{if } r = 1 \end{cases}$$

solves the Dirichlet problem on the disc with data f.

PROOF Pick $\epsilon > 0$. Choose $\delta > 0$ such that if $|s - t| < \delta$, then $|f(e^{is}) - f(e^{it})| < \epsilon$. Fix a point $e^{i\theta} \in \partial D$. We will first show that $\lim_{r \to 1^-} u(re^{i\theta}) = f(e^{i\theta}) = u(e^{i\theta})$. Now, for $0 < r < 1$,

$$|u(re^{i\theta}) - f(e^{i\theta})| = \left| \int_0^{2\pi} f(e^{i(\theta-t)}) P_r(t) \, dt - f(e^{i\theta}) \right|. \qquad (1.2.3.1)$$

Observe, using the sum from which we obtained the Poisson kernel, that

$$\int_0^{2\pi} |P_r(t)| \, dt = \int_0^{2\pi} P_r(t) \, dt = 1.$$

Thus we may rewrite (1.2.3.1) as

$$\int_0^{2\pi} \left[f(e^{i(\theta-t)}) - f(e^{i\theta}) \right] P_r(t) \, dt = \int_{|t| < \delta} \left[f(e^{i(\theta-t)}) - f(e^{i\theta}) \right] P_r(t) \, dt$$

$$+ \int_{\delta \le t \le 2\pi - \delta} \left[f(e^{i(\theta-t)}) - f(e^{i\theta}) \right] P_r(t) \, dt$$

$$\equiv I + II.$$

Now the term I does not exceed

$$\int_{|t|<\delta} |P_r(t)|\epsilon\, dt \le \epsilon \int_0^{2\pi} |P_r(t)|\, dt = \epsilon.$$

For the second, notice that $\delta < t < 2\pi - \delta$ implies that

$$|P_r(t)| = P_r(t) = \frac{1}{2\pi}\frac{1-r^2}{1-2r\cos t + r^2}$$

$$= \frac{1}{2\pi}\frac{1-r^2}{(1-2r\cos t + r^2\cos^2 t)+r^2(1-\cos^2 t)}$$

$$\le \frac{1}{2\pi}\frac{1-r^2}{(1-r\cos t)^2}$$

$$< \frac{1}{2\pi}\frac{1-r^2}{\delta^4/8}.$$

Thus

$$II \le 2\int_\delta^\pi 2\sup|f|\cdot\frac{1}{2\pi}\frac{1-r^2}{\delta^4/8}\, dt \to 0$$

as $r \to 1^-$. In fact, the proof shows that the convergence is uniform in θ. Putting together our estimates on I and II, we find that

$$\limsup_{r\to1^-}|u(re^{i\theta}) - f(e^{i\theta})| < \epsilon.$$

Since $\epsilon > 0$ was arbitrary, we see that

$$\limsup_{r\to1^-}|u(re^{i\theta}) - f(e^{i\theta})| = 0.$$

The proof is nearly complete. For $\theta \in \partial D$ and $\epsilon > 0$ fixed, choose $\delta > 0$ such that

(i) $|u(e^{i\theta}) - u(e^{i\psi})| < \epsilon/2$ when $|e^{i\theta} - e^{i\psi}| < \delta$;
(ii) $|u(re^{i\psi}) - u(e^{i\psi})| < \epsilon/2$ when $r > 1 - \delta, 0 \le \psi \le 2\pi$.

Let $z \in D$ satisfy $|z - e^{i\theta}| < \delta$. Then

$$|u(z) - u(e^{i\theta})| \le |u(z) - u(z/|z|)| + |u(z/|z|) - u(e^{i\theta})|$$

$$< \frac{\epsilon}{2} + \frac{\epsilon}{2},$$

where we have applied (ii) and (i) respectively. This shows that u is continuous at the boundary (it is obviously continuous elsewhere) and completes the proof. ∎

1.3 Lipschitz Spaces

Our first aim in this book is to study the boundary regularity for the Dirichlet problem: if the data f is "smooth," then will the solution of the Dirichlet problem be smooth *up to the boundary*? This is a venerable question in the theory of partial differential equations and will be a recurring theme throughout this book. In order to formulate the question precisely and give it a careful answer, we need suitable function spaces.

The most naive function spaces for studying the question formulated in the last paragraph are the C^k spaces, mentioned earlier. However, these spaces are not the most convenient for our study. The reason, which is of central importance, is as follows: We shall learn later, by a method of Hörmander [HO3], that the boundary regularity of the Dirichlet problem is equivalent to the boundedness of certain singular integral operators (see [STSI]) on the boundary. Singular integral operators, central to the understanding of many problems in analysis, are *not* bounded on the C^k spaces. (This fact explains the mysteriously imprecise formulation of regularity results in many books on partial differential equations. It also means that we shall have to work harder to get exact regularity results.)

Because of the remarks in the preceding paragraph, we now introduce the scale of Lipschitz spaces. They will be somewhat familiar, but there will be some new twists to which the reader should pay careful attention. A comprehensive study of these spaces appears in [KR2].

Now let $U \subseteq \mathbb{R}^N$ be an open set. Let $0 < \alpha < 1$. A function f on U is said to be *Lipschitz of order* α, and we write $f \in \Lambda_\alpha$, if

$$\sup_{\substack{h \neq 0 \\ x, x+h \in U}} \frac{|f(x+h) - f(x)|}{|h|^\alpha} + \|f\|_{L^\infty(U)} \equiv \|f\|_{\Lambda_\alpha(U)} < \infty.$$

We include the term $\|f\|_{L^\infty(U)}$ in this definition in order to guarantee that the Lipschitz norm is a true norm (without this term, constant functions would have "norm" zero and we would only have a semi-norm). In other contexts it is useful to use $\|f\|_{L^p(U)}$ rather than $\|f\|_{L^\infty(U)}$. See [KR3] for a discussion of these matters.

When $\alpha = 1$ the "first difference" definition of the space Λ_α makes sense, and it describes an important class of functions. *However, singular integral operators are not bounded on this space.* We set this space of functions apart by denoting it differently:

$$\|f\|_{\mathrm{Lip}_1(U)} \equiv \sup_{\substack{h \neq 0 \\ x, x+h \in U}} \frac{|f(x+h) - f(x)|}{|h|} + \|f\|_{L^\infty(U)} < \infty.$$

The space Lip_1 is important in geometric applications (see [FED]), but less so in the context of integral operators. Therefore we define

$$\|f\|_{\Lambda_1(U)} \equiv \sup_{\substack{h \neq 0 \\ x, x+h, x-h \in U}} \frac{|f(x+h) + f(x-h) - 2f(x)|}{|h|} + \|f\|_{L^\infty(U)} < \infty.$$

Inductively, if $0 < k \in \mathbf{Z}$ and $k < \alpha \leq k+1$, then we define a function f on U to be in Λ_α if f is bounded, $f \in C^1(U)$, and any first derivative $D_j f$ lies in $\Lambda_{\alpha-1}$. Equivalently, $f \in \Lambda_\alpha$ if and only if f is bounded and, for every nonnegative integer $\ell < \alpha$ and multiindex β of total order not exceeding ℓ we have $(\partial/\partial x)^\beta f$ exists, is continuous, and lies in $\Lambda_{\alpha-\ell}$.

The space Lip_k, $1 < k \in \mathbf{Z}$, is defined by induction in a similar fashion.

REMARK As an illustration of these ideas, observe that a function g is in $\Lambda_{5/2}(U)$ if g is bounded and the derivatives $\partial g/\partial x_j$, $\partial^2 g/\partial x_j \partial x_k$ exist and lie in $\Lambda_{1/2}$.

Prove as an exercise that if $\alpha' > \alpha$ then $\Lambda_{\alpha'} \subsetneqq \Lambda_\alpha$. Also prove that the Weierstrass nowhere-differentiable function

$$F(\theta) = \sum_{j=0}^{\infty} 2^{-j} e^{i2^j \theta}$$

is in $\Lambda_1(0, 2\pi)$ but not in $\text{Lip}_1(0, 2\pi)$. Construct an analogous example, for each positive integer k, of a function in $\Lambda_k \setminus \text{Lip}_k$.

If U is a bounded open set with smooth boundary and if $g \in \Lambda_\alpha(U)$ then does it follow that g extends to be in $\Lambda_\alpha(\bar{U})$? ∎

Let us now discuss the definition of C^k spaces in some detail. A function f on an open set $U \subseteq \mathbf{R}^N$ is said to be k-times continuously differentiable, written $f \in C^k(U)$, if all partial derivatives of f *up to and including order k* exist on U and are continuous. On $\mathbf{R}^1$, the function $f(x) = |x|$ lies in $C^0 \setminus C^1$. Examples to show that the higher order C^k spaces are distinct may be obtained by anti-differentiation. In fact, if we equip $C^k(U)$ with the norm

$$\|f\|_{C^k(U)} \equiv \|f\|_{L^\infty(U)} + \sum_{|\alpha| \leq k} \left\| \left(\frac{\partial^\alpha f}{\partial x^\alpha} \right) \right\|_{L^\infty(U)},$$

then elementary arguments show that $C^{k+1}(U)$ is contained in, but is nowhere dense in, $C^k(U)$.

It is natural to suspect that if all the kth order pure derivatives $(\partial/\partial x_j)^{\ell} f$ exist and are bounded, $0 \leq \ell \leq k$, then the function has all derivatives (including mixed ones) of order not exceeding k and they are bounded. In fact Mityagin and Semenov [MIS] showed this to be false in the strongest possible sense. *However, the analogous statement for Lipschitz spaces is true—see [KR2].*

Now suppose that U is a bounded open set in $\mathbb{R}^N$ with smooth boundary. We would like to talk about functions that are C^k on $\bar{U} = U \cup \partial U$. There are three ways to define this notion:

I. We say that a function f is in $C^k(\bar{U})$ if f and all its derivatives on U of order not exceeding k extend continuously to $\bar{U}$.

II. We say that a function f is in $C^k(\bar{U})$ if there is an open neighborhood W of $\bar{U}$ and a C^k function F on W such that $F\big|_{\bar{U}} = f$.

III. We say that a function f is in $C^k(\bar{U})$ if $f \in C^k(U)$ and for each $x_0 \in \partial U$ and each multiindex α such that $|\alpha| \leq k$ the limit

$$\lim_{U \ni x \to x_0} \frac{\partial^{\alpha}}{\partial x^{\alpha}} f(x)$$

exists.

We leave as an exercise for the reader to prove the equivalence of these definitions. Begin by using the implicit function theorem to map U locally to a boundary neighborhood of an upper half-space. See [HIR] for some help.

REMARK A basic regularity question for partial differential equations is as follows: consider the Laplace equation

$$\triangle u = f.$$

If $f \in \Lambda_{\alpha}(\mathbb{R}^N)$, then where (i.e. in what smoothness class) does the function u live (at least locally)?

In many texts on partial differential equations, the question is posed as "If $f \in C^k(\mathbb{R}^N)$ then where does u live?" The answer is generally given as "$u \in C^{k+2-\epsilon}_{\text{loc}}$ for any $\epsilon > 0$." Whenever a result in analysis is formulated in this fashion, it is safe to assume that either the most powerful techniques are not being used or (more typically) the results are being formulated in the language of the *incorrect spaces*. In fact, the latter situation obtains here. If one uses the Lipschitz spaces, then there is no ϵ-order loss of regularity: $f \in \Lambda_{\alpha}(\mathbb{R}^N)$ implies that u is locally in $\Lambda_{\alpha+2}(\mathbb{R}^N)$. Sharp results may also be obtained by using Sobolev spaces. We shall explore this matter in further detail as the book develops. ∎

1.4 Boundary Regularity for the Dirichlet Problem for the Laplacian on the Unit Disc

We begin this discussion by posing a question:

> **Question:** If we are given a "smooth" function f on the boundary of the unit disc D, then is the solution u to the Dirichlet problem for the Laplacian, with boundary data f, smooth up to closure of D? That is, if $f \in C^k(\partial D)$, then is $u \in C^k(\bar{D})$?

It turns out that the answer to this question is "no." But the reason is that we are using the wrong spaces. We can only get a sharp result if we use Lipschitz spaces. Thus we have:

> **Revised Question:** If we are given a "smooth" function f on the boundary of the unit disc D, then is the solution u to the Dirichlet problem for the Laplacian, with boundary data f, smooth up to the closure of D? That is, if $f \in \Lambda_\alpha(\partial D)$, then is $u \in \Lambda_\alpha(\bar{D})$?

We still restrict our detailed considerations to $\mathbb{R}^2$ for the moment. Also, it is convenient to work on the upper half-space $\mathcal{U} = \{(x, y) \in \mathbb{R}^2 : y > 0\}$. We think of the real line as the boundary of $\mathcal{U}$. By conformally mapping the disc to the upper half-space with the Cayley transformation $\phi(z) = i(1-z)/(1+z)$, we may calculate that the Poisson kernel for $\mathcal{U}$ is the function

$$P_y(x) = \frac{1}{\pi} \frac{y}{x^2 + y^2} \ .$$

For simplicity, we shall study Λ_α for $0 < \alpha < 2$ only. We shall see later that there are simple techniques for extending results from this range of α to all α. Now we have the following theorem.

THEOREM 1.4.1
Fix $0 < \alpha < 1$. If $f \in \Lambda_\alpha(\mathbb{R})$ then

$$u(x, y) = P_y f(x, y) \equiv \int_{\mathbb{R}} P_y(x - t) f(t) \, dt$$

lies in $\Lambda_\alpha(\bar{\mathcal{U}})$.

PROOF   Since

$$\int_{\mathbb{R}} |P_y(x - t)| \, dt = \int_{\mathbb{R}} P_y(x - t) \, dt = 1,$$

it follows that u is bounded by $\|f\|_{L^\infty}$.

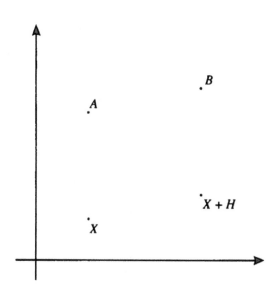

FIGURE 1.2

Now fix $X = (x_1, x_2) \in \mathcal{U}$. Fix also an $H = (h_1, h_2)$ such that $X + H \in \mathcal{U}$. We wish to show that

$$|u(X + H) - u(X)| \leq C|H|^\alpha.$$

Set $A = (x_1, x_2 + |H|), B = (x_1 + h_1, x_2 + h_2 + |H|)$. Clearly A, B lie in $\mathcal{U}$ because $X, X + H$ do. Refer to Figure 1.2. Then

$$|u(X + H) - u(X)| \leq |u(X) - u(A)| + |u(A) - u(B)| + |u(B) - u(X + H)|$$
$$\equiv I + II + III.$$

For the estimate of I we will use the following two facts:
Fact 1: The function u satisfies

$$\left| \frac{\partial^2}{\partial y^2} u(x, y) \right| \leq C y^{\alpha - 2}$$

for $(x, y) \in \mathcal{U}$.
Fact 2: The function u satisfies

$$\left| \frac{\partial}{\partial y} u(x, y) \right| \leq C y^{\alpha - 1}$$

for $(x, y) \in \mathcal{U}$.

Fact 2 follows from Fact 1, as we shall see below. Once we have Fact 2, the estimation of I proceeds as follows: Let $\gamma(t) = (x_1, x_2 + t), 0 \le t \le |H|$. Then, noting that $|\gamma(t)| \ge t$, we have

$$|u(X) - u(A)| = \left| \int_0^{|H|} \frac{d}{dt} \left(u(\gamma(t)) \right) dt \right|$$

$$\le \int_0^{|H|} \left| \frac{\partial u}{\partial y} \right|_{\gamma(t)} \right| dt$$

$$\le C \int_0^{|H|} |\gamma(t)|^{\alpha - 1} dt$$

$$\le C \int_0^{|H|} t^{\alpha - 1} dt$$

$$= C|H|^\alpha.$$

Thus, to complete the estimate on I, it remains to prove our two facts.

PROOF OF FACT 1   First we exploit the harmonicity of u to observe that

$$\left| \frac{\partial^2 u}{\partial y^2} \right| = \left| \frac{\partial^2 u}{\partial x^2} \right|.$$

Then

$$\left| \frac{\partial^2 u}{\partial y^2} \right| = \left| \frac{\partial^2 u}{\partial x^2} \right|$$

$$= \left| \frac{\partial^2}{\partial x^2} \int_{-\infty}^\infty P_y(x - t) f(t) \, dt \right|$$

$$= \left| \int_{-\infty}^\infty \frac{\partial^2}{\partial x^2} P_y(x - t) f(t) \, dt \right|$$

$$= \left| \int_{-\infty}^\infty \frac{\partial^2}{\partial t^2} P_y(x - t) f(t) \, dt \right|. \qquad (1.4.1.1)$$

Now, from the Fundamental Theorem of Calculus, we know that

$$\int_{-\infty}^\infty \frac{\partial^2}{\partial t^2} P_y(x - t) \, dt = 0.$$

As a result, line (1.4.1.1) equals

$$\left| \int_{-\infty}^{\infty} \frac{\partial^2}{\partial t^2} P_y(x-t)\left[f(t)-f(x)\right] dt \right|$$

$$\leq C \int_{-\infty}^{\infty} \left| \frac{\partial^2}{\partial t^2} P_y(x-t) \right| |x-t|^\alpha \, dt$$

$$= C \int_{-\infty}^{\infty} \left| \frac{2y\left(3(x-t)^2 - y^2\right)}{\left[(x-t)^2 + y^2\right]^3} \right| |x-t|^\alpha \, dt$$

$$= C \int_{-\infty}^{\infty} \left| \frac{y(3t^2 - y^2)}{(t^2+y^2)^3} \right| |t|^\alpha \, dt$$

$$\leq Cy \int_{-\infty}^{\infty} \frac{t^2+y^2}{(t^2+y^2)^3}(t^2+y^2)^{\alpha/2} \, dt$$

$$\leq Cy \int_{-\infty}^{\infty} \frac{1}{(t^2+y^2)^{2-\alpha/2}} \, dt$$

$$\overset{(t=ty)}{=} Cy \int_{-\infty}^{\infty} \frac{1}{(t^2 y^2 + y^2)^{2-\alpha/2}} y \, dt$$

$$= Cy^{\alpha-2} \int_{-\infty}^{\infty} \frac{1}{(t^2+1)^{2-\alpha/2}} \, dt$$

$$= Cy^{\alpha-2}.$$

This completes the proof of Fact 1.

PROOF OF FACT 2 First notice that, for any $x \in \mathbb{R}$,

$$\left| \frac{\partial u}{\partial y}(x,2) \right| = \frac{1}{\pi} \left| \left[\frac{\partial}{\partial y} \int_{-\infty}^{\infty} \frac{y}{(x-t)^2+y^2} f(t)\, dt \right] \right|_{y=2} \right|$$

$$\leq \frac{1}{\pi} \int_{-\infty}^{\infty} \left| \frac{\partial}{\partial y}\left[\frac{y}{(x-t)^2+y^2} \right] \right|_{y=2} |f(t)| \, dt$$

$$\leq \frac{1}{\pi} \int_{-\infty}^{\infty} \frac{|(x-t)^2+y^2-2y^2|}{\left[(x-t)^2+y^2\right]^2} |f(t)| \, dt \Big|_{y=2}$$

$$\leq \frac{1}{\pi} \int_{-\infty}^{\infty} \frac{|f(t)|}{(x-t)^2+y^2} \, dt \Big|_{y=2}$$

$$\leq C \cdot \|f\|_{L^\infty(\mathbb{R})}.$$

Now, if $y_0 \geq 2$, then from Fact 1 we may calculate that

$$
\left| \frac{\partial u}{\partial y}(x_0, y_0) \right| \leq \left| \int_2^{y_0} \frac{\partial^2 u}{\partial y^2}(x_0, y)\, dy + \frac{\partial u}{\partial y}(x_0, 2) \right|
$$

$$
\leq \int_2^{y_0} y^{\alpha-2}\, dy + \left| \frac{\partial u}{\partial y}(x_0, 2) \right|
$$

$$
\leq C_\alpha \left[y_0^{\alpha-1} + 2^{\alpha-1} \right] + C_2
$$

$$
\leq C' y_0^{\alpha-1}.
$$

A nearly identical argument shows that

$$
\left| \frac{\partial u}{\partial y}(x_0, y_0) \right| \leq C' y_0^{\alpha-1}
$$

when $0 \leq y_0 < 2$.

We have proved Facts 1 and 2 and therefore have completed our estimates of term I.

The estimate of III is just the same as the estimate for I and we shall say no more about it.

For the estimate of II, we write $A = (a_1, a_2)$ and $B = (b_1, b_2)$. Then

$$
|u(A) - u(B)| = |u(a_1, a_2) - u(b_1, b_2)|
$$

$$
\leq |u(a_1, a_2) - u(b_1, a_2)| + |u(b_1, a_2) - u(b_1, b_2)|
$$

$$
= II_1 + II_2.
$$

Assume for simplicity that $a_1 < b_1$ as shown in Figure 1.2. Now set $\eta(t) = (a_1 + t, a_2), 0 < t < b_1 - a_1$. Then

$$
II_1 = \int_0^{b_1-a_1} \frac{d}{dt}(u \circ \eta)(t)\, dt
$$

$$
= \left| \int_0^{b_1-a_1} \frac{d}{dt} \int_{-\infty}^{\infty} P_{a_2}(a_1 + t - s) f(s)\, ds\, dt \right|
$$

$$
= \int_0^{b_1-a_1} \int_{-\infty}^{\infty} \frac{d}{dt} P_{a_2}(a_1 + t - s) \left[f(s) - f(t) \right]\, ds\, dt.
$$

We leave it as an exercise for the reader to see that this last expression, in absolute value, does not exceed $C|H|^\alpha$.

The term II_2 is estimated in the same way that we estimated I. ∎

THEOREM 1.4.2
Fix $1 < \alpha < 2$. If $f \in \Lambda_\alpha(\mathbb{R})$ then

$$u(x,y) \equiv \int_{\mathbb{R}} P_y(x-t)f(t)\,dt$$

lies in $\Lambda_\alpha(\bar{\mathcal{U}})$.

DISCUSSION OF THE PROOF It follows from the first theorem that $u \in \Lambda_\beta$, all $0 < \beta < 1$. In particular, u is bounded. It remains to see that

$$\frac{\partial u}{\partial x} \in \Lambda_{\alpha-1} \quad \text{and} \quad \frac{\partial u}{\partial y} \in \Lambda_{\alpha-1}.$$

But

$$\frac{\partial u}{\partial x} = \frac{\partial}{\partial x}\int P_y(t)f(x-t)\,dt = \int P_y(t)f'(x-t)\,dt.$$

Now $f' \in \Lambda_{\alpha-1}$ by definition and we have already considered the case $0 < \alpha - 1 < 1$. Thus $\partial u/\partial x$ lies in $\Lambda_{\alpha-1}(\bar{\mathcal{U}})$.

To estimate $\partial u/\partial y$, we can instead estimate

$$\frac{\partial^2 u}{\partial y^2} = -\frac{\partial^2 u}{\partial x^2}$$

$$= -\frac{\partial^2}{\partial x^2}\int P_y(t)f(x-t)\,dt$$

$$= -\frac{\partial}{\partial x}\int P_y(t)f'(x-t)\,dt$$

$$= -\frac{\partial}{\partial x}\int P_y(x-t)f'(t)\,dt$$

$$= -\int \frac{\partial}{\partial x}P_y(x-t)f'(t)\,dt.$$

Using the ideas from the estimates on I in the proof of the last theorem, we see that

$$\left|\frac{\partial^2 u}{\partial y^2}\right| \le Cy^{\alpha-2}.$$

Likewise, because $\partial f/\partial x \in \Lambda_{\alpha-1}$, we may prove as in Facts 1 and 2 in the estimate of I in the proof of the last theorem that

$$\left|\frac{\partial^2 u}{\partial x\partial y}\right| \le Cy^{\alpha-2}.$$

But then our usual arguments show that $\partial u/\partial y \in \Lambda_{\alpha-1}$. We leave details to the reader.

The proof is now complete. ∎

THEOREM 1.4.3

If $f \in \Lambda_1(\mathbb{R})$, then

$$u(x, y) \equiv \int_{\mathbb{R}} P_y(x - t) f(t) \, dt$$

lies in $\Lambda_1(\bar{\mathcal{U}})$.

We break the proof up into a sequence of lemmas.

LEMMA 1.4.4

Fix $y > 0$. Take $(x_0, y_0) \in \mathcal{U}$. Then

$$u(x_0, y_0) = \int_0^y y' \frac{\partial^2}{\partial y'^2} u(x_0, y_0 + y') \, dy' - y \frac{\partial}{\partial y} u(x_0, y_0 + y) + u(x_0, y + y_0).$$

PROOF Note that when $y = 0$, then the right-hand side equals $0 - 0 + u(x_0, y_0)$. Also, the partial derivative with respect to y of the right side is identically zero. That completes the proof. ∎

LEMMA 1.4.5

If $f \in \Lambda_1(\mathbb{R})$, then

$$\left| \frac{\partial^2}{\partial y^2} u(x, y) \right| \leq C \cdot y^{-1}.$$

PROOF Notice that, because u is harmonic,

$$\left| \frac{\partial^2}{\partial y^2} u \right| = \left| \frac{\partial^2}{\partial x^2} u \right|$$

$$= \left| \frac{\partial^2}{\partial x^2} \int P_y(x - t) f(t) \, dt \right|$$

$$= \left| \int \left(\frac{\partial^2}{\partial x^2} P_y(x - t) \right) f(t) \, dt \right|$$

$$= \left| \int \left(\frac{\partial^2}{\partial t^2} P_y(x - t) \right) f(t) \, dt \right|$$

$$= \left| \int \left(\frac{\partial^2}{\partial t^2} P_y(t) \right) f(x - t) \, dt \right|$$

$$= \frac{1}{2} \left| \int \left(\frac{\partial^2}{\partial t^2} P_y(t) \right) [f(x + t) + f(x - t) - 2f(x)] \, dt \right|$$

(since $(\partial^2/\partial t^2)P_y(t)$ is even and has mean value zero). By hypothesis, the last line does not exceed

$$C \int \left| \frac{\partial^2}{\partial t^2} P_y(t) \right| |t| \, dt. \qquad (1.4.5.1)$$

But recall, as in the proof of 1.4.1, that

$$\left| \frac{\partial^2}{\partial t^2} P_y(t) \right| \leq C \cdot \frac{y}{(y^2 + t^2)^2} \cdot$$

Therefore we may estimate line (1.4.5.1) by

$$C \int \frac{y}{(y^2 + t^2)^2} |t| \, dt \overset{(t=ty)}{=} C \cdot \frac{1}{y} \int \frac{|t|}{(1 + t^2)^2} \, dt$$

$$= C \cdot \frac{1}{y} \, .$$

This is what we wanted to prove. ∎

COROLLARY 1.4.6
Our calculations have also proved that

$$\left| \frac{\partial^2}{\partial x^2} u(x, y) \right| \leq C \cdot y^{-1}.$$

REMARK   Similar calculations prove that

$$\left| \frac{\partial^3}{\partial x^2 \partial y} u(x, y) \right| \leq C \cdot y^{-2}, \qquad \left| \frac{\partial^3}{\partial x \partial y^2} u(x, y) \right| \leq C \cdot y^{-2},$$

$$\left| \frac{\partial^3}{\partial y^3} u(x, y) \right| \leq C \cdot y^{-2}, \qquad \left| \frac{\partial^2}{\partial x \partial y} u(x, y) \right| \leq C \cdot y^{-1}. \qquad ∎$$

If v is a function, $H = (h_1, h_2)$, $X = (x_1, x_2)$ and if $X, X + H \in \mathcal{U}$, then we define

$$\Delta_H^2 v(X) = v(X + H) + v(X - H) - 2v(X).$$

LEMMA 1.4.7
If all second derivatives of the function v exist, then we have

$$|\Delta_H^2 v(X)| \leq C \cdot |H|^2 \cdot \sup_{|t| \leq 1} |\nabla^2 v(X + tH)|.$$

PROOF   We apply the mean value theorem twice to the function $\phi(t) = v(X + tH)$. Thus

$$
\begin{aligned}
|\Delta_H^2 v(X)| &= |\phi(1) + \phi(-1) - 2\phi(0)| \\
&= |(\phi(1) - \phi(0)) - (\phi(0) - \phi(-1))| \\
&= |1 \cdot \phi'(\xi_1) - 1 \cdot \phi'(\xi_2)| \\
&= |\phi''(\xi_3) \cdot (\xi_1 - \xi_2)| \\
&\leq 2|\phi''(\xi_3)| \\
&\leq C|H|^2 \sup_{|t|\leq 1} |\nabla^2 v(X + tH)|.
\end{aligned}
$$

That proves the lemma.   ∎

FINAL ARGUMENT IN THE PROOF OF THEOREM 1.4.3   Fix $X = (x_0, y_0), H = (h_1, h_2)$. Then, by Lemma 1.4.4,

$$
\begin{aligned}
\Delta_H^2 u(x_0, y_0) &= \int_0^y y' \Delta_H^2 \left(\frac{\partial^2}{\partial y^2} u(x_0, y_0 + y') \right) dy' \\
&\quad - y \Delta_H^2 \left(\frac{\partial}{\partial y} u(x_0, y_0 + y) \right) + \Delta_H^2 u(x_0, y_0 + y) \\
&\equiv I + II + III,
\end{aligned}
$$

any $y > 0$. Now we must have that $h_2 < y_0$ or else $\Delta_H^2 u(x_0, y_0)$ makes no sense. Thus $y_0 + tH + y' \geq y'$, all $-1 \leq t \leq 1$.

Applying Lemma 1.4.5, we see that

$$
\begin{aligned}
|I| &\leq \int_0^y y' \cdot 4 \sup_{|t|\leq 1} \left| \frac{\partial^2}{\partial y^2} u(x_0 + th_1, y_0 + th_2 + y') \right| dy' \\
&\leq C \int_0^y y' \cdot (y')^{-1} dy' \\
&\leq C \cdot y.
\end{aligned}
$$

Next, Lemma 1.4.7 and the remark preceding it yield

$$
\begin{aligned}
|II| &\leq C \cdot y \cdot |H|^2 \sup_{|t|\leq 1} \left| \nabla^2 \frac{\partial}{\partial y} u(x_0 + th_1, y_0 + y + th_2) \right| \\
&\leq Cy|H|^2 \cdot (y)^{-2} \\
&= C|H|^2 \cdot y^{-1}.
\end{aligned}
$$

Finally, the same reasoning gives

$$|III| \leq |H|^2 \cdot \sup_{|t| \leq 1} |\nabla^2 u(x_0 + th_1, y_0 + th_2 + y)|$$

$$\leq C \cdot |H|^2 \cdot y^{-1}.$$

Now we take $y = 2|H|$. The estimates on I, II, III then combine to give

$$|\Delta_H^2 u(X)| \leq C|H|.$$

This is the desired estimate on u. ∎

REMARK   We have proved that if $f \in \Lambda_1(\mathbb{R})$ then $u = P_y f \in \Lambda_1(\mathcal{U})$ using the method of direct estimation. An often more convenient, and natural, methodology is to use interpolation of operators, as seen in the next theorem. ∎

THEOREM 1.4.8
Let $V \subseteq \mathbb{R}^m, W \subseteq \mathbb{R}^n$ be open with smooth boundary. Fix $0 < \alpha < \beta$ and assume that T is a linear operator such that $T : \Lambda_\alpha(V) \rightarrow \Lambda_\alpha(W)$ and $T : \Lambda_\beta(V) \rightarrow \Lambda_\beta(W)$. Then, for all $\alpha < \gamma < \beta$ we have $T : \Lambda_\gamma(V) \rightarrow \Lambda_\gamma(W)$.

Interpolation of operators, presented in the context of Lipschitz spaces, is discussed in detail in [KR2]. The subject of interpolation is discussed in a broader context in [STW] and [BGL]. Here is an application of the theorem:

Let $\alpha = 1/2, \beta = 3/2$, and let T be the Poisson integral operator from functions on $\mathbb{R}$ to functions on $\mathcal{U}$. We know that

$$T : \Lambda_\alpha(\mathbb{R}) \rightarrow \Lambda_\alpha(\mathcal{U})$$

and

$$T : \Lambda_\beta(\mathbb{R}) \rightarrow \Lambda_\beta(\mathcal{U}).$$

We may conclude from the theorem that

$$T : \Lambda_1(\mathbb{R}) \rightarrow \Lambda_1(\mathcal{U}).$$

Thus we have a neater way of seeing that Poisson integration is well behaved on Λ_1.

REMARK   Twenty years ago it was an open question whether, if T is a bounded linear operator on C^0 and on C^2, it follows that T is a bounded linear operator on C^1. The answer to this question is negative; details may be found in [MIS]. In fact, Λ_1 is the appropriate space that is intermediate to C^0 and C^2.

One might ask how Poisson integration is behaved on $\mathrm{Lip}_1(\mathbb{R})$. Set $f(x) = |x| \cdot \phi(x)$ where $\phi \in C_c^\infty(\mathbb{R})$, $\phi = 1$ near 0. One may calculate that

$$u(x,y) = P_y f(x) = \frac{y}{2} \cdot \ln\left[\frac{(1-x)^2 + y^2}{x^2 + y^2} \cdot \frac{(1+x)^2 + y^2}{x^2 + y^2}\right]$$

$$+ x\left[\arctan\frac{1-x}{y} - \arctan\frac{1+x}{y}\right]$$

$$+ \text{(smooth error)}.$$

Set $x = 0$. Then, for y small,

$$u(0,y) \approx \frac{y}{2} \cdot \ln\left[\frac{(1+y^2)^2}{y^4}\right] \approx y \cdot \ln\left(\frac{1+y^2}{y^2}\right)$$

$$\approx y \cdot \ln y \in \Lambda_1(\mathcal{U}) \setminus \mathrm{Lip}_1(\mathcal{U}).$$

Again, we see that the classical space Lip_1 does not suit our purposes, while Λ_1 does. We shall not encounter Lip_1 any more in this book. The space Λ_1 was invented by Zygmund (see [ZYG]), who called it the space of *smooth functions*. He denoted it by Λ_*.

Here is what we have proved so far: if ϕ is a piece of Dirichlet data for the disc that lies in $\Lambda_\alpha(\partial D), 0 < \alpha < 2$, then the solution u to the Dirichlet problem with that data is Λ_α up to the boundary. We did this by transferring the problem to the upper half-space by way of the Cayley transform and then using explicit calculations with the Poisson kernel for the half-space.

1.5   Regularity of the Dirichlet Problem on a Smoothly Bounded Domain and Conformal Mapping

We begin by giving a precise definition of a domain "with smooth boundary":

DEFINITION 1.5.1   Let $U \subseteq \mathbb{C}$ be a bounded domain. We say that U has smooth boundary if the boundary consists of finitely many curves and each of these is locally the graph of a C^∞ function.

In practice it is more convenient to have a different definition of domain with smooth boundary. A function ρ is called a *defining function* for U if ρ is defined in a neighborhood W of ∂U, $\nabla \rho \neq 0$ on ∂U, and $W \cap U = \{z \in W : \rho(z) < 0\}$. Now we say that U has smooth (or C^k) boundary if U has a defining function ρ that is smooth (or C^k). Yet a third definition of smooth boundary is that the boundary consists of finitely many curves γ_j, each of which is the trace of a smooth curve $\mathbf{r}(t)$ with nonvanishing gradient. We invite the reader to verify that these three definitions are equivalent.

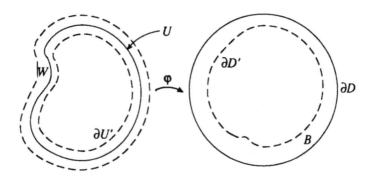

FIGURE 1.3

Our motivating question for the present section is as follows:

Let $\Omega \subseteq \mathbb{C}$ be a bounded domain with smooth boundary. Assume that $f \in \Lambda_\alpha(\partial\Omega)$. If $u \in C(\bar\Omega)$ satisfies (i) u is harmonic on Ω and (ii) $u|_{\partial\Omega} = f$, then does it follow that $u \in \Lambda_\alpha(\bar\Omega)$?

Here is a scheme for answering this question:

Step 1: Suppose at first that U is bounded and simply connected.

Step 2: By the Riemann mapping theorem, there is a conformal mapping ϕ : $U \to D$. Here D is the unit disc. We would like to reduce our problem to the Dirichlet problem on D for the data $f \circ \phi^{-1}$.

In order to carry out this program, we need to know that ϕ extends smoothly to the boundary. It is a classical result of Carathéodory [CAR] that if a simply connected domain U has boundary consisting of a Jordan curve, then any conformal map of the domain to the disc extends univalently and bicontinuously to the boundary. It is less well known that Painlevé, in his thesis [PAI], proved that when U has smooth boundary then the conformal mapping extends smoothly to the boundary. In fact, Painlevé's result long precedes that of Carathéodory.

We shall present here a modern approach to smoothness to the boundary for conformal mappings. These ideas come from [KER1]. See also [BKR] for a self-contained approach to these matters. Our purpose here is to tie the smoothness-to-the-boundary issue for mappings directly to the regularity theory of the Dirichlet problem for the Laplacian.

Refer to Figure 1.3. Let W be a collared neighborhood of ∂U. Set $\partial U' = \partial W \cap U$ and let $\partial D' = \phi(\partial U')$. Define B to be the region bounded by ∂D

and $\partial D'$. We solve the Dirichlet problem on B with boundary data

$$f(\zeta) = \begin{cases} 1 & \text{if } \zeta \in \partial D \\ 0 & \text{if } \zeta \in \partial D'. \end{cases}$$

Call the solution u.

Consider $v \equiv u \circ \phi : U \to \mathbb{R}$. Then, of course, v is still harmonic. By Carathéodory's theorem, v extends to ∂U, $\partial U'$, and

$$v = \begin{cases} 1 & \text{if } \zeta \in \partial U \\ 0 & \text{if } \zeta \in \partial U'. \end{cases}$$

Suppose that we knew that solutions of the Dirichlet problem on a smoothly bounded domain with C^∞ data are in fact C^∞ on the closure of the domain. Then, if we consider a first-order derivative $\mathcal{D}$ of v, we obtain

$$|\mathcal{D}v| = |\mathcal{D}(u \circ \phi)| = |\nabla u| \, |\nabla \phi| \le C.$$

It follows that

$$|\nabla \phi| \le \frac{C}{|\nabla u|} \, . \tag{1.5.2}$$

This will prove to be a useful estimate once we take advantage of the following lemma.

LEMMA 1.5.3 HOPF'S LEMMA
Let $\Omega \subset\subset \mathbb{R}^N$ have C^2 boundary. Let $u \in C(\bar{\Omega})$ with u harmonic and noncon-stant on Ω. Let $P \in \bar{\Omega}$ and assume that u takes a local minimum at P. Then the one-sided normal derivative satisfies

$$\frac{\partial u}{\partial \nu}(P) < 0.$$

PROOF   Suppose without loss of generality that $u > 0$ on Ω near P and that $u(P) = 0$. Let B_R be a ball that is internally tangent to $\bar{\Omega}$ at P. We may assume that the center of this ball is at the origin and that P has coordinates $(R, 0, \ldots, 0)$. Then, by Harnack's inequality (see [KR1]), we have for $0 < r < R$ that

$$u(r, 0, \ldots, 0) \ge c \cdot \frac{R^2 - r^2}{R^2 + r^2} \, ,$$

hence

$$\frac{u(r, 0, \ldots, 0) - u(R, 0, \ldots, 0)}{r - R} \le -c' < 0.$$

Therefore

$$\frac{\partial u}{\partial \nu}(P) \le -c' < 0.$$

This is the desired result.   ∎

Now let us return to the u from the Dirichlet problem that we considered prior to line (1.5.2). Hopf's lemma tells us that $|\nabla u| \geq c' > 0$ near ∂D. Thus, from (1.5.2), we conclude that

$$|\nabla \phi| \leq C. \tag{1.5.4}$$

Thus we have bounds on the first derivatives of ϕ.

To control the second derivatives, we calculate that

$$C \geq |\nabla^2 v| = |\nabla(\nabla v)| = |\nabla(\nabla(u \circ \phi))|$$
$$= |\nabla(\nabla u(\phi) \cdot \nabla \phi)| = |\left(\nabla^2 u \cdot [\nabla \phi]^2\right) + \left(\nabla u \cdot \nabla^2 \phi\right)|.$$

Here the reader should think of ∇ as representing a generic first derivative and ∇^2 a generic second derivative. We conclude that

$$|\nabla u|\,|\nabla^2 \phi| \leq C + |\nabla^2 u|\,|(\nabla \phi)^2| \leq C'.$$

Hence (again using Hopf's lemma),

$$|\nabla^2 \phi| \leq \frac{C'}{|\nabla u|} \leq C''.$$

In the same fashion, we may prove that $|\nabla^k \phi| \leq C_k$, any $k \in \{1, 2, \ldots\}$. This means (use the fundamental theorem of calculus) that $\phi \in C^\infty(\bar{\Omega})$.

We have arrived at the following situation: Smoothness to the boundary of conformal maps implies regularity of the Dirichlet problem on a smoothly bounded domain. Conversely, regularity of the Dirichlet problem can be used, together with Hopf's lemma, to prove the smoothness to the boundary of conformal mappings. We must find a way out of this impasse.

Our solution to the problem posed in the last paragraph will be to study the Dirichlet problem for a more general class of operators that is invariant under *smooth* changes of coordinates. We will study these operators by (i) localizing the problem and (ii) mapping the smooth domain under a diffeomorphism to an upper half-space. It will turn out that *elliptic operators* are invariant under these operations. We shall then use the calculus of pseudodifferential operators to prove local boundary regularity for elliptic operators.

There is an important point implicit in our discussion that deserves to be brought into the foreground. The Laplacian is invariant under conformal transformations (exercise). This observation was useful in setting up the discussion in the present section. But it turned out to be a point of view that is too narrow: we found ourselves in a situation of circular reasoning. We shall thus expand to a wider universe in which our operators are invariant under *diffeomorphisms*. This type of invariance will give us more flexibility and more power.

Let us conclude this section by exploring how the Laplacian behaves under a diffeomorphic change of coordinates. For simplicity, we restrict attention to

$\mathbb{R}^2$ with coordinates (x, y). Let

$$\phi(x, y) = (\phi_1(x, y), \phi_2(x, y)) \equiv (x', y')$$

be a diffeomorphism of $\mathbb{R}^2$. Let

$$\Delta = \frac{\partial^2}{\partial x^2} + \frac{\partial^2}{\partial y^2}.$$

In (x', y') coordinates, the operator Δ becomes

$$\phi_*(\Delta) = |\nabla\phi_1|^2 \frac{\partial^2}{\partial x'^2} + |\nabla\phi_2|^2 \frac{\partial^2}{\partial y'^2}$$

$$+ 2\left[\frac{\partial x'}{\partial x}\frac{\partial y'}{\partial x} + \frac{\partial x'}{\partial y}\frac{\partial y'}{\partial y}\right]\frac{\partial^2}{\partial x \partial y} + \text{(first-order terms)}.$$

In an effort to see what the new operator has in common with the old one, we introduce the notation

$$D = \sum a_\alpha \frac{\partial}{\partial x^\alpha},$$

where

$$\frac{\partial}{\partial x^\alpha} = \frac{\partial}{\partial x_1^{\alpha_1}}\frac{\partial}{\partial x_2^{\alpha_2}}\cdots\frac{\partial}{\partial x_n^{\alpha_n}}$$

is a differential monomial. Its "symbol" (for more on this, see the next two chapters) is defined to be

$$\sigma(D) = \sum a_\alpha(x)\xi^\alpha, \qquad \xi^\alpha = \xi_1^{\alpha_1}\xi_2^{\alpha_2}\cdots\xi_n^{\alpha_n}.$$

The symbol of the Laplacian $\Delta = (\partial^2/\partial x^2) + (\partial^2/\partial y^2)$ is

$$\sigma(\Delta) = \xi_1^2 + \xi_2^2.$$

Now associate to $\sigma(\Delta)$ a matrix $A_\Delta = (a_{ij})_{1 \leq i,j \leq 2}$, where $a_{ij} = a_{ij}(x)$ is the coefficient of $\xi_i\xi_j$ in the symbol. Thus

$$A_\Delta = \begin{pmatrix} 1 & 0 \\ 0 & 1 \end{pmatrix}.$$

The symbol of the transformed Laplacian (in the new coordinates) is

$$\sigma(\phi_*(\Delta)) = |\nabla\phi_1|^2\xi_1^2 + |\nabla\phi_2|^2\xi_2^2$$

$$+ 2\left[\frac{\partial x'}{\partial x}\frac{\partial y'}{\partial y} + \frac{\partial x'}{\partial y}\frac{\partial y'}{\partial y}\right]\xi_1\xi_2$$

$$+ \text{(lower order terms)}.$$

Then

$$A_{\sigma(\phi^*(\Delta))} = \begin{pmatrix} |\nabla\phi_1|^2 & \left[\frac{\partial x'}{\partial x}\frac{\partial y'}{\partial x} + \frac{\partial x'}{\partial y}\frac{\partial y'}{\partial y}\right] \\ \left[\frac{\partial x'}{\partial x}\frac{\partial y'}{\partial x} + \frac{\partial x'}{\partial y}\frac{\partial y'}{\partial y}\right] & |\nabla\phi_2|^2 \end{pmatrix}.$$

The matrix $A_{\sigma(\phi^*(\Delta))}$ is positive definite provided that the change of coordinates ϕ is a diffeomorphism (i.e., has nondegenerate Jacobian). It is this positive definiteness property of the symbol that will be crucial to the success of our attack on elliptic operators.

2

Review of Fourier Analysis

2.1   The Fourier Transform

A thorough treatment of Fourier analysis in Euclidean space may be found in
[STW] or [HOR4]. Here we give a sketch of the theory.

If $t, \xi \in \mathbb{R}^N$ then we let

$$t \cdot \xi \equiv t_1 \xi_1 + \cdots + t_N \xi_N.$$

We define the *Fourier transform* of an $f \in L^1(\mathbb{R}^N)$ by

$$\hat{f}(\xi) = \int f(t) e^{it \cdot \xi} \, dt.$$

Many references will insert a factor of 2π in the exponential or in the measure.
Others will insert a minus sign in the exponent. There is no agreement on this
matter. We have opted for this definition because of its simplicity. We note that
the significance of the exponentials $e^{it \cdot \xi}$ is that the only continuous multiplicative
homomorphisms of $\mathbb{R}^N$ into the circle group are the functions $\phi_\xi(t) = e^{it \cdot \xi}$.
(We leave this as an exercise for the reader. A thorough discussion appears
in [KAT] or [BAC].) These functions are called the *characters* of the additive
group $\mathbb{R}^N$.

Basic Properties of the Fourier Transform

PROPOSITION 2.1.1
If $f \in L^1(\mathbb{R}^N)$ then

$$\|\hat{f}\|_{L^\infty} \leq \|f\|_{L^1(\mathbb{R}^N)}.$$

PROOF   Observe that

$$|\hat{f}(\xi)| \leq \int |f(t)| \, dt. \qquad \blacksquare$$

PROPOSITION 2.1.2
If $f \in L^1(\mathbb{R}^N)$, f is differentiable, and $\partial f/\partial x_j \in L^1$, then

$$\widehat{\left(\frac{\partial f}{\partial x_j}\right)} = -i\xi_j \hat{f}(\xi).$$

PROOF Integrate by parts: if $f \in C_c^\infty$ then

$$\widehat{\left(\frac{\partial f}{\partial x_j}\right)} = \int \frac{\partial f}{\partial t_j} e^{it\cdot\xi} \, dt$$

$$= \int \cdots \int\int \frac{\partial f}{\partial t_j} e^{it\cdot\xi} \, dt_j \, dt_1 \ldots \widehat{dt_j} \ldots dt_N$$

$$= -\int \cdots \int f(t) \left(\frac{\partial}{\partial t_j} e^{it\cdot\xi}\right) dt_j \, dt_1 \ldots \widehat{dt_j} \ldots dt_N$$

$$= -i\xi_j \int \cdots \int f(t) e^{it\cdot\xi} \, dt.$$

The general case follows from a limiting argument. ∎

PROPOSITION 2.1.3
If $f \in L^1(\mathbb{R}^N)$ and $ix_j f \in L^1(\mathbb{R}^N)$, then

$$\widehat{(ix_j f)} = \frac{\partial}{\partial \xi_j} \hat{f}.$$

PROOF Integrate by parts. ∎

PROPOSITION 2.1.4 THE RIEMANN–LEBESGUE LEMMA
If $f \in L^1(\mathbb{R}^N)$, then

$$\lim_{\xi \to \infty} |\hat{f}(\xi)| = 0.$$

PROOF First assume that $f \in C_c^2(\mathbb{R}^N)$. We know that

$$\|\hat{f}\|_{L^\infty} \leq \|f\|_{L^1} \leq C$$

and

$$\left\|\xi_j^2 \hat{f}\right\|_{L^\infty} = \left\|\widehat{\left[\left(\frac{\partial^2}{\partial x_j^2}\right) f\right]}\right\|_{L^\infty} \leq \left\|\left(\frac{\partial^2}{\partial x_j^2}\right) f\right\|_{L^1} \leq C.$$

Then $(1 + |\xi|^2)\hat{f}$ is bounded. Therefore

$$|\hat{f}| \leq \frac{C}{1 + |\xi|^2} \overset{|\xi| \to \infty}{\longrightarrow} 0.$$

This proves the result for $f \in C_c^2$.

Now let $g \in L^1$ be arbitrary. By elementary measure theory, there is a function $\phi \in C_c(\mathbb{R}^N)$ such that $\|g - \phi\|_{L^1} < \epsilon/4$. It is then straightforward to construct a $\psi \in C_c^2$ such that $\|\phi - \psi\|_{L^1} < \epsilon/4$. It follows that $\|g - \psi\|_{L^1} < \epsilon/2$.

Choose M so large that when $|\xi| > M$ then $|\hat{\psi}(\xi)| < \epsilon/2$. Then, for $|\xi| > M$, we have

$$|\hat{g}(\xi)| = |(\widehat{g - \psi})(\xi) + \hat{\psi}(\xi)|$$
$$\leq |(\widehat{g - \psi})(\xi)| + |\hat{\psi}(\xi)|$$
$$\leq \|g - \psi\|_{L^1} + \frac{\epsilon}{2}$$
$$< \frac{\epsilon}{2} + \frac{\epsilon}{2} = \epsilon.$$

This proves the result. ∎

PROPOSITION 2.1.5
Let $f \in L^1(\mathbb{R}^N)$. Then $\hat{f}$ is uniformly continuous.

PROOF Apply the Lebesgue dominated convergence theorem and Proposition 2.1.4. ∎

Let $C_0(\mathbb{R}^N)$ denote the continuous functions on $\mathbb{R}^N$ that vanish at ∞. Equip this space with the supremum norm. Then the Fourier transform maps L^1 to C_0 continuously, with operator norm 1.

PROPOSITION 2.1.6
If $f \in L^1(\mathbb{R}^N)$, we let $\tilde{f}(x) = f(-x)$. Then $\hat{\tilde{f}}(\xi) = \tilde{\hat{f}}(\xi)$.

PROOF We calculate that

$$\hat{\tilde{f}}(\xi) = \int \tilde{f}(t)e^{it\cdot\xi}\,dt = \int f(-t)e^{it\cdot\xi}\,dt$$
$$= \int f(t)e^{-it\cdot\xi}\,dt = \tilde{\hat{f}}.$$ ∎

PROPOSITION 2.1.7
If ρ is a rotation of $\mathbb{R}^N$ then we define $\rho f(x) = f(\rho(x))$. Then $\widehat{\rho f} = \rho(\hat{f})$.

PROOF   Remembering that ρ is orthogonal, we calculate that

$$
\begin{aligned}
\widehat{\rho f}(\xi) \;&=\; \int (\rho f)(t) e^{i\xi\cdot t}\, dt = \int f(\rho(t)) e^{i\xi\cdot t}\, dt \\
\overset{(s=\rho(t))}{=}&\; \int f(s) e^{i\xi\cdot\rho^{-1}(s)}\, ds = \int f(s) e^{i\rho(\xi)\cdot s}\, ds \\
&=\; (\rho(\hat f))(\xi).
\end{aligned}
$$ ∎

PROPOSITION 2.1.8
We have

$$
\hat{\bar f} = \bar{\tilde{\hat f}}.
$$

PROOF   We calculate that

$$
\hat{\bar f}(\xi) = \int \bar f(t) e^{i\xi\cdot t}\, dt = \overline{\int f(t) e^{-i\xi\cdot t}\, dt} = \overline{\hat f(-\xi)} = \bar{\tilde{\hat f}}(\xi).
$$ ∎

PROPOSITION 2.1.9
If $\delta > 0$ and $f \in L^1(\mathbb{R}^N)$, then we set $\alpha_\delta f(x) = f(\delta x)$ and $\alpha^\delta f(x) = \delta^{-N} f(x/\delta)$. Then

$$
(\alpha_\delta f)\widehat{\,} = \alpha^\delta\left(\hat f\right)
$$

$$
\widehat{\alpha^\delta f} = \alpha_\delta \hat f.
$$

PROOF   We calculate that

$$
\begin{aligned}
(\alpha_\delta f)\widehat{\,} &= \int (\alpha_\delta f)(t) e^{it\cdot\xi}\, dt = \int f(\delta t) e^{it\cdot\xi}\, dt \\
&= \int f(t) e^{i(t/\delta)\cdot\xi} \delta^{-N}\, dt = \delta^{-N} \hat f(\xi/\delta) = \alpha^\delta(\hat f).
\end{aligned}
$$

That proves the first assertion. The proof of the second is similar.   ∎

If f, g are L^1 functions, then we define their *convolution* to be the function

$$
f * g(x) = \int f(x-t) g(t)\, dt = \int g(x-t) f(t)\, dt.
$$

It is a standard result of measure theory (see [RUD3]) that $f * g$ so defined is an L^1 function and $\|f * g\|_{L^1} \le \|f\|_{L^1} \|g\|_{L^1}$.

PROPOSITION 2.1.10
If $f, g \in L^1$, then

$$\widehat{f * g}(\xi) = \hat{f}(\xi) \cdot \hat{g}(\xi).$$

PROOF   We calculate that

$$\widehat{f * g}(\xi) = \int (f * g)(t) e^{i\xi \cdot t} \, dt = \int \int f(t - s) g(s) \, ds \, e^{i\xi \cdot t} \, dt$$

$$= \int \int f(t - s) e^{i\xi \cdot (t-s)} \, dt \, g(s) e^{i\xi \cdot s} \, ds = \hat{f}(\xi) \cdot \hat{g}(\xi).$$

The reader may justify the change in the order of integration.   ∎

PROPOSITION 2.1.11
If $f, g \in L^1$, then

$$\int \hat{f}(\xi) g(\xi) \, d\xi = \int f(\xi) \hat{g}(\xi) \, d\xi.$$

PROOF   This is a straightforward change in the order of integration.   ∎

The Inverse Fourier Transform

Our goal is to be able to recover f from $\hat{f}$. This program entails several technical difficulties. First, we need to know that the Fourier transform is univalent in order to have any hope of success. Second, we would like to say that

$$f(t) = c \cdot \int \hat{f}(\xi) e^{-it \cdot \xi} \, dt.$$

In general, however, the Fourier transform of an L^1 function is not integrable. Thus we need a family of *summability kernels* G_ϵ satisfying the following properties:

1. $G_\epsilon * f \rightarrow f$ as $\epsilon \rightarrow 0$;
2. $\widehat{G_\epsilon}(\xi) = e^{-\epsilon |\xi|^2}$;
3. $G_\epsilon * f$ and $\widehat{G_\epsilon} * f$ are both integrable.

It will be useful to prove formulas about $G_\epsilon * f$ and then pass to the limit as $\epsilon \rightarrow 0^+$.

LEMMA 2.1.12
We have

$$\int_{\mathbf{R}^N} e^{-|x|^2} \, dx = (\sqrt{\pi})^N.$$

PROOF It is enough to do the case $N = 1$. Set $I = \int_{-\infty}^{\infty} e^{-t^2}\, dt$. Then

$$I \cdot I = \int_{-\infty}^{\infty} e^{-s^2}\, ds \int_{-\infty}^{\infty} e^{-t^2}\, dt = \iint_{\mathbf{R}^2} e^{-|(s,t)|^2}\, ds\, dt$$

$$= \int_{0}^{2\pi} \int_{0}^{\infty} e^{-r^2} r\, dr\, d\theta = \pi.$$

Thus $I = \sqrt{\pi}$, as desired. ∎

REMARK Although this is the most common method for evaluating $\int e^{-|x|^2}\, dx$, several other approaches are provided in [HEI]. ∎

Now let us calculate the Fourier transform of $e^{-|x|^2}$. It suffices to treat the one-dimensional case because

$$\left(e^{-|x|^2}\right)^{\widehat{\ }} = \int_{\mathbf{R}^N} e^{-|x|^2} e^{ix\cdot\xi}\, dx = \int_{\mathbf{R}} e^{-x_1^2} e^{ix_1\xi_1}\, dx_1 \cdots \int_{\mathbf{R}} e^{-x_N^2} e^{ix_N\xi_N}\, dx_N.$$

Now when $N = 1$ we have

$$\int_{\mathbf{R}} e^{-x^2 + ix\xi}\, dx = \int_{\mathbf{R}} e^{(\xi/2 + ix)^2} e^{-\xi^2/4}\, dx$$

$$= e^{-\xi^2/4} \int_{\mathbf{R}} e^{(\xi/2 + ix)^2}\, dx$$

$$= e^{-\xi^2/4} \int_{\mathbf{R}} e^{(\xi/2 + ix/2)^2} \frac{1}{2}\, dx$$

$$= \frac{1}{2} e^{-\xi^2/4} \oint_{\Gamma} e^{(z/2)^2}\, dz. \qquad (2.1.12.1)$$

Here, for $\xi \in \mathbf{R}$ fixed, $\Gamma = \Gamma_\xi$ is the curve $t \mapsto \xi + it$. Let Γ_N be the part of the curve between $t = -N$ and $t = N$. Since $\int_\Gamma = \lim_{N\to\infty} \int_{\Gamma_N}$, it is enough for us to understand $\int_{\Gamma_N}$. Refer to Figure 2.1.

Now

$$\oint_{\Gamma_N \cup E_1^N \cup \tilde{\Gamma}_N \cup E_2^N} e^{(z^2/2)^2}\, dz = 0.$$

Therefore

$$\oint_{\Gamma_N} = -\left(\int_{\tilde{\Gamma}_N} + \int_{E_1^N} + \int_{E_2^N}\right).$$

But, as $N \to \infty$,

$$\oint_{E_1^N} + \oint_{E_2^N} \to 0.$$

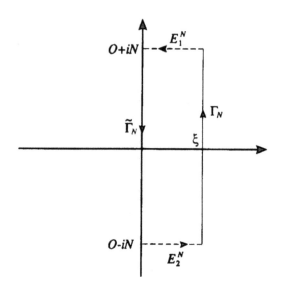

FIGURE 2.1

Thus

$$\lim_{N \to \infty} \oint_{\Gamma_N} = - \lim_{N \to \infty} \oint_{\tilde{\Gamma}_N} . \qquad (2.1.12.2)$$

Now we combine (2.1.12.1) and (2.1.12.2) to see that

$$\int_{\mathbf{R}} e^{-x^2 + ix\xi} \, dx = - \lim_{N \to \infty} \frac{1}{2} e^{-\xi^2/4} \int_{\tilde{\Gamma}_N} e^{(z/2)^2} \, dz$$

$$= \frac{1}{2} e^{-\xi^2/4} \int_{\mathbf{R}} e^{-t^2/4} \, dt$$

$$= e^{-\xi^2/4} \int_{\mathbf{R}} e^{-t^2} \, dt$$

$$= e^{-\xi^2/4} \sqrt{\pi}.$$

We conclude that, in $\mathbf{R}^1$,

$$(\widehat{e^{-x^2}})(\xi) = \sqrt{\pi} e^{-\xi^2/4},$$

and in $\mathbf{R}^N$ we have

$$(\widehat{e^{-|x|^2}})(\xi) = \sqrt{\pi}^N e^{-|\xi|^2/4}.$$

It is often convenient to scale this formula and write

$$(e^{-|x|^2/2})(\xi) = (2\pi)^{N/2} e^{-|\xi|^2/2}.$$

The function $G(x) = (2\pi)^{-N/2} e^{-|x|^2/2}$ is called the *Gauss–Weierstrass kernel*. It is a summability kernel (see [KAT]) for the Fourier transform. Observe that $\hat{G}(\xi) = e^{-|\xi|^2/2}$.

On $\mathbb{R}^N$ we define

$$G_\epsilon(x) = \alpha^{\sqrt{\epsilon}}(G)(x) = \epsilon^{-N/2}(2\pi)^{-N/2} e^{-|x|^2/(2\epsilon)}.$$

Then

$$\widehat{G_\epsilon}(\xi) = \left(\alpha^{\sqrt{\epsilon}}G\right)\widehat{\ }(\xi) = \alpha_{\sqrt{\epsilon}}\hat{G}(\xi) = e^{-\epsilon|\xi|^2/2},$$

$$\widehat{\widehat{G_\epsilon}}(\xi) = \left(e^{-\epsilon|\xi|^2/2}\right)\widehat{\ } = \left(\alpha_{\sqrt{\epsilon}}e^{-|\xi|^2/2}\right)\widehat{\ }$$

$$= \alpha^{\sqrt{\epsilon}}\left[(2\pi)^{N/2}e^{-|\xi|^2/2}\right]$$

$$= \epsilon^{-N/2}(2\pi)^{N/2}e^{-|\xi|^2/(2\epsilon)}.$$

Now assume that $f, \hat{f}$ are in L^1 and are continuous. We apply Proposition 2.1.11 with $g = \hat{G}_\epsilon \in L^1$. We obtain

$$\int f\widehat{\widehat{G_\epsilon}}(x)\,dx = \int \hat{f}(\xi)\widehat{G_\epsilon}(\xi)\,d\xi.$$

In other words,

$$\int f(x)\epsilon^{-N/2}(2\pi)^{N/2}e^{-|x|^2/(2\epsilon)}\,dx = \int \hat{f}(\xi)e^{-\epsilon|\xi|^2/2}\,d\xi. \qquad (2.1.13)$$

Now $e^{-\epsilon|\xi|^2/2} \to 1$ uniformly on compact sets. Thus $\int \hat{f}(\xi)e^{-\epsilon|\xi|^2}\,d\xi \to \int \hat{f}(\xi)\,d\xi$. That takes care of the right-hand side of (2.1.13).

Next observe that

$$\int G_\epsilon(x)\,dx = \int G_\epsilon(x)e^{i0\cdot\xi}\,dx = \widehat{G_\epsilon}(0) = 1.$$

Thus the left side of (2.1.13) equals

$$(2\pi)^N \int f(x)G_\epsilon(x)\,dx = (2\pi)^N \int f(0)G_\epsilon(x)\,dx$$

$$+ (2\pi)^N \int [f(x) - f(0)]G_\epsilon(x)\,dx$$

$$\to (2\pi)^N \cdot f(0).$$

Thus we have evaluated the limits of the left- and right-hand sides of (2.1.13). We have proved the following theorem.

THEOREM 2.1.14   THE FOURIER INVERSION FORMULA
If $f, \hat{f} \in L^1$ and both are continuous, then

$$f(0) = (2\pi)^{-N} \int \hat{f}(\xi)\, d\xi. \tag{2.1.14.1}$$

Of course there is nothing special about the point $0 \in \mathbb{R}^N$. We now exploit the compatibility of the Fourier transform with translations to obtain a more general formula. First, we define

$$(T_h f)(x) = f(x - h)$$

for any function f on $\mathbb{R}^N$ and any $h \in \mathbb{R}^N$. Then, by a change of variable in the integral,

$$\widehat{T_h f} = e^{ih\cdot\xi}\hat{f}(\xi).$$

Now we apply formula (2.1.14.1) in our theorem to $T_{-h}f$: The result is

$$(T_{-h}f)(0) = (2\pi)^{-N} \int (T_{-h}f)\widehat{\,}(\xi)\, d\xi$$

or

THEOREM 2.1.15
If $f, \hat{f} \in L^1$ then for any $h \in \mathbb{R}^N$ we have

$$f(h) = (2\pi)^{-N} \int \hat{f}(\xi)e^{-ih\cdot\xi}\, d\xi.$$

COROLLARY 2.1.16
The Fourier transform is univalent. That is, if $f, g \in L^1$ and $\hat{f} \equiv \hat{g}$ then $f = g$ almost everywhere.

PROOF   Since $f - g \in L^1$ and $\hat{f} - \hat{g} \equiv 0 \in L^1$, this is immediate from either the theorem or (2.1.13).   ∎

Since the Fourier transform is univalent, it is natural to ask whether it is surjective. We have

PROPOSITION 2.1.17
The operator

$$\widehat{} : L^1 \to C_0$$

is not onto.

PROOF   Seeking a contradiction, we suppose that the operator is in fact surjective. Then the open mapping principle guarantees that there is a constant $C > 0$ such that

$$\|f\|_{L^1} \le C\|\hat{f}\|_{L^\infty}.$$

On $\mathbf{R}^1$, let $g(\xi)$ be the characteristic function of the interval $[-1, 1]$. The inverse Fourier transform of g is a nonintegrable function. But then $\{G_{1/j} * g\}$ forms a sequence that is bounded in supremum norm but whose inverse Fourier transforms are unbounded in L^1 norm. That gives the desired contradiction. ∎

Plancherel's Formula

PROPOSITION 2.1.18   PLANCHEREL
If $f \in C_c^\infty(\mathbf{R}^N)$ then

$$\int |\hat{f}(\xi)|^2 \, d\xi = (2\pi)^N \int |f(x)|^2 \, dx.$$

PROOF   Define $g(x) = f * \tilde{\bar{f}} \in C_c^\infty(\mathbf{R}^N)$. Then

$$\hat{g} = \hat{f} \cdot \hat{\tilde{\bar{f}}} = \hat{f} \cdot \overline{\tilde{\hat{f}}} = \hat{f} \cdot \overline{\hat{f}} = |\hat{f}|^2. \qquad (2.1.18.1)$$

Now

$$g(0) = f * \tilde{\bar{f}}(0) = \int f(-t)\bar{f}(-t) \, dt = \int f(t)\bar{f}(t) \, dt = \int |f(t)|^2 \, dt.$$

By Fourier inversion and formula (2.1.18.1) we may now conclude that

$$\int |f(t)|^2 \, dt = g(0) = (2\pi)^{-N} \int \hat{g}(\xi) \, d\xi = (2\pi)^{-N} \int |\hat{f}(\xi)|^2 \, d\xi.$$

That is the desired formula. ∎

COROLLARY 2.1.19
If $f \in L^2(\mathbf{R}^N)$ then the Fourier transform of f can be defined in the following fashion: Let $f_j \in C_c^\infty$ satisfy $f_j \rightarrow f$ in the L^2 topology. It follows from the proposition that $\{\hat{f}_j\}$ is Cauchy in L^2. Let g be the L^2 limit of this latter sequence. We set $\hat{f} = g$.

It is easy to check that the definition of $\hat{f}$ given in the corollary is independent of the choice of sequence $f_j \in C_c^\infty$ and that

$$\int |\hat{f}(\xi)|^2 \, d\xi = (2\pi)^N \int |f(x)|^2 \, dx.$$

We now know that the Fourier transform $\mathcal{F}$ has the following mapping properties:

$$\mathcal{F}: L^1 \rightarrow L^\infty$$

$$\mathcal{F}: L^2 \rightarrow L^2.$$

The Riesz–Thorin interpolation theorem (see [STW]) now allows us to conclude that

$$\mathcal{F}: L^p \rightarrow L^{p'}, \qquad 1 \leq p \leq 2,$$

where $p' = p/(p-1)$. If $p > 2$ then $\mathcal{F}$ does not map L^p into any nice function space. The precise norm of $\mathcal{F}$ on L^p has been computed by Beckner [BEC].

Exercises: Restrict attention to dimension 1. Consider the Fourier transform $\mathcal{F}$ as a bounded linear operator on the Hilbert space $L^2(\mathbf{R}^N)$. Prove that the four roots of unity (suitably scaled) are eigenvalues of $\mathcal{F}$.

Prove that if $\rho(x)$ is a Hermite polynomial (see [STW], [WHW]), then the function $\rho(x)e^{-|x|^2/2}$ is an eigenfunction of $\mathcal{F}$. (Hint: $(ixf)\hat{} = (\hat{f})'$ and $\hat{f'} = -i\xi\hat{f}$.)

2.2 Schwartz Distributions

Thorough treatments of distribution theory may be found in [SCH], [HOR4], [TRE2]. Here we give a quick review.

We define the space of Schwartz functions:

$$S = \left\{\phi \in C^\infty(\mathbf{R}^N): \rho_{\alpha,\beta}(\phi) \equiv \sup_{x \in \mathbf{R}^N} \left|x^\alpha \left(\frac{\partial}{\partial x}\right)^\beta \phi(x)\right| < \infty,\right.$$

$$\left. \alpha = (\alpha_1,\ldots,\alpha_N), \beta = (\beta_1,\ldots,\beta_N)\right\}.$$

Observe that $e^{-|x|^2} \in S$ and $p(x) \cdot e^{-|x|^2} \in S$ for any polynomial p. Any derivative of a Schwartz function is still a Schwartz function. The Schwartz space is obviously a linear space.

It is worth noting that the space of C^∞ functions with compact support (which we have been denoting by C_c^∞) forms a proper subspace of S. Since as recently as 1930 there was some doubt as to whether C_c^∞ functions are genuine functions (see [OSG]), it may be worth seeing how to construct elements of this space.

Let the dimension N equal 1. Define

$$\lambda(x) = \begin{cases} e^{-1/|x|^2} & \text{if } x \geq 0 \\ 0 & \text{if } x < 0. \end{cases}$$

Then one checks, using l'Hôpital's Rule, that $\lambda \in C^\infty(\mathbb{R})$. Set

$$h(x) = \lambda(-x-1) \cdot \lambda(x+1) \in C_c^\infty(\mathbb{R}).$$

Moreover, if we define

$$g(x) = \int_{-\infty}^{x} h(t)\, dt,$$

then the function

$$f(x) = g(x+2) \cdot g(-x-2)$$

lies in C_c^∞ and is identically equal to a constant on $(-1,1)$. Thus we have constructed a standard "cutoff function" on $\mathbb{R}^1$. On $\mathbb{R}^N$, the function

$$F(x) \equiv f(x_1) \cdots f(x_N)$$

plays a similar role.

Exercise: [**The C^∞ Urysohn lemma**]   Let K and L be disjoint closed sets in $\mathbb{R}^N$. Prove that there is a C^∞ function ϕ on $\mathbb{R}^N$ such that $\phi \equiv 0$ on K and $\phi \equiv 1$ on L. (Details of this sort of construction may be found in [HIR].)

The Topology of the Space S

The functions $\rho_{\alpha,\beta}$ are seminorms on S. A neighborhood basis of 0 for the corresponding topology on S is given by the sets

$$N_{\epsilon,\ell,m} = \left\{ \phi : \sum_{\substack{|\alpha|\le\ell \\ |\beta|\le m}} \rho_{\alpha,\beta}(\phi) < \epsilon \right\}.$$

Exercise: The space S cannot be normed.

DEFINITION 2.2.1   *A **Schwartz distribution** α is a continuous linear functional on S. We write $\alpha \in S'$.*

Examples:

  1. If $f \in L^1$, then f induces a Schwartz distribution as follows:

$$S \ni \phi \longmapsto \int \phi f\, dx \in \mathbb{C}.$$

We see that this functional is continuous by noticing that

$$\left| \int \phi(x) f(x)\, dx \right| \le \sup |\phi| \cdot \|f\|_{L^1} = C \cdot \rho_{0,0}(\phi).$$

A similar argument shows that any finite Borel measure induces a distribution.

2. Differentiation is a distribution: On $\mathbb{R}^1$, for example, we have

$$S \ni \phi \mapsto \phi'(0)$$

satisfies

$$|\phi'(0)| \leq \sup_{x \in \mathbb{R}} |\phi'(x)| = \rho_{0,1}(\phi).$$

3. If $f \in L^p(\mathbb{R}^N)$, $1 \leq p \leq \infty$, then f induces a distribution:

$$T_f : S \ni \phi \mapsto \int \phi f \, dx \in \mathbb{C}.$$

To see that this functional is bounded, we first notice that

$$\left| \int \phi f \right| \leq \|f\|_{L^p} \cdot \|\phi\|_{L^{p'}}, \tag{2.2.2}$$

where $1/p + 1/p' = 1$. Now notice that

$$\left(1 + |x|^{N+1}\right) |\phi(x)| \leq C \left(\rho_{0,0}(\phi) + \rho_{N+1,0}(\phi)\right),$$

hence

$$|\phi(x)| \leq \frac{C}{1 + |x|^{N+1}} \left(\rho_{0,0}(\phi) + \rho_{N+1,0}(\phi)\right).$$

Finally,

$$\|\phi\|_{L^{p'}} \leq C \cdot \left[\int \left(\frac{1}{1+|x|^{N+1}}\right)^{p'} dx \right]^{1/p'} \cdot \left[\rho_{0,0}(\phi) + \rho_{N+1,0}(\phi)\right].$$

As a result, (2.2.2) tells us that

$$T_f(\phi) \leq C \|f\|_{L^p} \left(\rho_{0,0}(\phi) + \rho_{N+1,0}(\phi)\right).$$

Algebraic Properties of Distributions

(i) If $\alpha, \beta \in S'$ then $\alpha + \beta$ is defined by $(\alpha + \beta)(\phi) = \alpha(\phi) + \beta(\phi)$. Clearly $\alpha + \beta$ so defined is a Schwartz distribution.

(ii) If $\alpha \in S'$ and $c \in \mathbb{C}$ then $c\alpha$ is defined by $(c\alpha)(\phi) = c[\alpha(\phi)]$. We see that $c\alpha \in S'$.

(iii) If $\psi \in S$ and $\alpha \in S'$ then define $(\psi\alpha)(\phi) = \alpha(\psi\phi)$. It follows that $\psi\alpha$ is a distribution.

(iv) It is a theorem of Laurent Schwartz (see [SCH]) that there is no continuous operation of multiplication on S'. However, it is a matter of great interest, especially to mathematical physicists, to have such an operation.

Colombeau [CMB] has developed a substitute operation. We shall say no more about it here.

(v) Schwartz distributions may be differentiated as follows: If $\mu \in S'$ then $(\partial/\partial x)^\beta \mu \in S'$ is defined, for $\phi \in S$, by

$$\left[\left(\frac{\partial}{\partial x}\right)^\beta \mu\right](\phi) = (-1)^{|\beta|}\mu\left(\left(\frac{\partial}{\partial x}\right)^\beta \phi\right).$$

Observe that in case the distribution μ is induced by integration against a C_c^k function f, then the definition is compatible with what integration by parts would yield.

Let us differentiate the distribution induced by integration against the function $f(x) = |x|$ on $\mathbb{R}$. Now, for $\phi \in S$,

$$f'(\phi) \equiv -f(\phi')$$

$$= -\int_{-\infty}^{\infty} f\phi'\, dx$$

$$= -\int_0^{\infty} f(x)\phi'(x)\, dx - \int_{-\infty}^0 f(x)\phi'(x)\, dx$$

$$= -\int_0^{\infty} x\phi'(x)\, dx + \int_{-\infty}^0 x\phi'(x)\, dx$$

$$= -\left[x\phi(x)\right]_0^{\infty} + \int_0^{\infty} \phi(x)\, dx + \left[x\phi(x)\right]_{-\infty}^0 - \int_{-\infty}^0 \phi(x)\, dx$$

$$= \int_0^{\infty} \phi(x)\, dx - \int_{-\infty}^0 \phi(x)\, dx.$$

Thus f' consists of integration against $b(x) = -\chi_{(-\infty,0]} + \chi_{[0,\infty)}$. This function is often called the *Heaviside function*.

Exercise: Let $\Omega \subseteq \mathbb{R}^N$ be a smoothly bounded domain. Let ν be the unit outward normal vector field to $\partial\Omega$. Prove that $-\nu\chi_\Omega \in S'$. (Hint: Use Green's theorem. It will turn out that $(-\nu\chi_\Omega)(\phi) = \int_{\partial\Omega} \phi\, d\sigma$, where $d\sigma$ is area measure on the boundary.)

The Fourier Transform

The principal importance of the Schwartz distributions as opposed to other distribution theories (more on those below) is that they are well behaved under the Fourier transform. First we need a lemma:

LEMMA 2.2.3
If $f \in S$ then $\hat{f} \in S$.

PROOF   This is just an exercise with Propositions 2.1.2 and 2.1.3: the Fourier transform converts multiplication by monomials into differentiation and vice versa.   ∎

DEFINITION 2.2.4   If u is a Schwartz distribution, then we define a Schwartz distribution $\hat{u}$ by

$$\hat{u}(\phi) = u(\hat{\phi}).$$

By the lemma, the definition of $\hat{u}$ makes good sense. Moreover, by 2.2.5 below,

$$|\hat{u}(\phi)| = |u(\hat{\phi})| \leq \sum_{|\alpha|+|\beta| \leq M} \rho_{\alpha,\beta}(\hat{\phi})$$

for some $M > 0$ (by the definition of the topology on S). It is a straightforward exercise with 2.1.2 and 2.1.3 to see that the sum on the right is majorized by the sum

$$C \cdot \sum_{|\alpha|+|\beta| \leq M} \rho_{\alpha,\beta}(\phi).$$

In conclusion, the Fourier transform of a Schwartz distribution is also a Schwartz distribution.

Other Spaces of Distributions

Let $\mathcal{D} = C_c^\infty$ and $\mathcal{E} = C^\infty$. Clearly $\mathcal{D} \subseteq S \subseteq \mathcal{E}$. On each of the spaces $\mathcal{D}$ and $\mathcal{E}$ we use the semi-norms

$$\rho_{K,\alpha}(\phi) = \sup_K \left| \left(\frac{\partial}{\partial x} \right)^\alpha \phi \right|,$$

where $K \subseteq \mathbb{R}^N$ is a compact set and $\alpha = (\alpha_1, \dots, \alpha_N)$ is a multiindex. These induce a topology on $\mathcal{D}$ and $\mathcal{E}$ that turns them into topological vector spaces. The spaces $\mathcal{D}'$ and $\mathcal{E}'$ are defined to be the continuous linear functionals on $\mathcal{D}$ and $\mathcal{E}$ respectively. Trivially, $\mathcal{E}' \subseteq \mathcal{D}'$. The functional in $\mathbb{R}^1$ given by

$$\mu = \sum_{j=1}^\infty 2^j \delta_j,$$

where δ_j is the Dirac mass at j, is readily seen to be in $\mathcal{D}'$ but not in $\mathcal{E}'$.

The *support* of a distribution μ is defined to be the complement of the union of all open sets U such that $\mu(\phi) = 0$ for all elements of C_c^∞ that are supported in U. As an example, the support of the Dirac mass δ_0 is the origin: when δ_0 is applied to any testing function ϕ with support disjoint from 0 then the result is 0.

Exercise: Let $\mu \in \mathcal{D}'$. Then $\mu \in \mathcal{E}'$ if and only if μ has compact support. The elements of $\mathcal{E}'$ are sometimes referred to as the "compactly supported distributions."

PROPOSITION 2.2.5
A linear functional L on $\mathcal{S}$ is a Schwartz distribution (tempered distribution) if and only if there is a $C > 0$ and integers m and ℓ such that for all $\phi \in \mathcal{S}$ we have

$$|L(\phi)| \leq C \cdot \sum_{|\alpha| \leq \ell} \sum_{|\beta| \leq m} \rho_{\alpha,\beta}(\phi). \qquad (2.2.5.1)$$

SKETCH OF PROOF If an inequality like (2.2.5.1) holds, then clearly L is continuous.

For the converse, assume that L is continuous. Recall that a neighborhood basis of 0 in $\mathcal{S}$ is given by sets of the form

$$N_{\epsilon,\ell,m} = \left\{ \phi \in \mathcal{S} : \sum_{\substack{|\alpha| \leq \ell \\ |\beta| \leq m}} \rho_{\alpha,\beta}(\phi) < \epsilon \right\}.$$

Since L is continuous, the inverse image of an open set under L is open. Consider

$$L^{-1}\left(\{ z \in \mathbb{C} : |z| < 1 \} \right).$$

There exist ϵ, ℓ, m such that

$$N_{\epsilon,\ell,m} \subseteq L^{-1}\left(\{ z \in \mathbb{C} : |z| < 1 \} \right).$$

Thus

$$\sum_{\substack{|\alpha| \leq \ell \\ |\beta| \leq m}} \rho_{\alpha,\beta}(\phi) < \epsilon$$

implies that

$$|L(\phi)| < 1.$$

That is the required result, with $C = 1/\epsilon$. $\blacksquare$

Exercise: A similar result holds for $\mathcal{D}'$ and for $\mathcal{E}'$.

THEOREM 2.2.6   STRUCTURE THEOREM FOR $\mathcal{D}'$
If $u \in \mathcal{D}'$ then

$$u = \sum_{j=1}^{k} D^j \mu_j,$$

where μ_j is a finite Borel measure and each D^j is a differential monomial.

IDEA OF PROOF   For simplicity, restrict attention to $\mathbb{R}^1$. We know that the dual of the continuous functions with compact support is the space of finite Borel measures. In a natural fashion, the space of C^1 functions with compact support can be identified with a subspace of the set of ordered pairs of C_c functions: $f \leftrightarrow (f, f')$. Then every functional on C_c^1 extends, by the Hahn-Banach theorem, to a functional on $C_c \times C_c$. But such a functional will be given by a pair of measures. Combining this information with the definition of derivative of a distribution gives that an element of the dual of C_c^1 is of the form $\mu_1 + (\mu_2)'$. In a similar fashion, one can prove that an element of the dual of C_c^k must have the form $\mu_1 + (\mu_2)' + \cdots + (\mu_{k+1})^{(k)}$.

Finally, it is necessary to note that $\mathcal{D}'$ is nothing other than the countable union of the dual spaces $(C_c^k)'$.   ■

The theorem makes explicit the fact that an element of $\mathcal{D}'$ can depend on only finitely many derivatives of the testing function—that is, on finitely many of the norms $\rho_{\alpha,\beta}$.

We have already noted that the Schwartz distributions are the most convenient for Fourier transform theory. But the space $\mathcal{D}'$ is often more convenient in the theory of partial differential equations (because of the control on the support of testing functions). It will sometimes be necessary to pass back and forth between the two theories. In any given context, no confusion should result.

Exercise: Use the Paley–Wiener theorem (discussed in Section 4) or some other technique to prove that if $\phi \in \mathcal{D}$ then $\hat{\phi} \notin \mathcal{D}$. (This fact is often referred to as the Heisenberg uncertainty principle. In fact, it has a number of qualitative and quantitative formulations that are useful in quantum mechanics. See [FEG] for more on these matters.)

More on the Topology of $\mathcal{D}$ and $\mathcal{D}'$

We say that a sequence $\{\phi_j\} \subseteq \mathcal{D}$ converges to $\phi \in \mathcal{D}$ if

    1. all the functions ϕ_j have compact support in a single compact set K_0;

    2. $\rho_{K,\alpha}(\phi_j - \phi) \to 0$ for each compact set K and for every multiindex α.

The enemy here is the example of the "gliding hump": On $\mathbb{R}^1$, if ψ is a fixed C^∞ function and $\phi_j(x) \equiv \psi(x - j)$, then we do *not* want to say that the sequence $\{\phi_j\}$ converges to 0.

A functional μ on $\mathcal{D}$ is continuous if $\mu(\phi_j) \to \mu(\phi)$ whenever $\phi_j \to \phi$. This is equivalent to the already noted characterization that there exist a compact K and an $N > 0$ such that

$$|\mu(\phi)| \leq C \sum_{|\alpha| \leq N} \rho_{K,\alpha}(\phi)$$

for every testing function ϕ.

2.3 Convolution and Friedrichs Mollifiers

Recall that two integrable functions f and g are convolved as follows:

$$f * g = \int f(x - t)g(t)\, dt = \int g(x - t)f(t)\, dt.$$

In general, it is not possible to convolve two elements of $\mathcal{D}'$. However, we may successfully perform any of the following operations:

1. We may convolve an element $\mu \in \mathcal{D}'$ with an element $g \in D$.
2. We may convolve two distributions $\mu, \nu \in \mathcal{D}'$ *provided one of them is compactly supported.*
3. We may convolve $v_1, \ldots, v_k \in \mathcal{D}'$ *provided that all except possibly one is compactly supported.*

We shall now learn how to make sense of convolution. This is one of those topics in analysis (of which there are many) where understanding is best achieved by remembering the proof rather than the statements of the results.

DEFINITION 2.3.1 *We define the following convolutions:*

1. *If $\mu \in \mathcal{D}'$ and $g \in D$ then we define $(\mu * g)(\phi) = \mu(\tilde{g} * \phi)$, all $\phi \in D$.*
2. *If $\mu \in \mathcal{S}'$ and $g \in S$ then we define $(\mu * g)(\phi) = \mu(\tilde{g} * \phi)$, all $\phi \in S$.*
3. *If $\mu \in \mathcal{E}'$ and $g \in D$ then we define $(\mu * g)(\phi) = \mu(\tilde{g} * \phi)$, all $\phi \in \mathcal{E}$.*

Recall here that $\tilde{g}(x) \equiv g(-x)$. Observe in part (1) of the definition that $\tilde{g} * \phi \in D$, hence the definition makes sense. Similar remarks apply to parts (2) and (3) of the definition. In part (3), we must assume that $g \in \mathcal{D}$; otherwise $\tilde{g} * \phi$ does not necessarily make sense.

LEMMA 2.3.2
If $\alpha \in \mathcal{D}'$ and $g \in D$ then $\alpha * g$ is a function. What is more, if we let $T_h\phi(x) = \phi(x - h)$ then $(\alpha * g)(x) = \alpha(T_x\tilde{g})$.

PROOF   We calculate that

$$(\alpha * g)(\phi) = \alpha(\tilde{g} * \phi) = \alpha^x \left[\int \tilde{g}(x - t)\phi(t) \, dt \right].$$

Here the superscript on α denotes the variable in which α is acting. This last

$$= \alpha^x \left[\int (T_t\tilde{g})(x)\phi(t) \, dt \right] = \int \alpha^x (T_t\tilde{g})(x)\phi(t) \, dt. \quad \blacksquare$$

Next we introduce the concept of Friedrichs mollifiers. Let $\phi \in C_c^\infty$ be supported in the ball $B(0, 1)$. For convenience we assume that $\phi \geq 0$, although this is not crucial to the theory. Assume that $\int \phi(x) \, dx = 1$. Set $\phi_\epsilon(x) = \epsilon^{-N}\phi(x/\epsilon)$.

The family $\{\phi_\epsilon\}$ will be called a family of *Friedrichs mollifiers* in honor of K. O. Friedrichs. The use of such families to approximate a given function by smooth functions has become a pervasive technique in modern analysis. In functional analysis, such a family is sometimes called a weak approximation to the identity (for reasons that we are about to see). Observe that $\int \phi_\epsilon(x) \, dx = 1$ for every $\epsilon > 0$.

LEMMA 2.3.3
If $f \in L^p(\mathbf{R}^N), 1 \leq p \leq \infty$, then

$$\|f * \phi_\epsilon\|_{L^p} \leq \|f\|_{L^p}.$$

PROOF   The case $p = \infty$ is obvious, so we shall assume that $1 \leq p < \infty$. Then we may apply Jensen's inequality, with the unit mass measure $\phi_\epsilon(x) \, dx$, to see that

$$\|f * \phi_\epsilon\|_{L^p}^p = \int \left| \int f(x - t)\phi_\epsilon(t) \, dt \right|^p dx$$

$$\leq \int\int |f(x - t)|^p |\phi_\epsilon(t)| \, dt \, dx$$

$$= \int\int |f(x - t)|^p \, dx \, \phi_\epsilon(t) \, dt$$

$$= \int \|f\|_{L^p}^p \phi_\epsilon(t) \, dt$$

$$= \|f\|_{L^p}^p. \quad \blacksquare$$

REMARK   The function $f_\epsilon \equiv f * \phi_\epsilon$ is certainly C^∞ (just differentiate under the integral sign) but it is generally not compactly supported unless f is. $\blacksquare$

LEMMA 2.3.4
For $1 \le p < \infty$ we have

$$\lim_{\epsilon \to 0} \|f_\epsilon - f\|_{L^p} = 0.$$

PROOF   We will use the following claim: For $1 \le p < \infty$ we have

$$\lim_{h \to 0} \|T_h f - f\|_{L^p} = 0,$$

where $T_h f(x) = f(x - h)$. Assume the claim for the moment.
Now

$$
\begin{aligned}
\|f_\epsilon - f\|^p_{L^p} &= \left\| \int f(x - t)\phi_\epsilon(t)\, dt - \int f(x)\phi_\epsilon(t)\, dt \right\|^p_{L^p} \\
&= \left\| \int [f(x - t) - f(x)]\phi_\epsilon(t)\, dt \right\|^p_{L^p} \\
&= \left\| \int [T_t f(x) - f(x)]\phi_\epsilon(t)\, dt \right\|^p_{L^p} \\
&\le \int \|T_t f - f\|^p_{L^p} \phi_\epsilon(t)\, dt \\
&\stackrel{(t = \mu\epsilon)}{=} \int \|T_{\mu\epsilon} f - f\|^p_{L^p} \phi(\mu)\, d\mu.
\end{aligned}
$$

In the inequality here we have used Jensen's inequality. Now the claim and the Lebesgue dominated convergence theorem yield that $\|f_\epsilon - f\|_{L^p} \to 0$.

To prove the claim, first observe that if $\psi \in C_c$ then $\|T_h \psi - \psi\|_{\text{sup}} \to 0$ by uniform continuity. It follows that $\|T_h \psi - \psi\|_{L^p} \to 0$. Now if $f \in L^p$ is arbitrary and $\epsilon > 0$, then choose $\psi \in C_c$ such that $\|f - \psi\|_{L^p} < \epsilon/2$. Then

$$\limsup_{h \to 0} \|T_h f - f\|_{L^p} \le \limsup_{h \to 0} \|T_h(f - \psi)\|_{L^p} + \limsup_{h \to 0} \|T_h \psi - \psi\| \le \epsilon.$$

Since $\epsilon > 0$ was arbitrary, the claim follows.   ∎

LEMMA 2.3.5
If $f \in C_c$ then $f_\epsilon \to f$ uniformly.

PROOF   Let $\eta > 0$ and choose $\epsilon > 0$ such that if $|x - y| < \epsilon$ then $|f(x) - f(y)| < \eta$. Then

$$
\begin{aligned}
|f_\epsilon(x) - f(x)| &= \left| \int f(x - t)\phi_\epsilon(t)\, dt - f(x) \right| \\
&= \left| \int [f(x - t) - f(x)]\phi_\epsilon(t)\, dt \right|
\end{aligned}
$$

$$\leq \int_{|t|\leq\epsilon} |f(x-t) - f(x)||\phi_\epsilon(t)\,dt$$

$$\leq \int \eta\phi_\epsilon(t)\,dt$$

$$= \eta.$$

That completes the proof. ∎

Exercise: Is the last lemma true for a broader class of f? Prove that if $f \in C_c^k(\mathbf{R}^N)$, then $\|f_\epsilon - f\|_{C^k} \to 0$.

Now if $\alpha \in \mathcal{D}'$ and ϕ_ϵ is a family of Fredrichs mollifiers, then we define

$$\alpha_\epsilon(x) \equiv \alpha * \phi_\epsilon(x) = \alpha(\phi_\epsilon(x - \cdot)) = \alpha\left(T_x\tilde{\phi}_\epsilon\right).$$

LEMMA 2.3.6
Each α_ϵ is a C^∞ function. Moreover, $\alpha_\epsilon \to \alpha$ in the topology of $\mathcal{D}'$.

PROOF   For simplicity of notation, we restrict attention to dimension one. First let us see that α_ϵ is differentiable on $\mathbf{R}$. We calculate:

$$\frac{\alpha_\epsilon(x+h) - \alpha_\epsilon(x)}{h} = \frac{\alpha[\phi_\epsilon(x+h-\cdot)] - \alpha[\phi_\epsilon(x-\cdot)]}{h}$$

$$= \alpha\left(\frac{\phi_\epsilon(x+h-\cdot) - \phi_\epsilon(x-\cdot)}{h}\right). \qquad (2.3.6.1)$$

Observe that

$$\frac{\phi_\epsilon(x+h-\cdot) - \phi_\epsilon(x-\cdot)}{h} \to \phi_\epsilon'(x-\cdot)$$

in the topology of $\mathcal{D}$. Thus (2.3.6.1) implies that

$$\frac{\alpha_\epsilon(x+h) - \alpha_\epsilon(x)}{h} \to \alpha(\phi_\epsilon'(x-\cdot))$$

as $h \to 0$.

Now let us verify the convergence of α_ϵ to α. Fix a testing function $\psi \in \mathcal{D}$. Then

$$\alpha_\epsilon(\psi) = \int \alpha_\epsilon(x)\psi(x)\,dx = \int \alpha^s\left(\phi_\epsilon(x-s)\right)\psi(x)\,dx$$

$$= \alpha^s\left[\int \left(\phi_\epsilon(x-s)\right)\psi(x)\,dx\right] = \alpha^s\left[\int \left(\phi_\epsilon(x)\right)\psi(x+s)\,dx\right]$$

$$= \alpha^s\left[\int \phi_\epsilon(-x)\psi(s-x)\,dx\right]. \qquad (2.3.6.2)$$

Here a superscript on a distribution indicates the variable in which it is applied. We may assume that ϕ is even. Then, from (2.3.6.2),

$$\alpha_\epsilon(\psi) = \alpha(\psi_\epsilon). \qquad (2.3.6.3)$$

By the exercise preceding the lemma, $\psi_\epsilon \to \psi$ in every C^k norm. Therefore $\rho_{K,\alpha}(\psi_\epsilon - \psi) \to 0$ for every K and α. It follows that $\alpha_\epsilon(\psi) = \alpha(\psi_\epsilon) \to \alpha(\psi)$ as $\epsilon \to 0$. That completes the proof. ∎

PROPOSITION 2.3.7 SCHWARTZ

Let $A : C_c^\infty \to C^\infty$ be a linear, continuous operator that commutes with translations: $A(T_h(\phi)) = T_h(A(\phi))$. Then A is given by convolution with a distribution. That is, there is a distribution α such that

$$A(\phi) = \alpha * \phi$$

for all $\phi \in \mathcal{D}$.

PROOF Define α by

$$\alpha(\phi) = \left(A(\tilde{\phi})\right)(0).$$

Then, for all $\phi \in C_c^\infty$, we have

$$\alpha * \phi(x) = \alpha\left(T_x\tilde{\phi}\right) = \left(A\left(\widetilde{T_x\tilde{\phi}}\right)(0)\right)$$
$$= \left(A\left(T_{-x}\phi\right)\right)(0)$$
$$= \left(T_{-x}\left(A\phi\right)\right)(0)$$
$$= A\phi(x).$$

This is the desired result. ∎

As an application of the proposition, we will demonstrate that if $u, v \in \mathcal{D}'$ and if v has compact support, then we can define $u * v$ as a distribution. To see this, let $\phi \in \mathcal{D}$. Then

$$\phi \mapsto u * (v * \phi)$$

is a translation invariant operator from C_c^∞ to C^∞ that commutes with translations (notice that, because v is compactly supported, $v * \phi \in \mathcal{D}$). Thus this operator is given by convolution with a distribution α. We define $u * v = \alpha$.

We now assemble some remarks about when a distribution is a function and, more particularly, when it is a smooth function. First we note that if $\alpha \in \mathcal{D}'$ and α is compactly supported, then $\hat{\alpha}$ is a C^∞ function. Indeed,

$$\hat{\alpha}(\xi) = \alpha^x(e^{ix\cdot\xi}). \qquad (2.3.8)$$

To see this, let $\phi \in \mathcal{D}$. Then

$$\hat{\alpha}(\phi) \equiv \alpha(\hat{\phi}) = \alpha^{\xi}\left(\int \phi(x)e^{ix\cdot\xi}\,dx\right) = \int \phi(x)\alpha^{\xi}(e^{ix\cdot\xi})\,dx.$$

REMARK If $\alpha \in \mathcal{D}'$ is compactly supported and $\psi \in \mathcal{E}$, then we can define $\alpha(\psi)$ in the following manner: Let $\operatorname{supp}\alpha \subseteq K$ a compact set. Let $\Phi \in C_c^{\infty}$ be identically equal to 1 on K. Then we set $\alpha(\psi) = \alpha(\Phi \cdot \psi)$. ∎

Now let us use (2.3.8) to see that $\hat{\alpha}$ is C^{∞} when $\alpha \in \mathcal{D}'$ is compactly supported. For simplicity, we assume that the dimension is one. Then

$$\frac{d}{d\xi}\left(\hat{\alpha}(\xi)\right) = \lim_{h \to 0} \frac{\hat{\alpha}(\xi + h) - \hat{\alpha}(\xi)}{h}$$

$$= \lim_{h \to 0} \alpha^{x}\left[\frac{e^{ix\cdot(\xi+h)} - e^{ix\cdot\xi}}{h}\right]$$

$$= \alpha^{x}\left[ixe^{ix\xi}\right].$$

Notice that we may pass the limit inside the brackets because the Newton quotients converge in C^k for every k on the support of α. Thus we have shown that $\hat{\alpha}$ is differentiable. Iteration of this argument shows that $\hat{\alpha}$ is C^{∞}. We shall learn in the next section that, in fact, the Fourier transform of a compactly supported distribution is real analytic.

We conclude with a remark on how to identify a smooth distribution. The spirit of this remark will be a recurring theme throughout this book. We accomplish this identification by examining the decay of $\hat{\alpha}$ at infinity. Namely, let $\phi \in C_c^{\infty}$ be such that $\phi \equiv 1$ on a large compact set. We write $\hat{\alpha} = \phi\hat{\alpha}+(1-\phi)\hat{\alpha}$. Applying the inverse Fourier transform (denoted by $\check{}$), we see that

$$\alpha = (\phi\hat{\alpha})^{\vee} + \left((1 - \phi)\hat{\alpha}\right)^{\vee}.$$

Then, since $(\phi\hat{\alpha})$ is compactly supported, the first term is a C^{∞} function. We conclude that, in order to see whether α is C^{∞}, we must examine $((1 - \phi)\hat{\alpha})^{\vee}$. But this says, in effect, that we must examine the behavior of $\hat{\alpha}$ at infinity.

2.4   The Paley–Wiener Theorem

We begin by examining the so-called Fourier–Laplace transform. If $f \in C_c^{\infty}(\mathbb{R}^{N})$ and $\zeta = \xi + i\eta \in \mathbb{R}^{N} + i\mathbb{R}^{N} \approx \mathbb{C}^{N}$, then we define

$$\hat{f}(\zeta) = \int_{\mathbb{R}^{N}} f(x)e^{ix\cdot\zeta}\,dx.$$

More generally, if α is a compactly supported distribution, then we define its Fourier–Laplace transform to be

$$\hat{\alpha}^x(e^{ix\cdot\varsigma}).$$

Assume that the testing function f has support in the ball $B(0,A)$. Then

$$|\hat{f}(\varsigma)| = \left| \int_{\mathbf{R}^N} f(x) e^{ix\cdot\xi} e^{-x\cdot\eta}\, dx \right|$$

$$\leq \int_{\mathbf{R}^N} |f(x)| \left| e^{-x\cdot\eta} \right|\, dx$$

$$\leq \int_{\mathbf{R}^N} |f(x)| e^{A|\eta|}\, dx,$$

where we have used the support condition on f. Thus we see that

$$|\hat{f}(\varsigma)| \leq \|f\|_{L^1} e^{A|\operatorname{Im}\varsigma|}.$$

The Payley–Wiener theorem provides a converse to this estimate:

THEOREM 2.4.1
An entire function $U(\varsigma)$ is the Fourier–Laplace transform of a distribution with compact support in $B(0,A)$ if and only if there are positive constants C and K such that

$$|U(\varsigma)| \leq C \cdot (1 + |\varsigma|)^K e^{A|\operatorname{Im}\varsigma|}. \qquad (2.4.1.1)$$

Moreover, a distribution μ coincides with a function in $C_c^\infty(B(0,A))$ if and only if its Fourier–Laplace transform $U(\varsigma)$ is an entire analytic function and for every $K > 0$ there is a constant $C_K > 0$ such that

$$|U(\varsigma)| \leq C_K (1 + |\varsigma|)^{-K} e^{A|\operatorname{Im}\varsigma|}. \qquad (2.4.1.2)$$

PROOF First let us assume that $\alpha \in \mathcal{D}'$ with support in $B(0,A)$; we shall then prove that $U = \hat{\alpha}$ satisfies (2.4.1.1). Let $h \in C_c^\infty(\mathbf{R}^N)$ satisfy $h(t) \equiv 1$ when $|t| \leq 1/2$ and $h(t) \equiv 0$ when $|t| \geq 1$. For $\varsigma \in \mathbf{C}^N$ fixed and nonzero, we set

$$\phi_\varsigma(x) = e^{ix\cdot\varsigma} h\left(|\varsigma|(|x| - A)\right).$$

One checks that $\phi_\varsigma \in \mathcal{D}$ and $\phi_\varsigma(x) = e^{ix\cdot\varsigma}$ when $|x| \leq A + 1/(2|\varsigma|)$. Moreover, $\phi_\varsigma(x) = 0$ when $|\varsigma|(|x| - A) \geq 1$.
 Now for any $\phi \in \mathcal{D}$ we know that

$$|\alpha(\phi_\varsigma)| \leq C \sum_{|\beta| \leq K} \sup_{B(0,A)} |D^\beta \phi_\varsigma|$$

for some C, K. As a result,

$$|\hat{\alpha}(\zeta)| = |\alpha^x(e^{ix\cdot\zeta})| = |\alpha^x(\phi_\zeta(x))|$$

$$\leq C \sum_{|\beta|\leq K} \sup_{B(0,A)} |D^\beta \phi_\zeta|$$

$$\leq C(|\zeta|+1)^K \sup_{x\in B(0,A)} |e^{ix\cdot\zeta}|$$

$$\leq C(|\zeta|+1)^K e^{A|\text{Im}\,\zeta|}.$$

Next let us assume that α is a C_c^∞ function supported in $B(0,A)$. We shall prove (2.4.1.2). Now

$$|\hat{\alpha}(\zeta)| = \left|\int \alpha(x)e^{ix\cdot\zeta}\,dx\right|$$

$$\leq \int_{B(0,A)} |\alpha(x)|e^{A|\text{Im}\,\zeta|}\,dx$$

$$\leq \|\alpha\|_{L^1} e^{A|\text{Im}\,\zeta|}.$$

This is (2.4.1.2) with $K = 0$. To obtain (2.4.1.2) with $K = 1$ we write, for $\zeta_j \neq 0$,

$$|\hat{\alpha}(\zeta)| = \left|\int \alpha(x)e^{ix\cdot\zeta}\,dx\right|$$

$$= \left|\int \alpha(x)\frac{1}{i\zeta_j}\frac{\partial}{\partial x_j}e^{ix\cdot\zeta}\,dx\right|$$

$$= \left|\frac{1}{\zeta_j}\int \left(\frac{\partial}{\partial x_j}\alpha(x)\right)e^{ix\cdot\zeta}\,dx\right|$$

$$\leq \frac{1}{|\zeta_j|}\left\|\frac{\partial}{\partial x_j}\alpha\right\|_{L^1} e^{A|\text{Im}\,\zeta|}.$$

We can iterate this argument to obtain (2.4.1.2) for every K.

Next we prove that if (2.4.1.2) holds then α is a C_c^∞ function with support in $B(0,A)$. We write

$$|\alpha(x)| = \left|(2\pi)^{-N}\int U(\xi)e^{-ix\cdot\xi}\,d\xi\right|$$

$$= \left|(2\pi)^{-N}\int U(\xi+i\eta)e^{-ix\cdot(\xi+i\eta)}\,d\xi\right|.$$

If $\zeta = \xi + i\eta$ then from (2.4.1.2) we can write

$$|\alpha(x)| \leq C \int (1 + |\zeta|)^{-K} e^{A|\text{Im}\,\zeta|} e^{x \cdot \eta}\, d\xi$$

$$= Ce^{A|\eta| + x\eta} \int (1 + |\zeta|)^{-K}\, d\xi$$

$$\leq Ce^{A|\eta| + x \cdot \eta}$$

provided that $K \geq N + 1$.

Now for $x \in \mathbb{R}^N$ fixed we take $\eta = -tx$, where t is a positive real number. Then

$$|\alpha(x)| \leq Ce^{At|x| - t|x|^2} = Ce^{t|x|(A - |x|)}.$$

If $|x| > A$ then, as $t \to +\infty$, the right side of this inequality tends to 0. Thus $\alpha(x) = 0$. But this simply means that supp $\alpha \subseteq B(0, A)$.

We leave it as an exercise to prove that (2.4.1.1) implies that U is the Fourier–Laplace transform of a distribution supported in $B(0, A)$. ∎

3

Pseudodifferential Operators

3.1 Introduction to Pseudodifferential Operators

Consider the partial differential equation $\Delta u = f$. We wish to study the existence and regularity properties of solutions to this equation and equations like it. It turns out that, in practice, existence follows from a suitable *a priori* regularity estimate (to be defined below). Therefore we shall concentrate for now on regularity.

The *a priori* regularity problem is as follows: If $u \in C_c^\infty(\mathbb{R}^N)$ and if

$$\Delta u = f, \tag{3.1.1}$$

then how may we estimate u in terms of f? Taking the Fourier transform of both sides of (3.1.1) yields

$$(\Delta u)\widehat{\ } = \hat{f}$$

or

$$-\sum_j |\xi_j|^2 \hat{u}(\xi) = \hat{f}(\xi).$$

Arguing formally, we may solve this equation for $\hat{u}$:

$$\hat{u}(\xi) = -\frac{1}{|\xi|^2}\hat{f}(\xi). \tag{3.1.2}$$

Suppose for specificity that we are working in $\mathbb{R}^2$. Then $-1/|\xi|^2$ has a nonintegrable singularity and we find that equation (3.1.2) does not provide useful information.

The problem of studying existence and regularity for linear partial differential operators with constant coefficients was studied systematically in the 1950's by

Ehrenpreiss and Malgrange, among others. The approach of Ehrenpreiss was to write

$$u(x) = c \cdot \int \hat{u}(\xi) e^{-ix\cdot\xi} \, d\xi = \int -\frac{1}{|\xi|^2} \hat{f}(\xi) e^{-ix\cdot\xi} \, d\xi.$$

Using Cauchy theory, he was able to relate this last integral to

$$\int -\frac{1}{|\xi + i\eta|^2} \hat{f}(\xi + i\eta) e^{-ix\cdot(\xi+i\eta)} \, d\xi.$$

In this way he avoided the singularity at $\xi = 0$ of the right-hand side of (3.1.2).

Malgrange's method, by contrast, was to first study (3.1.1) for those f such that $\hat{f}$ vanishes to some finite order at 0 and then to apply some functional analysis.

We have already noticed that, for the study of C^∞ regularity, the behavior of the Fourier transform on the finite part of space is of no interest. Thus the philosophy of pseudodifferential operator theory is to replace the Fourier multiplier $1/|\xi|^2$ by the multiplier $(1 - \phi(\xi))/|\xi|^2$, where $\phi \in C_c^\infty(\mathbb{R}^N)$ is identically equal to 1 near the origin. Thus we define

$$(\widehat{Pg})(\xi) = -\frac{1 - \phi(\xi)}{|\xi|^2} \hat{g}(\xi)$$

for any $g \in C_c^\infty$. Equivalently,

$$Pg = \left(-\frac{1 - \phi(\xi)}{|\xi|^2} \hat{g}(\xi) \right)^{\vee}.$$

Now we look at $u - Pf$, where f is the function on the right of (3.1.1):

$$\begin{aligned}(\widehat{u - Pf}) &= \hat{u} - \widehat{Pf} \\ &= -\frac{1}{|\xi|^2} \hat{f} + \frac{1 - \phi(\xi)}{|\xi|^2} \hat{f} \\ &= -\frac{\phi(\xi)}{|\xi|^2} \hat{f}.\end{aligned}$$

Then $u - Pf$ is a distribution whose Fourier transform has compact support, that is, $u - Pf$ is C^∞. So studying the regularity of u is equivalent to studying the regularity of Pf. This is precisely what we mean when we say that P is a parametrix for the partial differential operator Δ. And the point is that P has symbol $-(1 - \phi)/|\xi|^2$, which is free of singularities.

Now let L be a partial differential operator with (smooth) variable coefficients:

$$L = \sum_\alpha a_\alpha(x) \left(\frac{\partial}{\partial x} \right)^\alpha.$$

The classical approach to studying such operators was to reduce to the constant coefficient case by "freezing coefficients": Fix a point $x_0 \in \mathbb{R}^N$ and write

$$L = \sum_\alpha a_\alpha(x_0) \left(\frac{\partial}{\partial x}\right)^\alpha + \sum_\alpha \left(a_\alpha(x) - a_\alpha(x_0)\right) \left(\frac{\partial}{\partial x}\right)^\alpha.$$

For a reasonable class of operators (elliptic), the second term turns out to be negligible because it has small coefficients. The principal term, the first, has constant coefficients.

The idea of freezing the coefficients is closely related to the idea of passing to the symbol of the operator L. We set

$$\ell(x,\xi) = \sum_\alpha a_\alpha(x)(-i\xi)^\alpha.$$

(This definition of symbol is slightly different from that in Section 1.5 because we need to make peace with the Fourier transform.) The motivation is that if $\phi \in \mathcal{D}$ and *if L has constant coefficients* then

$$L\phi = c \int e^{-ix\cdot\xi} \ell\hat{\phi} \, d\xi.$$

However, even in the variable coefficient case we might hope that a **parametrix** for L is given by

$$\phi \longmapsto \left(\frac{1}{\ell(x,\xi)}\hat{\phi}\right)^{\smile}.$$

Assume for simplicity that $\ell(x,\xi)$ vanishes only at $\xi = 0$ (in fact, this is exactly what happens in the elliptic case). Let $\Phi \in C_c^\infty$ satisfy $\Phi(\xi) \equiv 1$ when $|\xi| \leq 1$ and $\Phi(\xi) \equiv 0$ when $|\xi| \geq 2$. Set

$$m(x,\xi) = \left(1 - \Phi(\xi)\right) \frac{1}{\ell(x,\xi)}.$$

We hope that m, acting as a Fourier multiplier by

$$T : f \longmapsto \left((1 - \Phi(\xi)) \frac{1}{\ell(x,\xi)}\hat{f}(x,\xi)\right)^{\smile},$$

gives an approximate right inverse for L. More precisely, we hope that equations of the following form hold:

$$T \circ L = \text{id} + (\text{negligible error term})$$

$$L \circ T = \text{id} + (\text{negligible error term}).$$

In the constant coefficient case, composition of operators corresponds to multiplication of symbols so that we would have

$$((L \circ T)f)\widehat{} = \ell(\xi) \cdot \left(\frac{1 - \Phi(\xi)}{\ell(\xi)}\right) \hat{f}(\xi)$$

$$= (1 - \Phi(\xi))\hat{f}(\xi).$$

In the variable coefficient case, we hope for an equation such as this with the addition of an error.

A calculus of pseudodifferential operators is a collection of integral operators that contains all elliptic partial differential operators and their parametrices and such that the collection is closed under composition and the taking of adjoints. Once the calculus is in place, then, when one is given a partial or pseudodifferential operator, one can instantly write down a parametrix and obtain estimates. Pioneers in the development of pseudodifferential operators were Mikhlin ([MIK1], [MIK2]) and Calderón and Zygmund [CZ2].

One of the classical approaches to developing a calculus of operators finds it roots in the work of Hadamard [HAD] and Riesz [RIE] and Calderón and Zygmund [CZ1]. To explain this approach, we introduce two types of integral operators.

The first are based on the classical Calderón–Zygmund singular integral kernels. Such a kernel is defined to be a function of the form

$$K(x) = \frac{\Omega(x)}{|x|^N},$$

where Ω is a smooth function on $\mathbb{R}^N \setminus \{0\}$ that is homogeneous of degree zero (i.e., $\Omega(\lambda x) = \Omega(x)$ for all $\lambda > 0$). Then it can be shown (see [STSI]) that the Cauchy principal value integral

$$T_K f(x) \equiv \lim_{\epsilon \to 0^+} \int_{|t| > \epsilon} f(x - t) K(t)\, dt$$

converges for almost every x when $f \in L^p$ and that T is a bounded operator from L^p to $L^p, 1 < p < \infty$.

The second type of operator is called a Riesz potential. The Riesz potential of order α has kernel

$$k_\alpha(x) \equiv c_{N,\alpha} \cdot \frac{1}{|x|^{N-\alpha}}, \qquad 0 < \alpha < N,$$

where $c_{N,\alpha}$ is a positive constant that will be of no interest here. The Riesz potentials are sometimes called *fractional integration operators* because the Fourier multiplier corresponding to k_α is $c'_{N,\alpha}|\xi|^{-\alpha}$. If we think about the fact that multiplication on the Fourier transform side by $(-|\xi|^2)^\alpha, \alpha > 0$, corresponds to applying a power of the Laplacian—that is, it corresponds to differentiation—

then it is reasonable that a Fourier multiplier $|\xi|^\beta$ with $\beta < 0$ should correspond to integration of some order.

Now the classical idea of creating a calculus is to consider the smallest algebra generated by the singular integral operators and the Riesz potentials. Unfortunately, it is not the case that the composition of two singular integrals is a singular integral, nor is it the case that the composition of a singular integral and a fractional integral is (in any simple fashion) an operator of one of the component types. Thus, while this calculus could be used to solve some problems, it is rather clumsy.

Here is a second, and rather old, attempt at a calculus of pseudodifferential operators:

DEFINITION 3.1.1 *A function $p(x,\xi)$ is said to be a **symbol** of order m if p is C^∞, has compact support in the x variable, and is homogeneous of degree m in ξ when ξ is large. That is, we assume that there is an $M > 0$ such that if $|\xi| > M$ and $\lambda > 1$ then*

$$p(x, \lambda\xi) = \lambda^m p(x, \xi).$$

It is possible to show that symbols so defined, and the corresponding operators

$$T_p f \equiv \int \hat{f}(\xi) p(x, \xi) e^{-ix \cdot \xi} \, d\xi,$$

form an algebra in a suitable sense. These may be used to study elliptic operators effectively.

But the definition of symbol that we have just given is needlessly restrictive. For instance, the symbol of even a constant coefficient partial differential operator is not generally homogeneous and we would have to deal with only the top order terms. It was realized in the mid-1960s that homogeneity was superfluous to the intended applications. The correct point of view is to control the decay of derivatives of the symbol at infinity. In the next section we shall introduce the Kohn–Nirenberg approach to pseudodifferential operators.

3.2   A Formal Treatment of Pseudodifferential Operators

Now we give a careful treatment of an algebra of pseudodifferential operators. We begin with the definition of the symbol classes.

DEFINITION 3.2.1   KOHN–NIRENBERG [KON1] *Let $m \in \mathbb{R}$. We say that a smooth function $\sigma(x, \xi)$ on $\mathbb{R}^N \times \mathbb{R}^N$ is a **symbol** of order m if there is a compact set $K \subseteq \mathbb{R}^N$ such that $\operatorname{supp} \sigma \subseteq K \times \mathbb{R}^N$ and, for any pair of multiindices α, β,* *there is a constant $C_{\alpha,\beta}$ such that*

$$\left| D_\xi^\alpha D_x^\beta \sigma(x,\xi) \right| \le C_{\alpha,\beta} \left(1 + |\xi| \right)^{m-|\alpha|}. \tag{3.2.1.1}$$

We write $\sigma \in S^m$.

As a simple example, if $\Phi \in C_c^\infty(\mathbb{R}^N)$, $\Phi \equiv 1$ near the origin, define

$$\sigma(x,\xi) = \Phi(x)(1 - \Phi(\xi))(1 + |\xi|^2)^{m/2}.$$

Then σ is a symbol of order m. We leave it as an exercise for the reader to verify condition (3.2.1.1).

For our purposes, namely the local boundary regularity of the Dirichlet problem, the Kohn–Nirenberg calculus will be sufficient. We shall study this calculus in detail. However, we should mention that there are several more general calculi that have become important. Perhaps the most commonly used calculus is the Hörmander calculus [HOR2]. Its symbols are defined as follows:

DEFINITION 3.2.2 *Let $m \in \mathbb{R}$ and $0 \le \rho,\ \delta \le 1$. We say that a smooth function $\sigma(x,\xi)$ lies in the symbol class $S_{\rho,\delta}^m$ if*

$$\left| D_\xi^\alpha D_x^\beta \sigma(x,\xi) \right| \le C_{\alpha,\beta}(1 + |\xi|)^{m-\rho|\alpha|+\delta|\beta|}.$$

The Kohn–Nirenberg symbols are special cases of the Hörmander symbols with $\rho = 1$ and $\delta = 0$ and with the added convenience of restricting the x support to be compact. Hörmander's calculus is important for the study of the $\bar\partial$-Neumann problem (treated in our Chapter 8). In that context symbols of class $S_{1/2,1/2}^1$ arise naturally.

Even more general classes of operators, which are spatially inhomogenous and nonisotropic in the phase variable ξ, have been developed. Basic references are [BEF2], [BEA1], and [HOR5]. Pseudodifferential operators with "rough symbols" have been studied by Meyer [MEY] and others.

The significance of the index m in the notation S^m is that it tells us how the corresponding pseudodifferential operator acts on certain function spaces. While one may formulate results for C^k spaces, Lipschitz spaces, and other classes of functions, we find it most convenient at first to work with the Sobolev spaces.

DEFINITION 3.2.3 *If $\phi \in \mathcal{D}$ then we define the norm*

$$\|\phi\|_{H^s} = \|\phi\|_s \equiv \left(\int |\hat\phi(\xi)|^2 \left(1 + |\xi|^2 \right)^s d\xi \right)^{1/2}.$$

We let $H^s(\mathbb{R}^N)$ be the closure of $\mathcal{D}$ with respect to $\|\ \ \|_s$.

In the case that s is a nonnegative integer,

$$(1 + |\xi|^{2s}) \approx (1 + |\xi|^2)^s \approx \sum_{|\alpha| \leq 2s} |\xi|^\alpha \approx \left(\sum_{|\alpha| \leq s} |\xi|^\alpha \right)^2$$

for $|\xi|$ large. Therefore

$$\phi \in H^s \quad \text{if and only if} \quad \hat{\phi} \cdot \left(\sum_{|\alpha| \leq s} |\xi|^\alpha \right) \in L^2.$$

This last condition means that $\hat{\phi}\xi^\alpha \in L^2$ for all multiindices α with $|\alpha| \leq s$. That is,

$$\left(\frac{\partial}{\partial x} \right)^\alpha \phi \in L^2 \quad \forall \alpha \text{ such that } |\alpha| \leq s.$$

Thus we have

PROPOSITION 3.2.4
If s is a nonnegative integer then

$$H^s = \left\{ f \in L^2 : \frac{\partial^\alpha}{\partial x^\alpha} f \in L^2 \text{ for all } \alpha \text{ with } |\alpha| \leq s \right\}.$$

Here derivatives are interpreted in the sense of distributions.

Notice in passing that if $s > r$ then $H^s \subseteq H^r$ because

$$|\hat{\phi}|^2 (1 + |\xi|^2)^r \leq |\hat{\phi}|^2 (1 + |\xi|^2)^s.$$

The Sobolev spaces turn out to be easy to work with because they are modeled on L^2—indeed, each H^s is canonically isomorphic as a Hilbert space to L^2 (exercise). But they are important because they can be related to more classical spaces of smooth functions. That is the content of the Sobolev imbedding theorem:

THEOREM 3.2.5   SOBOLEV
Let $s > N/2$. If $f \in H^s(\mathbf{R}^N)$, then f can be corrected on a set of measure zero to be continuous.
  More generally, if $k \in \{0, 1, 2, \ldots\}$ and if $f \in H^s$, $s > N/2 + k$, then f can be corrected on a set of measure zero to be C^k.

PROOF   For the first part of the theorem, let $f \in H^s$. By definition, there exist $\phi_j \in \mathcal{D}$ such that $\|\phi_j - f\|_{H^s} \to 0$. Then

$$\|\phi_j - f\|_{L^2} = \|\phi_j - f\|_0 \leq \|\phi_j - f\|_s^2 \to 0. \tag{3.2.5.1}$$

Our plan is to show that $\{\phi_j\}$ is an equibounded, equicontinuous family of functions. Then the Ascoli–Arzelá theorem [RUD1] will imply that there is a subsequence converging uniformly on compact sets to a (continuous) function g. But (3.2.5.1) guarantees that a subsequence of this subsequence converges pointwise to the function f. So $f = g$ almost everywhere and the required assertion follows.

To see that $\{\phi_j\}$ is equibounded, we calculate that

$$|\phi_j(x)| = c \cdot \left| \int e^{-ix\cdot\xi} \hat{\phi}_j(\xi) \, d\xi \right|$$

$$\leq c \cdot \int |\hat{\phi}_j(\xi)| (1 + |\xi|^2)^{s/2} (1 + |\xi|^2)^{-s/2} \, d\xi$$

$$\leq c \cdot \left(\int |\hat{\phi}_j(\xi)|^2 (1 + |\xi|^2)^s \, d\xi \right)^{1/2} \cdot \left(\int (1 + |\xi|^2)^{-s} \, d\xi \right)^{1/2}.$$

Using polar coordinates, we may see easily that, for $s > N/2$,

$$\int (1 + |\xi|^2)^{-s} \, d\xi < \infty.$$

Therefore

$$|\phi_j(x)| \leq C \|\phi_j\|_{H^s} \leq C'$$

and $\{\phi_j\}$ is equibounded.

To see that $\{\phi_j\}$ is equicontinuous, we write

$$|\phi_j(x) - \phi_j(y)| = c \left| \int \hat{\phi}_j(\xi) \left(e^{-ix\cdot\xi} - e^{-iy\cdot\xi} \right) d\xi \right|.$$

Observe that $|e^{-ix\cdot\xi} - e^{-iy\cdot\xi}| \leq 2$ and, by the mean value theorem,

$$|e^{-ix\cdot\xi} - e^{-iy\cdot\xi}| \leq |x - y| \, |\xi|.$$

Then, for any $0 < \epsilon < 1$,

$$|e^{-ix\cdot\xi} - e^{-iy\cdot\xi}| = |e^{-ix\cdot\xi} - e^{-iy\cdot\xi}|^{1-\epsilon} |e^{-ix\cdot\xi} - e^{-iy\cdot\xi}|^{\epsilon} \leq 2^{1-\epsilon} |x - y|^{\epsilon} |\xi|^{\epsilon}.$$

Therefore

$$|\phi_j(x) - \phi_j(y)| \leq C \int |\hat{\phi}_j(\xi)| |x - y|^{\epsilon} |\xi|^{\epsilon} \, d\xi$$

$$\leq C |x - y|^{\epsilon} \int |\hat{\phi}_j(\xi)| (1 + |\xi|^2)^{\epsilon/2} \, d\xi$$

$$\leq C |x - y|^{\epsilon} \|\phi_j\|_{H^s} \left(\int (1 + |\xi|^2)^{-s+\epsilon} \, d\xi \right)^{1/2}.$$

If we select $0 < \epsilon < 1$ such that $-s + \epsilon < -N/2$, then we find that $\int (1 + |\xi|^2)^{-s+\epsilon} d\xi$ is finite. It follows that the sequence $\{\phi_j\}$ is equicontinuous and we are done.

The second assertion may be derived from the first by a simple inductive argument. We leave the details as an exercise. ∎

REMARKS

1. If $s = N/2$, then the first part of the theorem is false (exercise).
2. The theorem may be interpreted as saying that $H^s \subseteq C^k$ for $s > k + N/2$. In other words, the identity provides a continuous imbedding of H^s into C^k. A converse is also true. Namely, if $H^s \subseteq C^k$ for some nonnegative integer k then $s > k + N/2$.

To see this, notice that the hypotheses $u_j \to u$ in H^s and $u_j \to v$ in C^k imply that $u = v$. Therefore the inclusion of H^s into C^k is a closed map. It is therefore continuous by the closed graph theorem. Thus there is a constant C such that

$$\|f\|_{C^k} \leq C \|f\|_{H^s}.$$

For $x \in \mathbf{R}^N$ fixed and α a multiindex with $|\alpha| \leq k$, the tempered distribution e_x^α defined by

$$e_x^\alpha(\phi) = \left(\frac{\partial^\alpha}{\partial x^\alpha} \right) \phi(x)$$

is bounded in $(C^k)^*$ with bound independent of x and α (but depending on k). Hence $\{e_x^\alpha\}$ form a bounded set in $(H^s)^* \equiv H^{-s}$ (this point is discussed in detail in Lemma 3.2.9 below). As a result, for $|\alpha| \leq k$ we have that

$$\|e_x^\alpha\|_{H^{-s}} = \left(\int |(\widehat{e_x^\alpha})(\xi)|^2 (1 + |\xi|^2)^{-s} d\xi \right)^{1/2}$$

$$= \left(\int |(-i\xi)^\alpha e^{ix\cdot\xi}|^2 (1 + |\xi|^2)^{-s} d\xi \right)^{1/2}$$

$$\leq C \left(\int (1 + |\xi|^2)^{-s+|\alpha|} d\xi \right)^{1/2}$$

is finite, independent of x and α. But this can only happen if $2(k - s) < -N$, that is, if $s > k + N/2$. ∎

Exercise: Imitate the proof of the Sobolev theorem to prove Rellich's lemma: If $s > r$, then the inclusion map $i : H^s \to H^r$ is a compact operator.

THEOREM 3.2.6
Let $p \in S^m$ and define the associated pseudodifferential operator $P = Op(p) = T_p$ by

$$P(\phi) = \int \hat{\phi}(\xi) p(x, \xi) e^{-ix \cdot \xi} \, d\xi.$$

Then

$$P : H^s \to H^{s-m}$$

continuously.

REMARKS   Notice that if $m > 0$, then we lose smoothness under P. Likewise, if $m < 0$ then P is essentially a fractional integration operator and we gain smoothness. We say that the pseudodifferential operator T_p has *order* m precisely when its symbol is of order m.

Observe also that in the constant coefficient case (which is misleadingly simple), we would have $p(x, \xi) = p(\xi)$ and the proof of the theorem would be as follows:

$$\|P(\phi)\|_{s-m}^2 = \int \left| (\widehat{P(\phi)})(\xi) \right|^2 (1 + |\xi|^2)^{s-m} \, d\xi$$

$$= \int |p(\xi) \hat{\phi}(\xi)|^2 (1 + |\xi|^2)^{s-m} \, d\xi$$

$$\leq c \int |\hat{\phi}(\xi)|^2 (1 + |\xi|^2)^s \, d\xi$$

$$= c\|\phi\|_s^2. \qquad \blacksquare$$

To prove the theorem in full generality is more difficult. We shall break it up into several lemmas.

LEMMA 3.2.7
For any complex numbers a, b we have

$$\frac{1 + |a|}{1 + |b|} \leq 1 + |a - b|.$$

PROOF   We have

$$1 + |a| \leq 1 + |a - b| + |b|$$

$$\leq 1 + |a - b| + |b| + |b| \, |a - b|$$

$$= (1 + |a - b|)(1 + |b|). \qquad \blacksquare$$

LEMMA 3.2.8
If $p \in S^m$ then, for any multiindex α and integer $k > 0$, we have

$$\left| \mathcal{F}_x \left(D_x^\alpha p(x, \xi) \right) (\eta) \right| \leq C_{k,\alpha} \frac{(1 + |\xi|)^m}{(1 + |\eta|)^k} \,.$$

Here $\mathcal{F}_x$ denotes the Fourier transform in the x variable.

PROOF If α is any multiindex and γ is any multiindex such that $|\gamma| = k$, then

$$|\eta^\gamma| \left| \mathcal{F}_x \left(D_x^\alpha p(x, \xi) \right) \right| = \left| \mathcal{F}_x \left(D_x^\gamma D_x^\alpha p(x, \xi) \right) (\eta) \right|$$
$$\leq \| D_x^{\alpha + \gamma} p(x, \xi) \|_{L^1(x)} \leq C_{k,\alpha} \cdot (1 + |\xi|)^m \,.$$

As a result,

$$\left(|\eta^\gamma| + 1 \right) \left| \mathcal{F}_x \left(D_x^\alpha p(x, \xi) \right) \right| \leq \left(C_{0,\alpha} + C_{k,\alpha} \right) \cdot (1 + |\xi|)^m \,.$$

This is what we wished to prove. ∎

LEMMA 3.2.9
We have that

$$\left(H^s \right)^* = H^{-s}.$$

PROOF   Observe that

$$H^s = \left\{ g : \hat{g} \in L^2 \left((1 + |\xi|^2)^s \, d\xi \right) \right\} \,.$$

But then H^s and H^{-s} are clearly dual to each other by way of the pairing

$$\langle f, g \rangle = \int \hat{f}(\xi) \hat{g}(\xi) \, d\xi. \qquad\qquad ∎$$

The upshot of the last lemma is that, in order to estimate the H^s norm of a function (or Schwartz distribution) ϕ, it is enough to prove an inequality of the form

$$\left| \int \phi(x) \bar{\psi}(x) \, dx \right| \leq C \|\psi\|_{H^{-s}}$$

for every $\psi \in \mathcal{D}$.

PROOF OF THEOREM 3.2.6   Fix $\phi \in \mathcal{D}$. Let $p \in S^m$ and let $P = \text{Op}(p)$. Then

$$P\phi(x) = \int e^{-ix \cdot \xi} p(x, \xi) \hat{\phi}(\xi) \, d\xi.$$

Define

$$\hat{S}_x(\lambda, \xi) = \int e^{ix \cdot \lambda} p(x, \xi) \, dx.$$

This function is well defined since p is compactly supported in x. Then

$$\widehat{P\phi}(\eta) = \int\int e^{-ix \cdot \xi} p(x, \xi) \hat{\phi}(\xi) \, d\xi \, e^{i\eta \cdot x} \, dx$$

$$= \int\int p(x, \xi) \hat{\phi}(\xi) e^{ix \cdot (\eta - \xi)} \, dx \, d\xi$$

$$= \int \hat{S}_x(\eta - \xi, \xi) \hat{\phi}(\xi) \, d\xi.$$

We want to estimate $\|P\phi\|_{s-m}$. By the remarks following Lemma 3.2.9, it is enough to show that, for $\psi \in \mathcal{D}$,

$$\left| \int P\phi(x) \bar{\psi}(x) \, dx \right| \leq C \|\phi\|_{H^s} \|\psi\|_{H^{m-s}}.$$

We have

$$\left| \int P\phi(x) \bar{\psi}(x) \, dx \right| = \left| \int \widehat{P\phi}(\xi) \hat{\bar{\psi}}(\xi) \, d\xi \right|$$

$$= \left| \int \left(\int \hat{S}_x(\xi - \eta, \eta) \hat{\phi}(\eta) \, d\eta \right) \hat{\bar{\psi}}(\xi) \, d\xi \right|$$

$$= \int\int \hat{S}_x(\xi - \eta, \eta)(1 + |\eta|)^{-s}(1 + |\xi|)^{s-m}$$

$$\times \hat{\bar{\psi}}(\xi)(1 + |\xi|)^{m-s} \hat{\phi}(\eta)(1 + |\eta|)^s \, d\eta \, d\xi.$$

Define

$$K(\xi, \eta) = \left| \hat{S}_x(\xi - \eta, \eta)(1 + |\eta|)^{-s}(1 + |\xi|)^{s-m} \right|.$$

We claim that

$$\int |K(\xi, \eta)| \, d\xi \leq C$$

and

$$\int |K(\xi, \eta)| \, d\eta \leq C.$$

Assume the claim for the moment. Then

$$\left| \int P\phi(x)\bar{\psi}(x)\,dx \right| \leq \iint K(\xi,\eta)|\hat{\bar{\psi}}(\xi)|(1+|\xi|)^{m-s}|\hat{\phi}(\eta)|(1+|\eta|)^s\,d\eta\,d\xi$$

$$\leq C\left(\iint K(\xi,\eta)(1+|\xi|^2)^{m-s}|\hat{\bar{\psi}}(\xi)|^2\,d\xi\,d\eta \right)^{1/2}$$

$$\times \left(\iint K(\xi,\eta)(1+|\eta|^2)^s|\hat{\phi}(\eta)|^2\,d\xi\,d\eta \right)^{1/2}$$

$$\leq C\left(\int |\hat{\bar{\psi}}(\xi)|^2(1+|\xi|^2)^{m-s}\,d\xi \right)^{1/2}$$

$$\times \left(\int |\hat{\phi}(\eta)|^2(1+|\eta|^2)^s\,d\eta \right)^{1/2}$$

$$= C\|\psi\|_{H^{m-s}} \cdot \|\phi\|_{H^s}.$$

That is the desired estimate. It remains to prove the claim.
By Lemma 3.2.8 we know that

$$\left| \hat{S}_x(\zeta,\xi) \right| \leq C_k(1+|\xi|)^m \cdot (1+|\zeta|)^{-k}.$$

But now, by Lemma 3.2.7, we have

$$|K(\xi,\eta)| \equiv \left| \hat{S}_x(\xi-\eta,\eta)(1+|\eta|)^{-s}(1+|\xi|)^{s-m} \right|$$

$$\leq C_k(1+|\eta|)^m(1+|\xi-\eta|)^{-k}(1+|\eta|)^{-s}(1+|\xi|)^{s-m}$$

$$= C_k \left(\frac{1+|\eta|}{1+|\xi|} \right)^{m-s} \cdot (1+|\xi-\eta|)^{-k}$$

$$\leq C_k(1+|\xi-\eta|)^{m-s}(1+|\xi-\eta|)^{-k}.$$

We may specify k as we please, so we choose it so large that $m-s-k \leq -N-1$.
Then the claim is obvious and the theorem is proved. ∎

3.3   The Calculus of Pseudodifferential Operators

The three central facts about our pseudodifferential operators are these:

1. If $p \in S^m$ then $T_p : H^s \to H^{s-m}$.

2. If $p \in S^m$ then $(T_p)^*$ is "essentially" $T_{\bar{p}}$. In particular, the symbol of $(T_p)^*$ lies in S^m.

3. If $p \in S^m, q \in S^n$, then $T_p \circ T_q$ is "essentially" T_{pq}. In particular, the symbol of $T_p \circ T_q$ lies in S^{m+n}.

We have already proved (1); in this section we shall give precise formulations to (2) and (3) and we shall prove them.

We begin with (2), and for motivation consider a simple example. Let $A = a(x)(\partial/\partial x_1)$. Let us calculate A^*. If $\phi, \psi \in \mathcal{D}$ then

$$\langle A^*\phi, \psi \rangle_{L^2} = \langle \phi, A\psi \rangle_{L^2}$$

$$= \int \phi(x) \left(\overline{a(x)\frac{\partial \psi}{\partial x_1}(x)} \right) dx$$

$$= -\int \frac{\partial}{\partial x_1} \left(\bar{a}(x)\phi(x) \right) \bar{\psi}(x)\, dx$$

$$= \int \left(-\bar{a}(x)\frac{\partial}{\partial x_1} - \frac{\partial \bar{a}}{\partial x_1}(x) \right) \phi(x) \cdot \bar{\psi}(x)\, dx.$$

Then

$$A^* = -\bar{a}(x)\frac{\partial}{\partial x_1} - \frac{\partial \bar{a}}{\partial x_1}(x)$$

$$= \mathrm{Op}\left(-\bar{a}(x)(-i\xi_1) - \frac{\partial \bar{a}}{\partial x_1}(x) \right)$$

$$= \mathrm{Op}\left(i\xi_1 \bar{a}(x) - \frac{\partial \bar{a}}{\partial x_1}(x) \right).$$

Thus we see in this example that the "principal part" of the adjoint operator (that is, the term with the highest degree monomial in ξ of the symbol of A^*) is $i\xi_1 \bar{a}(x)$, and this is just the conjugate of the symbol of A.

In general it turns out that the symbol of A^* for a general pseudodifferential operator A is given by the asymptotic expansion

$$\sum_\alpha D_x^\alpha \left(\frac{\partial}{\partial \xi} \right)^\alpha \overline{\sigma(A)} \frac{1}{\alpha!} .$$

Here $D_x^\alpha \equiv (i\partial/\partial x)^\alpha$. We shall learn more about asymptotic expansions later. The basic idea of an asymptotic expansion is that, in a given application, the asymptotic expansion may be written in more precise form as

$$\sum_{|\alpha| \leq k} D_x^\alpha \left(\frac{\partial}{\partial \xi}\right)^\alpha \overline{\sigma(A)} \frac{1}{\alpha!} + \mathcal{E}_k.$$

One selects k so large that the error term $\mathcal{E}_k$ is negligible.

If we apply this asymptotic expansion to the operator $a(x)\partial/\partial x_1$ that was just considered, it yields that

$$\sigma(A^*) = i\xi_1 \bar{a}(x) - \frac{\partial \bar{a}}{\partial x_1}(x),$$

which is just what we calculated by hand.

Now let us look at an example to motivate how compositions of pseudodifferential operators will behave. Let the dimension N be 1 and let

$$A = a(x)\frac{d}{dx} \qquad \text{and} \qquad B = b(x)\frac{d}{dx}.$$

Then $\sigma(A) = a(x)(-i\xi)$ and $\sigma(B) = b(x)(-i\xi)$. Moreover, if $\phi \in \mathcal{D}$ then

$$(A \circ B)(\phi) = \left(a(x)\frac{d}{dx}\right)\left(b(x)\frac{d\phi}{dx}\right)$$

$$= \left(a(x)\frac{db}{dx}(x)\frac{d}{dx} + a(x)b(x)\frac{d^2}{dx^2}\right)\phi.$$

Thus we see that

$$\sigma(A \circ B) = a(x)\frac{db}{dx}(x)(-i\xi) + a(x)b(x)(-i\xi)^2.$$

Notice that the principal part of the symbol of $A \circ B$ is

$$a(x)b(x)(-i\xi)^2 = \sigma(A) \cdot \sigma(B).$$

In general, the Kohn–Nirenberg formula says (in $\mathbb{R}^N$) that

$$\sigma(A \circ B) = \sum_\alpha \frac{1}{\alpha!}\left(\frac{\partial}{\partial \xi}\right)^\alpha (\sigma(A)) \cdot D_x^\alpha(\sigma(B)). \qquad (3.3.1)$$

Recall that the *commutator*, or bracket, of two operators is

$$[A, B] \equiv AB - BA.$$

Here juxtaposition of operators denotes composition. A corollary of the Kohn–Nirenberg formula is that

$$\sigma([A, B]) = \sum_{|\alpha|>0} \frac{(\partial/\partial\xi)^\alpha \sigma(A) D_x^\alpha \sigma(B) - (\partial/\partial\xi)^\alpha \sigma(B) D_x^\alpha \sigma(A)}{\alpha!}$$

(notice here that the $\alpha = 0$ term cancels out) so that $\sigma([A, B])$ has order strictly less than $(\text{order}(A) + \text{order}(B))$. This phenomenon is illustrated concretely in $\mathbb{R}^1$ by the operators $A = a(x)d/dx, B = b(x)d/dx$. One calculates that

$$AB - BA = \left(a(x)\frac{db}{dx}(x) - b(x)\frac{da}{dx}(x) \right) \frac{d}{dx} \, ,$$

which has order *one* instead of two.

Our final key result in the development of pseudodifferential operators is the asymptotic expansion for a symbol. We shall first have to digress a bit on the subject of asymptotic expansions.

Let f be a C^∞ function defined in a neighborhood of 0 in $\mathbb{R}$. Then

$$f(x) \sim \sum_0^\infty \frac{1}{n!} \frac{d^n f}{dx^n}(0) \, x^n. \tag{3.3.2}$$

We are certainly not asserting that the Taylor expansion of an arbitrary C^∞ function converges back to the function, or even that it converges at all (generically just the opposite is true).

This formal expression (3.3.2) means instead the following: Given an $N > 0$ there exists an $M > 0$ such that whenever $m > M$ and x is small then the partial sum S_m satisfies

$$|f(x) - S_m| < C|x|^N.$$

Now we present a notion of asymptotic expansion that is related to this one, but is specially adapted to the theory of pseudodifferential operators:

DEFINITION 3.3.3 *Let $\{a_j\}$ be symbols in $\cup_m S^m$. We say that another symbol a satisfies*

$$a \sim \sum_j a_j$$

if for every $L \in \mathbb{R}^+$ there is an $M \in \mathbb{Z}^+$ such that

$$a - \sum_{j=1}^M a_j \in S^{-L}.$$

DEFINITION 3.3.4 *Let* $K \subset\subset \mathbb{R}^N$ *be a fixed compact set. Let* Ψ_K *be the set of symbols with x-support in* K. *If* $p \in \Psi_K$, *then we will think of the corresponding pseudodifferential operator* P *as*

$$P : C_c^\infty(K) \to C_c^\infty(K).$$

(This makes sense because $P\phi(x) = \int e^{-ix\cdot\xi} p(x,\xi)\hat{\phi}(\xi)\, d\xi.$)

Now our main result is

THEOREM 3.3.5
Fix a compact set K *and pick* $p \in S^m \cap \Psi_K$. *Let* $P = Op(p)$. *Then* P^* *has symbol in* $S^m \cap \Psi_K$ *given by*

$$\sigma(P^*) \sim \sum_\alpha D_x^\alpha \left(\frac{\partial}{\partial\xi}\right)^\alpha \overline{p(x,\xi)} \cdot \frac{1}{\alpha!}\ .$$

We will prove this theorem in stages. There is a technical difficulty that arises almost immediately: Recall that if an operator T is given by integration against a kernel $K(x,y)$, then the roles of x and y are essentially symmetric. If we attempt to calculate the adjoint of T by formal reasoning, there is no difficulty in seeing that T^* is given by integration against the kernel $\overline{K(y,x)}$. However, at the symbol level matters are different. Namely, in our symbols $p(x,\xi)$, the role of x and ξ is not symmetric. If we attempt to calculate the symbol of $Op(p)$ by a formal calculation, then this lack of symmetry serves as an obstruction.

It was Hörmander who determined a device for dealing with the problem just described. We shall now describe his method. We introduce a new class of symbols $r(x,\xi,y)$. Such a smooth function on $\mathbb{R}^N \times \mathbb{R}^N \times \mathbb{R}^N$ is said to be in the symbol class T^m if there is a compact set K such that

$$\operatorname*{supp}_x r(x,\xi,y) \subseteq K$$

and

$$\operatorname*{supp}_y r(x,\xi,y) \subseteq K$$

and, for any multiindices α, β, γ, there is a constant $C_{\alpha,\beta,\gamma}$ such that

$$\left| \left(\frac{\partial}{\partial\xi}\right)^\alpha \left(\frac{\partial}{\partial x}\right)^\beta \left(\frac{\partial}{\partial y}\right)^\gamma r(x,\xi,y) \right| \leq C_{\alpha,\beta,\gamma} |\xi|^{m-|\alpha|}.$$

The corresponding operator R is defined by

$$R\phi(x) = \iint e^{i(y-x)\cdot\xi} r(x,\xi,y)\phi(y)\, dy\, d\xi. \tag{3.3.6}$$

Notice that the integral is not absolutely convergent and must therefore be interpreted as an iterated integral.

PROPOSITION 3.3.7
Let $r \in T^m$ have x- and y-supports contained in a compact set K. Then the operator R defined as in (3.3.6) defines a pseudodifferential operator of Kohn–Nirenberg type with symbol $p \in \Psi_K$ having an asymptotic expansion

$$p(x,\xi) \sim \sum_{\alpha} \frac{1}{\alpha!} \partial_\xi^\alpha D_y^\alpha r(x,\xi,y)\Big|_{y=x}.$$

PROOF We calculate that

$$\int e^{iy\cdot\xi} r(x,\xi,y)\phi(y)\, dy = \widehat{\left(r(x,\xi,\cdot)\phi(\cdot)\right)}$$

$$= \left(\hat{r}_3(x,y,\cdot) * \hat{\phi}(\cdot)\right)(\xi).$$

Here $\hat{r}_3$ indicates that we have taken the Fourier transform of r in the third variable. By the definition of $R\phi$ we have

$$R\phi(x) = \iint e^{i(-x+y)\cdot\xi}\, r(x,\xi,y)\phi(y)\, dy\, d\xi$$

$$= \int e^{-ix\cdot\xi}\left[\hat{r}_3(x,\xi,\cdot) * \hat{\phi}(\cdot)\right](\xi)\, d\xi$$

$$= \iint \hat{r}_3(x,\xi,\xi-\eta)\hat{\phi}(\eta)\, d\eta e^{-ix\cdot\xi}\, d\xi$$

$$= \iint \hat{r}_3(x,\xi,\xi-\eta)e^{-ix\cdot(\xi-\eta)}\, d\xi\hat{\phi}(\eta)e^{-ix\cdot\eta}\, d\eta$$

$$\equiv \int p(x,y)\hat{\phi}(\eta)e^{-ix\cdot\eta}\, d\eta.$$

Here

$$p(x,\eta) \equiv \int \hat{r}_3(x,\xi,\xi-\eta)e^{-ix(\xi-\eta)}\, d\xi$$

$$= \int e^{-ix\cdot\xi}\hat{r}_3(x,\xi+\eta,\xi)\, d\xi.$$

Now if we expand the function $\hat{r}_3(x,\eta+\cdot,\xi)$ in a Taylor expansion in powers of ξ, it is immediate that p has the claimed asymptotic expansion. In particular, one sees that $p \in S^m$. In detail, we have

$$\hat{r}_3(x,\eta+\xi,\xi) = \sum_{|\alpha|<k} \partial_\eta^\alpha \hat{r}_3(x,\eta,\xi)\frac{\xi^\alpha}{\alpha!} + \mathcal{R}.$$

Thus (dropping the ubiquitous c from the Fourier integrals),

$$p(x,\eta) = \sum_{|\alpha|<k} \int e^{-ix\cdot\xi} \partial_\eta^\alpha \hat{r}_3(x,\eta,\xi) \frac{\xi^\alpha}{\alpha!}\, d\xi + \int \mathcal{R}\, d\xi$$

$$= \sum_{|\alpha|<k} \frac{1}{\alpha!} \partial_\eta^\alpha D_y^\alpha r(x,\eta,y)\Big|_{y=x} + \int \mathcal{R}\, d\xi.$$

The rest is formal checking. ∎

PROOF OF THEOREM 3.3.5   Let $p \in \Psi_K \cap S^m$ and choose $\phi, \psi \in \mathcal{D}$. Then, with P the pseudodifferential operator corresponding to the symbol p, we have

$$\langle \phi, P^*\psi \rangle \equiv \langle P\phi, \psi \rangle$$

$$= \int \left[\int e^{-ix\cdot\xi} p(x,\xi)\hat{\phi}(\xi)\, d\xi \right] \bar{\psi}(x)\, dx$$

$$= \iiint e^{-i(x-y)\cdot\xi} \phi(y)\, dy\, p(x,\xi)\, d\xi \bar{\psi}(x)\, dx.$$

Let us suppose for the moment that p is compactly supported in ξ. With this extra hypothesis the integral is absolutely convergent and we may write

$$\langle \phi, P^*\psi \rangle = \int \phi(y) \left[\iint e^{i(x-y)\cdot\xi} \overline{p(x,\xi)\psi(x)}\, d\xi\, dx \right] dy. \qquad (3.3.5.1)$$

Thus we have

$$P^*\psi(y) = \iint e^{i(x-y)\cdot\xi} \overline{p(x,\xi)}\psi(x)\, d\xi\, dx.$$

Now let $\rho \in C_c^\infty$ be a real-valued function such that $\rho \equiv 1$ on K. Set

$$r(x,\xi,y) = \rho(x) \cdot p(y,\xi).$$

Then

$$P^*\psi(y) = \iint e^{i(x-y)\cdot\xi} \overline{p(x,\xi)}\rho(y)\psi(x)\, d\xi\, dx$$

$$= \int e^{i(x-y)\cdot\xi} r(y,\xi,x)\psi(x)\, d\xi\, dx$$

$$\equiv R\psi(y),$$

where we define R by means of the multiple symbol r. (Note that the roles of x and y here have unfortunately been reversed.)

By Proposition 3.3.7, P^* is then a classical pseudodifferential operator with symbol p^* whose asymptotic expansion is

$$p^*(x,\xi) \sim \sum_\alpha \frac{1}{\alpha!} \partial_\xi^\alpha D_y^\alpha \left[\rho(x)\overline{p(y,\xi)} \right] \Bigg|_{y=x}$$

$$\sim \sum_\alpha \frac{1}{\alpha!} \partial_\xi^\alpha D_x^\alpha \overline{p(x,\xi)}.$$

We have used here the fact that $\rho \equiv 1$ on K. The theorem is thus proved with the extra hypothesis of compact support of the symbol in ξ.

To remove the extra hypothesis, let $\phi \in C_c^\infty$ satisfy $\phi \equiv 1$ if $|\xi| \leq 1$ and $\phi \equiv 0$ if $|\xi| \geq 2$. Let

$$p_j(x,\xi) = \phi(\xi/j) \cdot p(x,\xi).$$

Observe that $p_j \to p$ in the C^k topology on compact sets for any k. Also, by the special case of the theorem already proved,

$$(\mathrm{Op}(p_j))^* \sim \sum_\alpha \frac{1}{\alpha!} \partial_\xi^\alpha D_x^\alpha \overline{p_j(x,\xi)}$$

$$\sim \sum_\alpha \frac{1}{\alpha!} \partial_\xi^\alpha D_x^\alpha \left[\overline{\phi(\xi/j)p(x,\xi)} \right] .$$

The proof is completed now by letting $j \to \infty$. ∎

THEOREM 3.3.8   KOHN–NIRENBERG
Let $p \in \Psi_K \cap S^m, q \in \Psi_K \cap S^n$. Let P,Q denote the pseudodifferential operators associated with p,q respectively. Then $P \circ Q = \mathrm{Op}(\sigma)$ where

1. *$\sigma \in \Psi_K \cap S^{m+n}$;*
2. *$\sigma \sim \sum_\alpha \frac{1}{\alpha!} \partial_\xi^\alpha p(x,\xi) D_x^\alpha q(x,\xi).$*

PROOF   We may shorten the proof by using the following trick: write $Q = (Q^*)^*$ and recall that Q^* is defined by

$$Q^*\phi(y) = \iint e^{i(x-v)\cdot\xi} \phi(x)\overline{q(x,\xi)}\, dx\, d\xi$$

$$= \left(\int e^{ix\cdot\xi} \phi(x)\overline{q(x,\xi)}\, dx \right)^{\vee} (y).$$

Here we have used (3.3.5.1).
Then

$$Q\phi(x) = \left(\int e^{iy\cdot\xi} \phi(y)\overline{q^*(y,\xi)}\, dy \right)^{\vee} (x), \qquad (3.3.8.1)$$

where q^* is the symbol of Q^*. (Note that q^* is *not* $\bar{q}$; however, we do know that $\bar{q}$ is the principal part of q^*.) Then, using (3.3.8.1), we may calculate that

$$(P \circ Q)(\phi)(x) = \int e^{-ix\cdot\xi} p(x,\xi)(\widehat{Q\phi})(\xi)\, d\xi$$

$$= \iint e^{-ix\cdot\xi} p(x,\xi) e^{iy\cdot\xi} \overline{q^*(y,\xi)} \phi(y)\, dy\, d\xi$$

$$= \iint e^{-i(x-y)\cdot\xi} \left[p(x,\xi)\overline{q^*(y,\xi)} \right] \phi(y)\, dy\, d\xi.$$

Set $\tilde{q} = \overline{q^*}$. Define

$$r(x,\xi,y) = p(x,\xi) \cdot \tilde{q}(y,\xi).$$

One verifies directly that $r \in T^{n+m}$. We leave this as an exercise. Thus R, the associated operator, equals $P \circ Q$. By Proposition 3.3.7 there is a classical symbol σ such that $R = \mathrm{Op}(\sigma)$ and

$$\sigma(x,\xi) \sim \sum_\alpha \frac{1}{\alpha!} \partial_\xi^\alpha D_y^\alpha r(x,\xi,y)\Big|_{y=x}.$$

Developing this last line we obtain

$$\sigma(x,\xi) \sim \sum_\alpha \frac{1}{\alpha!} \partial_\xi^\alpha D_y^\alpha \left(p(x,\xi)\tilde{q}(y,\xi) \right)\Big|_{y=x}$$

$$\sim \sum_\alpha \frac{1}{\alpha!} \partial_\xi^\alpha \left[p(x,\xi) D_y^\alpha \tilde{q}(y,\xi) \right]\Big|_{y=x}$$

$$\sim \sum_\alpha \frac{1}{\alpha!} \partial_\xi^\alpha \left[p(x,\xi) D_x^\alpha \tilde{q}(x,\xi) \right]$$

$$\sim \sum_\alpha \frac{1}{\alpha!} \sum_{\alpha^1+\alpha^2=\alpha} \frac{\alpha!}{\alpha^1!\alpha^2!} \left[\partial_\xi^{\alpha^1} p(x,\xi) \right] \left[\partial_\xi^{\alpha^2} D_x^\alpha \tilde{q}(x,\xi) \right]$$

$$\sim \sum_\alpha \sum_{\alpha^1+\alpha^2=\alpha} \left[\partial_\xi^{\alpha^1} p(x,\xi) \right] \left[\partial_\xi^{\alpha^2} D_x^{\alpha^2} D_x^{\alpha^1} \tilde{q}(x,\xi) \right] \frac{1}{\alpha^1!\alpha^2!}$$

$$\sim \sum_{\alpha^1,\alpha^2} \frac{1}{\alpha^1!} \left[\partial_\xi^{\alpha^1} p(x,\xi) \right] \frac{1}{\alpha^2!} \left[\partial_\xi^{\alpha^2} D_x^{\alpha^2} D_x^{\alpha^1} \tilde{q}(x,\xi) \right]$$

$$\sim \sum_{\alpha^1} \frac{1}{\alpha^1!} \partial_\xi^{\alpha^1} p(x,\xi) D_x^{\alpha^1} \left[\sum_{\alpha^2} \frac{1}{\alpha^2!} \partial_\xi^{\alpha^2} D_x^{\alpha^2} \tilde{q}(x,\xi) \right]$$

$$\sim \sum_{\alpha^1} \frac{1}{\alpha^1!} \partial_\xi^{\alpha^1} p(x,\xi) D_x^{\alpha^1} q(x,\xi).$$

Here we have used the fact that the expression inside the brackets is just the asymptotic expansion for the symbol of $(Q^*)^*$. That completes the proof. ∎

The next proposition is a useful device for building pseudodifferential operators. Before we can state it we need a piece of terminology: we say that two pseudodifferential operators P and Q are equal *up to a smoothing operator* if $P - Q \in S^k$ for all $k < 0$. In this circumstance we write $P \sim Q$.

PROPOSITION 3.3.9
Let $p_j, j = 0, 1, 2, \ldots$, be symbols of order m_j, $m_j \searrow -\infty$. Then there is a symbol $p \in S^{m_0}$, unique modulo smoothing operators, such that

$$p \sim \sum_0^\infty p_j.$$

PROOF Let $\psi : \mathbb{R}^n \to [0, 1]$ be a C^∞ function such that $\psi \equiv 0$ when $|x| \leq 1$ and $\psi \equiv 1$ when $|x| \geq 2$. Let $1 < t_1 < t_2 < \cdots$ be a sequence of positive numbers that increases to infinity. We will specify these numbers later. Define

$$p(x, \xi) = \sum_{j=0}^\infty \psi(\xi/t_j) p_j(x, \xi).$$

Note that for every fixed x, ξ the sum is finite, for $\psi(\xi/t_j) = 0$ as soon as $t_j > |\xi|$. Thus p is a well-defined C^∞ function.

Our goal is to choose the t_j's so that p has the correct asymptotic expansion. We claim that there exist $\{t_j\}$ such that

$$\left| D_x^\beta D_\xi^\alpha \left(\psi(\xi/t_j) p_j(x, \xi) \right) \right| \leq 2^{-j} (1 + |\xi|)^{m_j - |\alpha|}.$$

Assume the claim for the moment. Then for any multiindices α, β we have

$$|D_x^\beta D_\xi^\alpha p(x, \xi)| \leq \sum_{j=0}^\infty \left| D_x^\beta D_\xi^\alpha \left(\psi(\xi/t_j) p_j(x, \xi) \right) \right|$$

$$\leq \sum_{j=0}^\infty 2^{-j} (1 + |\xi|)^{m_j - |\alpha|}$$

$$\leq C \cdot (1 + |\xi|)^{m_0 - |\alpha|}.$$

It follows that $p \in S^{m_0}$. Now we want to show that p has the right asymptotic expansion. Let $0 < k \in \mathbf{Z}$ be fixed. We will show that

$$p - \sum_{j=0}^{k-1} p_j$$

lives in S^{m_k}. We have

$$p - \sum_{j=0}^{k-1} p_j = \left[p(x,\xi) - \sum_{j=0}^{k-1} \psi(\xi/t_j) p_j(x,\xi) \right]$$

$$- \sum_{j=0}^{k-1} \left(1 - \psi(\xi/t_j) \right) p_j(x,\xi)$$

$$\equiv q(x,\xi) + s(x,\xi).$$

It follows directly from our construction that $q(x,\xi) \in S^{m_k}$. Since $[1 - \psi(\xi/t_j)]$ has compact support in $B(0, 2t_1)$ for every j, it follows that $s(x,\xi) \in S^{-\infty}$. Then

$$p - \sum_{j=0}^{k-1} p_j \in S^{m_k}$$

as we asserted.

We wish to see that p is unique modulo smoothing terms. Suppose that $q \in S^{m_0}$ and $q \sim \sum_{j=0}^{\infty} p_j$. Then

$$p - q = \left(p - \sum_{j<k} p_j \right) - \left(q - \sum_{j<k} p_j \right)$$

$$\in S^{m_k}$$

for any k. That establishes the uniqueness.

It remains to prove the claim. First observe that, for $|\alpha| = j$,

$$\left| D_\xi^\alpha \psi(\xi/t_\alpha) \right| = \frac{1}{t_j^{|\alpha|}} \left| (D_\xi^\alpha \psi)(\xi/t_j) \right|$$

$$\leq \frac{1}{t_j^{|\alpha|}} \left| \sup_{|\xi| \leq 2t_j} \left\{ (D^\alpha \psi) \cdot (1 + |\xi|)^{|\alpha|} \right\} \right| (1 + |\xi|)^{-|\alpha|}$$

$$\leq C_j (1 + |\xi|)^{-j},$$

with C independent of j. Therefore

$$\left|D_\xi^\alpha\left(\psi(\xi/t_j)p_j(x,\xi)\right)\right| = \left|\sum_{\gamma\le\alpha}\binom{\alpha}{\gamma}D_\xi^\gamma\psi(\xi/t_j)D_\xi^{\alpha-\gamma}p_j(x,\xi)\right|$$

$$\le \sum_{\gamma\le\alpha}\binom{\alpha}{\gamma}C_\gamma(1+|\xi|)^{-|\gamma|}C_{\gamma,\alpha}(1+|\xi|)^{m_j-(|\alpha|-|\gamma|)}$$

$$\le C_j(1+|\xi|)^{m_j-|\alpha|}.$$

Consequently,

$$\left|D_x^\beta D_\xi^\alpha\left(\psi(\xi/t_j)p_j(x,\xi)\right)\right| \le C_{j,\alpha,\beta}(1+|\xi|)^{m_j-|\alpha|}$$

$$\le C_j(1+|\xi|)^{m_j-|\alpha|}$$

for every $j \ge |\alpha| + |\beta|$ (here we have set $C_j = \max\{C_{j,\alpha,\beta} : |\alpha| + |\beta| \le j\}$).

Now recall that $\psi(\xi) = 0$ if $|\xi| \le 1$. Then $\psi(\xi/t_j) \ne 0$ implies that $|\xi| \ge t_j$. Thus we choose t_j so large that $t_j > t_{j-1}$ and

$$|\xi| \ge t_j \quad \text{implies} \quad C_j(1+|\xi|)^{m_j-m_{j-1}} \le 2^{-j}.$$

Then it follows that

$$\left|D_x^\beta D_\xi^\alpha\left(\psi(\xi/t_j)p_j(x,\xi)\right)\right| \le 2^{-j}(1+|\xi|)^{m_{j-1}-|\alpha|},$$

which establishes the claim and finishes the proof of the proposition. ∎

4

Elliptic Operators

4.1 Some Fundamental Properties of Partial Differential Operators

We begin this chapter by discussing some general properties that it is desirable for a partial differential operator to have. We will consider *why* these properties are desirable and illustrate with examples. We follow this discussion by introducing an important, and easily recognizable, class of partial differential operators that enjoy these desirable properties: the elliptic operators.

Our first topic of discussion is locality and pseudolocality. Let $T : C_c^\infty \to C^\infty$. We say that T is *local* if whenever a testing function ϕ vanishes on an open set U then $T\phi$ also vanishes on U. The most important examples of local operators are differential operators. In fact, the converse is true as well:

THEOREM 4.1.1 PEETRE
If

$$T : C_c^\infty \to C^\infty$$

is a linear operator that is local then T is a partial differential operator.

PROOF See [HEL]. ∎

The calculation of a derivative at a point involves only the values of the function at points nearby. Thus the notion of locality is well suited to differentiation. In particular, it means that if T is local and $\phi = \psi$ on an open set then $T\phi = T\psi$ on that open set. For the purposes of studying regularity for differential operators, literal equality is too restrictive and not actually necessary. Therefore we make the following definitions:

DEFINITION 4.1.2 *Let $\alpha \in \mathcal{D}'$ and $U \subseteq \mathbb{R}^N$ an open set. If there is a C^∞*

function f on U such that for all $\phi \in \mathcal{D}$ that are supported in U we have

$$\alpha(\phi) = \int f(x)\phi(x)\,dx,$$

then we say that α is C^∞ on U.

The *singular support* of a distribution α is defined to be the complement of the union of all the open sets on which α is C^∞.

DEFINITION 4.1.3   *A linear operator* $T : \mathcal{D}' \to \mathcal{D}'$ *is said to be **pseudolocal** if whenever U is an open set and $\alpha \in \mathcal{D}'$ is C^∞ on U then $T\alpha$ is C^∞ on U.*

Now we have

THEOREM 4.1.4
If T is a pseudodifferential operator, then T is pseudolocal.

The theorem may be restated as "the singular support of Tu is contained in the singular support of u for every distribution u."

A sort of converse to this theorem was proved by R. Beals in [BEA2-4]. That is, in some sense the only pseudolocal operators are pseudodifferential. We shall not treat that result in detail here.

The proof of the theorem will proceed in stages. First, we need to define how a pseudodifferential operator operates on a distribution. Let P be a pseudodifferential operator and let $u \in \mathcal{D}'$ have compact support. We want Pu to be a distribution. For any testing function ϕ, we set

$$\langle Pu, \phi \rangle = \langle u, {}^t P\phi \rangle.$$

Here ${}^t P$ is the transpose of P which we define by

$$\langle {}^t P\phi, \psi \rangle = \langle \phi, P\psi \rangle,$$

for $\phi, \psi \in \mathcal{D}$.

To illustrate the definition, suppose that u is given by integration against an L^1 function f. Suppose also that the pseudodifferential operator P is given by integration against the kernel $K(x, x - y)$ with $K(\cdot, y) \in L^1$ and $K(x, \cdot) \in L^1$. Then P has symbol $\widehat{K}_2(x, \xi)$. We see that

$$
\begin{aligned}
(Pu)(\phi) &= \int \left[\int K(x, x - y) f(y)\,dy \right] \phi(x)\,dx \\
&= \int \left[\int K(x, x - y) \phi(x)\,dx \right] f(y)\,dy \\
&= \int {}^t P\phi(y) f(y)\,dy.
\end{aligned}
$$

This calculation is consistent with the last definition.

PROOF OF THEOREM 4.1.4   Let U be an open set on which the distribution u is C^∞. Fix $x \in U$. Let $\phi \in C_c^\infty(U)$ satisfy $\phi \equiv 1$ in a neighborhood of x. Finally, let $\psi \in C_c^\infty(U)$ satisfy $\psi \equiv 1$ on the support of ϕ. Then we have $\psi u \in \mathcal{D}$, hence $\phi P \psi u \in C_c^\infty$. (Note here that we are using implicitly the fact that if $\sigma(P) \in S^m$, then P maps H^s to H^{s-m}, hence, by the Sobolev imbedding theorem, P maps C_c^∞ to C^∞.)

We wish to find the symbol of $\phi P \psi$. This is where the calculus of pseudodifferential operators will come in handy. Let us write our operator as

$$T = \mathcal{M}_\phi \circ P \circ \mathcal{M}_\psi,$$

where the symbol $\mathcal{M}$ denotes a multiplication operator. Of course $\sigma(\mathcal{M}_\phi) = \phi(x)$ and $\sigma(\mathcal{M}_\psi) = \psi(x)$. Hence

$$\sigma(T) = \phi(x)\sigma\big(P \circ \mathcal{M}_\psi\big)$$

$$\sim \phi(x)\left(\sum_\alpha \frac{1}{\alpha!}\partial_\xi^\alpha p D_x^\alpha \psi\right).$$

In the last equality we have used the Kohn–Nirenberg formula. But, on the support of ϕ, any derivative of ψ of order at least 1 vanishes. As a result,

$$\sigma(T) \sim \phi(x)\sigma(P)(x, \xi).$$

Therefore

$$\sigma(T) - \phi(x)\sigma(P) \in S^{-k} \qquad \forall\, k \geq 0.$$

In other words,

$$T - \mathcal{M}_\phi \circ P : H^s \to H^{s+k}$$

for all $k \geq 0$ and every s. By Sobolev's theorem,

$$T - \mathcal{M}_\phi \circ P : H^s \to C^\infty.$$

But $\mathcal{D}' = \bigcup_s H^s$ so our distribution u lies in some H^s.

We conclude that

$$\big(T - \mathcal{M}_\phi P\big)u \in C^\infty(U).$$

Since $\phi P \psi u$ is C^∞, we may conclude that $\phi P u \in C^\infty$. But $\phi \equiv 1$ in a neighborhood of x. Therefore Pu is C^∞ near x. Since x was an arbitrary point of U, we are done. ∎

DEFINITION 4.1.5   *A linear operator $T : \mathcal{D}' \to \mathcal{D}'$ is called **hypoelliptic** if* sing supp $u \subseteq$ sing supp Tu.

Notice that the containment defining hypoellipticity is just the opposite from that defining pseudolocality. A good, but simple, example of a hypoelliptic operator on the real line is d/dx. For if

$$\frac{d}{dx} u = f$$

and f is smooth on an open set U, then u is also smooth on U. The reason, of course, is that we may recover u from f *on this open set* by integration. A more interesting example of a hypoelliptic operator is the Laplacian Δ on $\mathbf{R}^N$, $N \geq 2$. Thus the equation $\Delta u = f$ entails u being smooth wherever f is. This is proved, in analogy with the one-variable case, by constructing a right inverse (or at least a parametrix) for the operator Δ. If the right inverse or parametrix is a pseudodifferential operator, then the last theorem will tell us that Δ is hypoelliptic.

We now introduce an important class of pseudodifferential operators which are hypoelliptic and enjoy several other appealing regularity properties. These are the elliptic operators:

DEFINITION 4.1.6 *We say that a symbol $p \in S^m$ is **elliptic** on an open set $U \subseteq \mathbf{R}^N$ if there exists a continuous function $c(x) > 0$ on U such that*

$$|p(x,\xi)| \geq c(x)|\xi|^m$$

for ξ large. A partial differential operator or, more generally, a pseudodifferential operator L is elliptic precisely when its symbol $\sigma(L)$ is elliptic.

Example 1
Let Δ be the Laplacian. Then $\sigma(\Delta) = \sum_{j=1}^{N}(-i\xi_j)^2 = -\sum_{j=1}^{N}|\xi_j|^2$. It follows that

$$|\sigma(\Delta)| = |\xi|^2 \geq 1 \cdot |\xi|^2$$

on all of space. Hence Δ is elliptic.   □

Example 2
Now let

$$\delta_{ij} = \begin{cases} 1 & \text{if } i = j \\ 0 & \text{if } i \neq j. \end{cases}$$

Then

$$\Delta = \sum_{i,j}(\delta_{ij})\frac{\partial^2}{\partial x_i \partial x_j} \ .$$

If ϵ_{ij}, $i, j = 1, \ldots, N$ are C^∞ functions having small C^0 norms, then the operator

$$\sum_{i,j} (\delta_{ij} + \epsilon_{ij}(x)) \frac{\partial^2}{\partial x_i \partial x_j}$$

is elliptic.   ◻

Example 3

Let (a_{ij}) be a positive definite $N \times N$ matrix of constants. Let $b_1, \ldots, b_N$ be scalars. Then the partial differential operator

$$\sum_{i,j} a_{ij} \frac{\partial^2}{\partial x_i \partial x_j} + \sum_j b_j \frac{\partial}{\partial x_j}$$

is elliptic.   ◻

If P is a partial differential operator that is elliptic of order m (usually m is positive, but it is not necessary to assume this), and if $Q = \sum a_\alpha(x) \partial^\alpha$ is a partial differential operator with continuous coefficients of strictly lower order, then $P + Q$ is still elliptic, no matter what Q is. To see this, let $\sigma(P) = p(x, \xi)$ with $|p(x, \xi)| \geq c(x)|\xi|^m$ for $|\xi|$ large. Then

$$|\sigma(P + Q)| \geq |\sigma(P)| - |\sigma(Q)|$$

$$\geq c(x)|\xi|^m - \left| \sum_{|\alpha|<m} a_\alpha(x)(-i\xi)^\alpha \right|$$

$$\geq c(x)|\xi|^m - b(x)|\xi|^{m-1}$$

$$\geq \frac{c(x)}{2} |\xi|^m$$

for ξ large and x in a bounded set. As a result of this calulation, we can say that a partial differential operator is elliptic if and only if its principal symbol $\sum_{|\alpha|=m} a_\alpha(x) \partial^\alpha$ is elliptic.

A natural question to ask is: must an elliptic operator be of even order? If the dimension is at least two, and the order is at least two, then the answer is yes. We leave the easy verification as an exercise.

In the case of dimension two and order one, let us consider the example

$$P = \frac{\partial}{\partial x} + i \frac{\partial}{\partial y} .$$

Then

$$\sigma(P) = -i\xi_1 + i(-i\xi_2) = -i\xi_1 + \xi_2.$$

Thus

$$|\sigma(P)| \geq |\xi|.$$

In dimension one, consider an operator

$$P = \sum_{j=1}^{m} a_j(x) \frac{d^j}{dx^j}.$$

Then

$$\sigma(P) = \sum_{j=1}^{m} a_j(x)(-i\xi)^m$$

so that

$$|\sigma(P)| \approx |a_m(x)| \cdot |\xi|^m$$

for ξ large. Then P is elliptic if and only if a_m is nowhere vanishing.

Our next main goal is to prove existence and regularity theorems for elliptic partial differential operators.

4.2 Regularity for Elliptic Operators

We begin with some terminology.

DEFINITION 4.2.1 *A pseudodifferential operator is said to be* ***smoothing*** *if its symbol p lies in $S^{-\infty} = \cap_m S^m$.*

DEFINITION 4.2.2 *If P is a pseudodifferential operator, then a left (resp. right)* ***parametrix*** *for P on an open set U is a pseudodifferential operator Q such that there is a $\psi \in C_c^\infty$, $\psi \equiv 1$ on U, with $QP - \psi I$ (resp. $PQ - \psi I$) smoothing.*

The next proposition is fundamental to our regularity theory.

PROPOSITION 4.2.3
If P is an elliptic pseudodifferential operator of order m and if L is a relatively compact open set in the (x variable) domain of $\sigma(P)$, then there exists a pseudodifferential operator of order $-m$ that is a two-sided parametrix for P on L.

PROOF By hypothesis we have

$$|\sigma(P)| \equiv |p(x,\xi)| \geq a(x)|\xi|^m$$

for $x \in \operatorname{supp} p(\cdot, \xi)$, ξ large with $a(x) > 0$ a continuous function. Select a relatively compact open set $L \subseteq \operatorname{supp} p(\cdot, \xi)$ so that there is a constant $c_0 > 0$ with $a(x) > c_0$ on L. We also select $K_0 > 0$ such that, for $|\xi| \geq K_0$ and $x \in L$, it holds that $|p(x, \xi)| \geq c_0 |\xi|^m$. Let ψ be a C_c^∞ function with support in the (x variable) domain of $\sigma(P)$ such that $\psi \equiv 1$ on L. Also choose a function $\phi \in C^\infty$ such that $\phi(\xi) = 1$ for $|\xi| \geq K_0 + 2$ and $\phi(\xi) = 0$ for $|\xi| \leq K_0 + 1$. Then set

$$q_0(x, \xi) = \psi(x) \frac{\phi(\xi)}{p(x, \xi)}$$

and

$$Q_0 = \operatorname{Op}(q_0).$$

Observe that $\phi \equiv 0$ on a neighborhood of the zeroes of $p(x, \xi)$; hence q_0 is C^∞. Furthermore, q_0 has compact support in x. Finally, $q_0 \in S^{-m}$. To see this, first notice that

$$|q_0(x, \xi)| \leq C \sup |\psi| \sup |\phi| (1 + |\xi|)^{-m}.$$

Moreover,

$$\left| \frac{\partial}{\partial \xi_j} q_0(x, \xi) \right| = |\psi(x)| \cdot \left| \frac{p(x, \xi)(\partial \phi / \partial \xi_j)(\xi) - \phi(\xi)(\partial p(x, \xi) / \partial \xi_j)}{p^2(x, \xi)} \right|$$

$$\leq C \cdot \left[\frac{(1 + |\xi|)^m \cdot |(\partial \phi / \partial \xi_j)(\xi)|}{(1 + |\xi|)^{2m}} + \frac{(1 + |\xi|)^{m-1}}{(1 + |\xi|)^{2m}} \right]$$

for $|\xi|$ large. But this is

$$\leq C \cdot \left[\left| \frac{\partial \phi}{\partial \xi_j}(\xi) \right| (1 + |\xi|)^{-m} + (1 + |\xi|)^{-m-1} \right].$$

However, $\partial \phi / \partial \xi_j$ is compactly supported, so it follows that

$$\left| \frac{\partial}{\partial \xi_j} q_0(x, \xi) \right| \leq C \cdot (1 + |\xi|)^{-m-1}.$$

Arguing in a similar fashion, one can show that

$$|\partial_\xi^\alpha q_0(x, \xi)| \leq C_\alpha (1 + |\xi|)^{-m-|\alpha|}.$$

Since x derivatives are harmless, we conclude that $q_0 \in S^{-m}$.

Now consider $Q_0 \circ P$. By the Kohn–Nirenberg formula,

$$\sigma(Q_0 \circ P) \sim \sum_\alpha \frac{1}{\alpha!} \partial_\xi^\alpha q_0(x,\xi) D_x^\alpha p(x,\xi).$$

That is,

$$\sigma(Q_0 \circ P) = q_0(x,\xi) p(x,\xi) + \tilde{r}_{-1}(x,\xi). \qquad (4.2.3.1)$$

Notice that $q_0(x,\xi) p(x,\xi) = \psi(x)\phi(\xi)$ and $\tilde{r}_{-1}(x,\xi) \in S^{-1}$. Now set $\sigma(Q_0 \circ P) = \psi(x) + r_{-1}(x,\xi)$, where r_{-1} is *defined* by this equation and (4.2.3.1). The equation that defines r_{-1} shows that we may suppose that r_{-1} has compact support in the x variable. Observe also that $r_{-1} \in S^{-1}$.

Define

$$q_1(x,\xi) = \tilde{\psi}(x) \cdot \frac{\phi(\xi)}{p(x,\xi)} \cdot (-r_{-1}(x,\xi)),$$

where $\tilde{\psi}$ is a C_c^∞ function that is identically equal to 1 on the x-support of r_{-1}.

Consider $Q_0 + Q_1 \equiv \mathrm{Op}(q_0 + q_1)$. We calculate $(Q_0 + Q_1) \circ P$. By the Kohn–Nirenberg formula,

$$\sigma\big((Q_0 + Q_1) \circ P\big) = \sigma(Q_0 \circ P) + \sigma(Q_1 \circ P)$$
$$= \psi(x) + \tilde{\psi}(x) r_{-1} - \tilde{\psi}(x)\phi(\xi) r_{-1} + \tilde{r}_{-2}.$$

Notice that $\tilde{r}_{-2} \in S^{-2}$. Since $\tilde{\psi}(x)(1 - \phi(\xi))$ is compactly supported in both x and ξ, it follows that $\tilde{\psi}(x)(1 - \phi(\xi)) r_{-1}$ is smoothing. Therefore we may write

$$\sigma\big((Q_0 + Q_1) \circ P\big) = \psi(x) + r_{-2}$$

with $r_{-2} \in S^{-2}$.

Now suppose inductively that we have constructed $q_0, \ldots, q_{k-1}$ such that, setting $Q_j = \mathrm{Op}(q_j)$ for $j = 0, 1, \ldots, k-1$, we have

$$\sigma\big((Q_0 + \cdots + Q_{k-1}) \circ P\big) = \psi(x) + r_{-k}$$

with $r_{-k} \in S^{-k}$. Define

$$q_k(x,\xi) = \tilde{\psi}(x) \frac{\phi(\xi)}{p(x,\xi)} \cdot (-r_{-k}(x,\xi)).$$

Let Q be the pseudodifferential operator having symbol q, where

$$q \sim \sum_{k=0}^\infty q_k(x,\xi)$$

(here we are using 3.3.9). Then

$$\sigma(Q \circ P) = \psi(x) + s(x, \xi),$$

with $s(x, \xi) \in S^{-\infty}$, i.e., $Q \circ P - \psi I$ is smoothing. We also will write $Q \circ P = \psi I + S$. Thus Q is a left parametrix for P.

A similar construction yields a right parametrix $\tilde{Q}$ for P. We write $P \circ \tilde{Q} = \psi I + \tilde{S}$.

We will now show that $\sigma(Q - \tilde{Q}) \in S^{-\infty}$ on a slightly smaller open set $W \subset\subset L$. By adjusting notation, one can of course arrange (after the proof) for the equations $Q \circ P = \psi I + S, P \circ \tilde{Q} = \psi I + \tilde{S}$, and $\sigma(Q - \tilde{Q}) \in S^{-\infty}$ to all be valid on the original open set L. We now interpolate an open set $W \subset\subset V \subset\subset L$. Let $\rho \in C_c^\infty(V)$ satisfy $\rho \equiv 1$ on W. Let $\mu \in C_c^\infty(L)$ satisfy $\mu \equiv 1$ on V. In what follows we will use continually the fact that the composition of *any* pseudodifferential operator with a smoothing operator is still smoothing. The identity of our smoothing operators may change from line to line, but we will denote them all by S or $\tilde{S}$. Now

$$\begin{aligned}
\rho Q \mu &= \rho Q (P \circ \tilde{Q} - \tilde{S})\mu \\
&= \rho Q \circ P(\tilde{Q} - \tilde{S})\mu \\
&= \rho(I + S) \circ (\tilde{Q} - \tilde{S})\mu \\
&= \rho(\tilde{Q} - \tilde{S} + S\tilde{Q} - S\tilde{S})\mu \\
&= \rho\tilde{Q}\mu + \tilde{S}\mu.
\end{aligned}$$

Thus $Q - \tilde{Q}$ is smoothing on W when applied to functions in $C_c^\infty(V)$. We conclude that $Q - \tilde{Q}$ is smoothing and we are done. ∎

Now we can present our basic interior regularity result for elliptic partial differential (in fact, even pseudodifferential) operators.

THEOREM 4.2.4
Let $U \subseteq \mathbb{R}^N$ be an open set. Let P be a pseudodifferential operator that is elliptic of order $m > 0$ on U. If $f \in H_{loc}^s$ and u is a solution of the equation $Pu = f$, then $u \in H_{loc}^{s+m}$.

PROOF The hard work has already been done. Since the theorem is local, we can suppose that u and f have compact support. (It is important to develop some intuition about this: the point is to see that we can consider ϕu rather than u and ρf rather than f, where ϕ, ρ are C_c^∞ cutoff functions. This amounts to commuting ϕ past P and ρ past the parametrix, noticing that this process gives rise to error terms of lower order, and then thinking about how the error terms would affect the parametrix.) Let $\psi \in C_c^\infty$ satisfy $\psi \equiv 1$ on supp u. Let Q be a left parametrix for P:

$$S \equiv Q \circ P - \psi I$$

is smoothing. Then

$$u = \psi u = (Q \circ P - S)u = Qf - Su.$$

Since $f \in H^s$ and Q is of order $-m$, it follows that $Qf \in H^{s+m}$; also, Su is smooth. Then $Qf - Su \in H^{s+m}$, that is, $u \in H^{s+m}$. ∎

Now that we have our basic regularity result in place we will use it, and some functional analysis, to prove our basic existence result. We would be remiss not to begin by mentioning the paramount discovery of Hans Lewy in 1956 [LEW2]: not all partial differential operators are locally solvable. In fact, we shall discuss Lewy's example in detail in Chapter 9.

Although we do not attempt to formulate the most general local solvability result, we present a result that will suffice in our applications. The most important ideas connected with local solvability of partial differential operators appear in [NTR], [BEF1], and the references therein.

THEOREM 4.2.5
Let P be an elliptic partial differential operator on a domain Ω. If $f \in \mathcal{D}'(\Omega)$ and $x_0 \in \Omega$, then there exists a $u \in D'(\Omega)$ such that $Pu = f$ in a neighborhood of x_0.

PROOF We may assume, by the usual arguments, that f has compact support—that is, $f \in \mathcal{E}'$. Then $f \in H^s$ for some s. Thus we may take the partial differential operator P to have compact support in a neighborhood K of x_0. Let $\psi \in C_c^\infty(\Omega)$ satisfy $\psi \equiv 1$ on K. Let Q be a right parametrix for P such that

$$P \circ Q - \psi I \equiv S \qquad (4.2.5.1)$$

and S is smoothing. By looking at the left side of this equation we see that $\sigma(S)$ has x-support also lying in supp ψ. Thus

$$S : H^s \to C_c^\infty(\text{supp } \psi).$$

Then $S : H^s \to H^s$ is compact by Rellich's lemma. Therefore the equation

$$(S + I)u = f \qquad (4.2.5.2)$$

can be solved if $f \in \mathcal{N}^{\perp}$, where

$$\mathcal{N} = \{g \in H^s : (S^* + I)g = 0\};$$

here the inner product is with respect to the Hilbert space H^s. Notice that, since S^* is also smoothing (why?) and $g = -S^* g$ for $g \in \mathcal{N}$, we know that $\mathcal{N}$ is a finite-dimensional space of C^{∞} functions.

Let $g_1, \ldots, g_M$ be a basis of $\mathcal{N}$. We claim that there exists a neighborhood U of x_0 such that $g_1, \ldots, g_M$ are linearly independent as functionals on $C_c^{\infty}(\Omega \backslash U)$. To see this, suppose that U does not exist. Let U_n be neighborhoods of x_0 such that $U_n \searrow x_0$. For each n, let $c_{n1}, \ldots, c_{nM}$ be constants (not all zero) such that

$$G_n \equiv \sum_{j=1}^{M} c_{nj} g_j = 0$$

as functionals on $C_c^{\infty}(\Omega \backslash U_n)$. By linearity, we may suppose that each G_n has norm 1 in $\mathcal{N}$ for all n (note in passing that $G_n \neq 0$ for every n). Since $\mathcal{N}$ is finite dimensional, there is a convergent subsequence G_{n_m} such that $G_{n_m} \to G \in \mathcal{N}$ and G has norm 1. However, viewed as a functional on $C_c^{\infty}(\Omega)$, G has support $\{x_0\}$. Since G is C^{∞}, we have a contradiction.

Thus we have proved the existence of U. We may assume that $U \subset\subset K$. Now we can find a C^{∞} function h with support in $\Omega \backslash U$ such that $f + h$ is perpendicular to $\mathcal{N}$ in the $H^s(\Omega)$ topology. By (4.2.5.2), there is a distribution u such that

$$(I + S)u = h + f = f \quad \text{on} \quad U.$$

By (4.2.5.1), we see that $P(Qu) = f$ on U. Therefore Qu solves the differential equation. ∎

Exercise: The proofs of the last two theorems, suitably adapted, show that the local solution exhibits a gain in smoothness, in the Sobolev topology, of order m.

It is also the case that elliptic partial differential equations exhibit a gain of order m in the Lipschitz topology. We shall present no details for this assertion. Given the machinery that we have developed, it is only necessary to check that if $p \in S^m$ then $\text{Op}(p) : \Lambda_\alpha^{loc} \to \Lambda_{\alpha-m}^{loc}$ for any $\alpha > 0$. This material is discussed in [TAY].

Exercise: Prove that the elliptic operator P in Theorem 4.2.5 is surjective in a suitable sense by showing that its adjoint is injective.

4.3 Change of Coordinates

It is a straightforward calculation to see, as we have indicated earlier, that an elliptic partial differential operator remains elliptic under a smooth change of coordinates. In fact pseudodifferential operators behave rather nicely under change of coordinates, and that makes them a powerful tool. They are vital, for instance, in the proof of the Atiyah–Singer index theorem because of this invariance and because they can be smoothly deformed more readily than classical partial differential operators. We shall treat coordinate changes in the present section.

Let $U, \tilde{U}$ be open sets in $\mathbf{R}^N$ and $\Phi : \tilde{U} \to U$ a C^∞ diffeomorphism. If $\tilde{x} \in \tilde{U}$ then set $x = \Phi(\tilde{x})$. Suppose that P is a pseudodifferential operator with x-support contained in U. Define, for $\phi \in \mathcal{D}(U)$,

$$\tilde{\phi}(x) = \phi \circ \Phi(\tilde{x})$$

and

$$\tilde{P}(\tilde{\phi})(\tilde{x}) \equiv P\phi(x).$$

Then $\tilde{P}$ operates on elements of $\mathcal{D}(\tilde{U})$ and we would like to determine which properties of P are preserved under the transformation $P \mapsto \tilde{P}$.

As an example, let us consider the transformation

$$\Phi(\tilde{x}) = a\tilde{x} + b,$$

where a is an invertible $N \times N$ matrix and b is an N-vector. Then

$$
\begin{aligned}
\tilde{P}\tilde{\phi}(\tilde{x}) \quad &\equiv \quad P\phi(x) \\
&= \quad \int e^{-ix\cdot\xi} p(x,\xi)\hat{\phi}(\xi)\, d\xi \\
&= \quad \iint e^{-i(x-y)\cdot\xi} p(x,\xi)\phi(y)\, dy\, d\xi \\
&= \quad \int e^{-i(a\tilde{x}+b-a\tilde{y}-b)\cdot\xi} p(a\tilde{x}+b,\xi)\phi(a\tilde{y}+b) \,|\det a|\, d\tilde{y}\, d\xi \\
&= \quad \iint e^{-i(\tilde{x}-\tilde{y})\cdot{}^t a\xi} p(a\tilde{x}+b,\xi)\tilde{\phi}(\tilde{y}) \,|\det a|\, d\tilde{y}\, d\xi \\
\overset{({}^t a\xi=\eta)}{=} \quad &\iint e^{-i(\tilde{x}-\tilde{y})\cdot\eta} p\left(a\tilde{x}+b, ({}^t a)^{-1}\eta\right)\tilde{\phi}(\tilde{y})\, d\tilde{y}\, d\eta \\
&= \quad \iint e^{-i\tilde{x}\cdot\eta} p(a\tilde{x}+b, ({}^t a)^{-1}\eta)\hat{\tilde{\phi}}(\eta)\, d\eta.
\end{aligned}
$$

Thus we see that $\tilde{P}$ is a pseudodifferential operator with symbol

$$\sigma(\tilde{P})(\tilde{x}, \xi) = p\left(a\tilde{x} + b, \left({}^{t}a\right)^{-1}\xi\right).$$

Notice that the $\tilde{x}$-support of $\sigma(\tilde{P})$ is a subset of the inverse image under Φ of the x-support of p. Thus the support is compact in $\tilde{U}$.

It turns out that for general Φ we can only expect the principal symbol of a pseudodifferential operator to transform nicely.

DEFINITION 4.3.1   *Let $\sigma = \sigma(x, \xi)$ be a symbol of order m. We call a symbol σ_L the **leading** (or **principal**) **symbol** of σ if*

(i)   *σ_L is homogeneous of degree m for $|\xi|$ large;*

(ii)   *$\sigma_L(x, \xi) - \sigma(x, \xi) \in S^{m-\epsilon}$ for some $\epsilon > 0$.*

Now let P be a pseudodifferential operator of order m, elliptic on an open set $U' \subset\subset U$. Let $\Phi : \tilde{U} \to U$ be any C^{∞} diffeomorphism. As before, define $\tilde{P}(\tilde{\phi})(\tilde{x}) = P\phi(x)$ for $\tilde{\phi} \in C_c^{\infty}(\tilde{U})$, $\tilde{\phi}(\tilde{x}) = \phi(\Phi(x))$.

THEOREM 4.3.2
There exists an open set $\tilde{V} \subset\subset \tilde{U}$ such that $\tilde{P}$ is defined by

$$\tilde{P}\tilde{\phi}(\tilde{x}) = (P\phi)(x)$$

for $\tilde{\phi} \in C_c(\tilde{V})$. Then

$$\sigma_L(\tilde{P})(\tilde{x}, \eta) = p\left(\Phi(\tilde{x}), \left({}^{t}(\mathrm{Jac}\,\Phi(\tilde{x}))\right)^{-1}\eta\right).$$

In particular, if P is elliptic on $V \equiv \Phi(\tilde{V})$, then $\tilde{P}$ is elliptic on $\tilde{V}$.

PROOF   Fix $\tilde{x} \in \tilde{U}$. Consider

$$\Phi(\tilde{x}) - \Phi(\tilde{y}) = \int_0^1 \frac{d}{dt} \Phi\left(t\tilde{x} + (1-t)\tilde{y}\right) dt$$

$$= \int_0^1 \mathrm{Jac}\,\Phi\left(\tilde{x} + (1-t)\tilde{y}\right) \cdot (\tilde{x} - \tilde{y})\, dt$$

$$\equiv H(\tilde{x}, \tilde{y}) \cdot (\tilde{x} - \tilde{y}).$$

Observe that $H(\tilde{x}, \tilde{x}) = \mathrm{Jac}\,\Phi(\tilde{x})$, which is invertible. Since the Jacobian is a continuous function of its argument, it follows that $H(\tilde{x}, \tilde{y})$ is invertible if $\tilde{y}$ is close to $\tilde{x}$. Now we choose $\tilde{V}$ a neighborhood of $\tilde{x}$ so small that $|\det H(\tilde{x}, \tilde{y})| \geq c > 0$ on $\tilde{V} \times \tilde{V}$. Then, for $x \in \tilde{V}$ and supp $\phi \subseteq \tilde{V}$ we have

$$\tilde{P}\tilde{\phi}(\tilde{x}) = P\phi(x) = \iint e^{-i(x-y)\cdot\xi} p(x, \xi)\phi(y)\, dy\, d\xi$$

$$= \iint e^{-i\left(\Phi(\tilde{x}) - \Phi(\tilde{y})\right)\cdot\xi} p(\Phi(\tilde{x}), \xi)\tilde{\phi}(\tilde{y}) \left|\frac{\partial \Phi}{\partial \tilde{y}}\right| d\tilde{y}\, d\xi.$$

Here we have used the notation $|\partial\Phi/\partial\tilde{y}|$ to denote $|\det \operatorname{Jac} \Phi(\tilde{y})|$. This last

$$= \iint e^{-i\left(H(\tilde{x},\tilde{y})\cdot(\tilde{x}-\tilde{y})\right)\cdot\xi} p(\Phi(\tilde{x}), \xi)\tilde{\phi}(\tilde{y}) \left|\frac{\partial \Phi}{\partial \tilde{y}}\right| d\tilde{y}\, d\xi$$

$$= \iint e^{-i(\tilde{x}-\tilde{y})\cdot({}^t H(\tilde{x},\tilde{y})\xi)} p(\Phi(\tilde{x}), \xi)\tilde{\phi}(\tilde{y}) \left|\frac{\partial \Phi}{\partial \tilde{y}}\right| d\tilde{y}\, d\xi$$

$$\overset{({}^t H(\tilde{x},\tilde{y})\xi=\eta)}{=} \iint e^{-i(\tilde{x}-\tilde{y})\cdot\eta} p\left(\Phi(\tilde{x}), ({}^t H(\tilde{x}, \tilde{y}))^{-1}\eta\right)\tilde{\phi}(\tilde{y})$$

$$\times \left|\frac{\partial \Phi}{\partial \tilde{y}}\right| \cdot \left|\det \left({}^t H(\tilde{x}, \tilde{y})\right)^{-1}\right| d\tilde{y}\, d\xi.$$

Now set

$$r(\tilde{x}, \eta, \tilde{y}) = p\left(\Phi(\tilde{x}), \left({}^t H(\tilde{x}, \tilde{y})\right)^{-1}\eta\right) \left|\frac{\partial \Phi}{\partial \tilde{y}}\right| \cdot \left|\det \left({}^t H(\tilde{x}, \tilde{y})\right)^{-1}\right|.$$

Then r is a Hörmander symbol and, if $\tilde{p}$ is the classical symbol of $\tilde{P}$, then

$$\tilde{p}(\tilde{x}, \eta) = \sum_{\alpha \geq 0} \frac{1}{\alpha!}\partial_\eta^\alpha D_{\tilde{y}}^\alpha r(\tilde{x}, \eta, \tilde{y})\Big|_{\tilde{y}=\tilde{x}}$$

$$= p\left(\Phi(\tilde{x}), \left({}^t H(\tilde{x}, \tilde{y})\right)^{-1}\eta\right) + \text{higher order terms}.$$

This completes the proof. ∎

4.4 Restriction Theorems for Sobolev Spaces

Let $S = \{(x_1, \ldots, x_{N-1}, 0)\} \subseteq \mathbb{R}^N$. Then S is a hypersurface, and is the boundary of $\{(x_1, \ldots, x_N) \in \mathbb{R}^N : x_N > 0\}$. It is the simplest example of the type of geometric object that arises as the boundary of a domain. It is natural to want to be able to restrict a function f defined on a neighborhood of S, or on one side of S, to S. If f is continuous on a neighborhood of S, then the restriction of f to S is trivially and unambiguously defined, simply because a continuous function is well defined at every point.

If instead $f \in H^{N/2+\epsilon}$ then we may apply the Sobolev imbedding theorem to correct f on a set of measure zero (in a unique manner) to obtain a continuous function. The corrected f may then be restricted. Restriction *prior* to the correction on the set of measure zero is *prima facie* ambiguous—a Sobolev space "function" is really an equivalence class of functions any pair of which agrees up to a set of measure zero. Unfortunately this discussion is flawed: the set S itself has measure zero. Thus two different corrections of f may have different restrictions to S.

In this section we wish to develop a notion of calculating the restriction or "trace" of a Sobolev space function on a hypersurface. We want the following characteristics to hold: (i) the restriction operation is successful on H^r for $r \ll N/2$; (ii) the restriction operation should work naturally in the context of Sobolev classes and not rely on the Sobolev imbedding theorem.

THEOREM 4.4.1
Identify S with $\mathbf{R}^{N-1}$ in a natural way. Let $s > 1/2$. Then the mapping

$$C_c^\infty \ni \phi \longmapsto \phi\big|_{\mathbf{R}^{N-1}}$$

extends to a bounded linear operator from $H^s(\mathbf{R}^N)$ to $H^{s-1/2}(\mathbf{R}^{N-1})$. That is, there exists a constant $C = C(s) > 0$ such that

$$\left\| \phi\big|_S \right\|_{H^{s-1/2}(\mathbf{R}^{N-1})} \le C(s) \|\phi\|_{H^s(\mathbf{R}^N)}.$$

REMARK Since $H^s(\mathbf{R}^N)$ and $H^{s-1/2}(\mathbf{R}^{N-1})$ are defined to be the closures of $C_c^\infty(\mathbf{R}^N)$ and $C_c^\infty(\mathbf{R}^{N-1})$ respectively, we may use the theorem to conclude the following: If T is any $(N-1)$-dimensional affine subspace of $\mathbf{R}^N$, then a function $f \in H^s(\mathbf{R}^N)$ has a well-defined trace on T. Conversely, we shall see that if $g \in H^{s-1/2}(T), s > 1/2$, then there is a function $\tilde{g} \in H^s(\mathbf{R}^N)$ such that $\tilde{g}$ has trace g on T. We leave it as an exercise for the reader to apply the implicit function theorem to see that these results are still valid if T is a sufficiently smooth hypersurface (not necessarily affine).

It is a bit awkward to state the theorem as we have (that is, as an *a priori* estimate on C_c^∞ functions). As an exercise, the reader should attempt to reformulate the theorem directly in terms of the H^s spaces to see that in fact the statement of Theorem 4.4.1 is as simple as it can be made. ∎

PROOF OF THEOREM 4.4.1   We introduce the notation $(x', x_N) = (x_1, \ldots, x_N)$ for an element of $\mathbf{R}^N$. If $u \in C_c^\infty(\mathbf{R}^N)$ then we will use the notation $u^r(x')$

to denote $u(x', 0)$. Now we have

$$\|u^\tau\|^2_{H^{s-1/2}(\mathbf{R}^{N-1})} = \int_{\mathbf{R}^{N-1}} |\widehat{u^\tau}(\xi')|^2 (1 + |\xi'|^2)^{s-1/2} \, d\xi'$$

$$= \int_{\mathbf{R}^{N-1}} \left[-\int_0^\infty \frac{\partial}{\partial x_N} |\widehat{u}_1(\xi', x_N)|^2 (1 + |\xi'|^2)^{s-1/2} \, dx_N \right] d\xi'.$$

Here $\widehat{u}_1$ denotes the partial Fourier transform in the variable x'. The product rule yields that, if D is a first derivative, then $D(|h|^2) \le 2|h| \cdot |Dh|$. Therefore the last line does not exceed

$$2 \int_{\mathbf{R}^{N-1}} \int_0^\infty |\widehat{u}_1(\xi', x_N)| \cdot \left| \frac{\partial}{\partial x_N} \widehat{u}_1(\xi', x_N) \right| \cdot (1 + |\xi'|^2)^{s-1/2} \, dx_N \, d\xi'.$$

Since $2\alpha\beta \le \alpha^2 + \beta^2$, the last line is

$$\le \int_{\mathbf{R}^{N-1}} \int_0^\infty \left| \frac{\partial}{\partial x_N} \widehat{u}_1(\xi', x_N) \right|^2 dx_N (1 + |\xi'|^2)^{s-1} \, d\xi'$$

$$+ \int_{\mathbf{R}^{N-1}} \int_0^\infty |\widehat{u}_1(\xi', x_N)|^2 \, dx_N (1 + |\xi'|^2)^s \, d\xi'$$

$$\equiv I + II.$$

Now apply Plancherel's theorem to term II in the x_N variable. The result is

$$II \le C \int_{\mathbf{R}^{N-1}} \int_{\mathbf{R}} |\widehat{u}(\xi', \xi_N)|^2 \, d\xi_N (1 + |\xi'|^2)^s \, d\xi'$$

$$\le C \|u\|^2_{H^s(\mathbf{R}^N)}.$$

Plancherel's theorem applied, in the x_N variable, to the term I yields

$$I \le C \int_{\mathbf{R}^{N-1}} \int_{\mathbf{R}} \left| \frac{\partial}{\partial x_N} \widehat{u}_1(\xi', x_N) \right|^2 dx_N (1 + |\xi'|^2)^{s-1} \, d\xi'$$

$$\le C \int_{\mathbf{R}^{N-1}} \int_{\mathbf{R}} |\widehat{u}(\xi', \xi_N)|^2 \cdot |\xi_N|^2 \, d\xi_N (1 + |\xi'|^2)^{s-1} \, d\xi'$$

$$\le C \int_{\mathbf{R}^N} |\widehat{u}(\xi)|^2 \cdot (1 + |\xi|^2)^s \, d\xi$$

$$= C \|u\|_{H^s(\mathbf{R}^N)}. \qquad \blacksquare$$

As an immediate corollary we have:

COROLLARY 4.4.2
Let $u \in H^s(\mathbf{R}^N)$ and let α be a multiindex such that $s > |\alpha| + 1/2$. Then $D^\alpha u$ has trace in $H^{s-|\alpha|-1/2}(\mathbf{R}^{N-1})$.

Now we present a converse to the theorem. Again set $S = \{(x', 0) \in \mathbf{R}^N\}$.

THEOREM 4.4.3
Assume that $\phi_0, \ldots, \phi_k$ are defined on S, each $\phi_j \in H^{s-j-1/2}(\mathbf{R}^{N-1})$ with $s > 1/2 + k$. Then there exists a function $f \in H^s(\mathbf{R}^N)$ such that $D_N^j f$ has trace ϕ_j on $S, j = 0, \ldots, k$. Moreover,

$$\|f\|_{H^s(\mathbf{R}^N)} \le C_{s,k} \sum_{j=0}^{k} \|\phi_j\|^2_{H^{s-j-1/2}(\mathbf{R}^{N-1})}.$$

PROOF   Let $h \in C_c^\infty(\mathbf{R}), h \equiv 1$ in a neighborhood of the origin, $0 \le h \le 1$. We define

$$\widehat{u_1}(\xi', x_N) = \sum_{j=0}^{k} \frac{1}{j!}(-ix_N)^j h\left(x_N(1 + |\xi'|^2)^{1/2}\right)\widehat{\phi_j}(\xi')$$

and

$$f(x) = \int e^{-ix'\cdot\xi'} \widehat{u_1}(\xi', x_N)\, d\xi'.$$

This is the function f that we seek. For if m is any integer, $0 \le m \le k$, then

$$D_N^m f(x', 0) = \int e^{-ix'\cdot\xi'} D_N^m \widehat{u_1}(x', x_N)\big|_{x_N=0}\, d\xi'$$

$$= \int e^{-ix'\cdot\xi'} i^m \sum_{j=0}^{k} \frac{1}{j!} \sum_{\ell=0}^{m} \binom{m}{\ell} \frac{\partial^\ell}{\partial x_N^\ell}(-ix_N)^j\bigg|_{x_N=0}$$

$$\times \frac{\partial^{m-\ell}}{\partial x_N^{m-\ell}}[h(x_N(1 + |\xi'|^2)^{1/2})]\bigg|_{x_N=0} \cdot \widehat{\phi_j}(\xi')\, d\xi'$$

$$= \int e^{-ix'\cdot\xi'} i^m \sum_{j=0}^{k} \frac{1}{j!} \binom{m}{j} j!(-i)^j$$

$$\times \frac{\partial^{m-j}}{\partial x_N^{m-j}}[h(x_N(1 + |\xi'|^2)^{1/2})]\bigg|_{x_N=0} \cdot \widehat{\phi_j}(\xi')\, d\xi'$$

$$= \int e^{-ix'\cdot\xi'} \widehat{\phi_m}(\xi')\, d\xi'$$

$$= \phi_m(x').$$

This verifies our first assertion.

Now we need to check that f has the right Sobolev norm. We have

$$\|f\|_{H^2}^2 = \int_{\mathbf{R}^N} |\hat{f}(\xi)|^2 (1 + |\xi|^2)^s \, d\xi$$

$$= \int_{\mathbf{R}^N} \left| \widehat{\left(\hat{u}_1(\xi', \cdot) \right)_2}(\xi_N) \right|^2 (1 + |\xi|^2)^s \, d\xi$$

$$= \int_{\mathbf{R}^N} \left| \sum_{j=0}^k \frac{1}{j!} \frac{\partial^j}{\partial \xi_N^j} \hat{h}\left(\frac{\xi_N}{(1 + |\xi'|^2)^{1/2}} \right) \frac{1}{(1 + |\xi'|^2)^{1/2}} \hat{\phi}_j(\xi') \right|^2$$

$$\times (1 + |\xi|^2)^s \, d\xi.$$

Now we use the fact that $|\sum_{j=0}^k \zeta_j|^2 \le C \cdot (\sum_{j=0}^k |\zeta_j|^2)$, where the constant C depends only on k. The result is that the last line is

$$\le C \int_{\mathbf{R}^N} \sum_{j=0}^k \frac{1}{(j!)^2} \left| \frac{\partial^j}{\partial \xi_N^j} \hat{h}\left(\frac{\xi_N}{(1 + |\xi'|^2)^{1/2}} \right) \right|^2$$

$$\times \frac{1}{1 + |\xi'|^2} \cdot |\widehat{\phi_j}(\xi)|^2 (1 + |\xi|^2)^s \, d\xi$$

$$\le C \int_{\mathbf{R}^N} \sum_{j=0}^k \frac{1}{(j!)^2} \left| \hat{h}^{(j)} \left(\frac{\xi_N}{(1 + |\xi'|^2)^{1/2}} \right) \right|^2$$

$$\times \frac{1}{(1 + |\xi'|^2)^{j+1}} \cdot |\widehat{\phi_j}(\xi)|^2 (1 + |\xi|^2)^s \, d\xi. \qquad (4.4.3.1)$$

Now $d\xi = d\xi' d\xi_N$. We make the change of variables

$$\xi_N \mapsto \xi_N (1 + |\xi'|^2)^{1/2}.$$

Then

$$1 + |\xi|^2 = 1 + |\xi'|^2 + |\xi_N|^2 \mapsto 1 + |\xi'|^2 + |\xi_N|^2 (1 + |\xi'|^2)$$

$$= (1 + |\xi'|^2)(1 + |\xi_N|^2).$$

Then (4.4.3.1) is

$$\leq C \sum_{j=0}^{k} \frac{1}{(j!)^2} \int_{\mathbf{R}^{N-1}} \int_{\mathbf{R}} |\hat{h}^{(j)}(\xi_N)|^2 \frac{1}{(1+|\xi'|^2)^{j+1}} (1+|\xi'|^2)^{1/2}$$

$$\times |\widehat{\phi_j}(\xi')|^2 \left[(1+|\xi'|^2)(1+|\xi_N|^2)\right]^s d\xi' d\xi_N$$

$$= C \sum_{j=0}^{k} \frac{1}{(j!)^2} \int_{\mathbf{R}^{N-1}} |\hat{\phi}_j(\xi')|^2 (1+|\xi'|^2)^{s-j-1/2} d\xi'$$

$$\times \int_{\mathbf{R}} |\hat{h}^{(j)}(\xi_N)|^2 (1+|\xi_N|^2)^s d\xi_N$$

$$\leq C \sum_{j=0}^{k} \frac{1}{(j!)^2} \|\phi_j\|_{H^{s-j-1/2}(\mathbf{R}^{N-1})} \cdot \|h\|^2_{H^{s+k}(\mathbf{R})}$$

$$\leq C_{s,k} \sum_{j=0}^{k} \|\phi_j\|^2_{H^{s-j-1/2}(\mathbf{R}^{N-1})}.$$

That completes the proof of our trace theorem. ∎

REMARK   The extension $\hat{u}_1$ was constructed by a scheme based on ideas that go back at least to A. P. Caldéron—see [STSI] and references therein. ∎

5

Elliptic Boundary Value Problems

5.1 The Constant Coefficient Case

We begin our study of boundary value problems by considering $\Omega = \mathbb{R}^N_+, \partial\Omega = \{(x',0) : x' \in \mathbb{R}^{N-1}\}$. We will study the problem

$$\begin{cases} P(D)u = f & \text{on } \mathbb{R}^N_+ \\ B_j(D)u = g_j & \text{on } \partial\mathbb{R}^N_+, \quad j = 1,\ldots,k. \end{cases} \quad (*)$$

Here P will be an elliptic operator. At first both P and the B_j will have constant coefficients. Our aim is to determine what conditions on the operators P, B_j will make this a well-posed and solvable boundary value problem. We shall assume that P is homogeneous of degree $m > 0$ and that

$$|P(\xi)| \geq C|\xi|^m.$$

Examples: Let

$$P(D) = \Delta = \frac{\partial}{\partial x_1^2} + \frac{\partial}{\partial x_2^2}.$$

Example 1
First consider $\Omega = \mathbb{R}^2_+$. Let us discuss the boundary value problem

$$\begin{cases} \Delta u = 0 \\ u\big|_{\partial\Omega} = 0 \\ \frac{\partial u}{\partial x_1}\big|_{\partial\Omega} = 1. \end{cases}$$

This system has no solution because the second boundary condition is inconsistent with the first. The issue here turns out to be one of transversality of boundary conditions involving derivatives. See the next example. ☐

Example 2

We repair the first example by making the second boundary condition transverse: the problem

$$\left\{ \begin{array}{l} \Delta u = 0 \\ u\big|_{\partial\Omega} = 0 \\ \frac{\partial u}{\partial x_2}\big|_{\partial\Omega} = 1 \end{array} \right.$$

has the unique solution $u(x_1, x_2) = x_2$.   ☐

Example 3

Let $\Omega = \mathbb{R}^2_+$. Consider the boundary value problem

$$\left\{ \begin{array}{l} \Delta u = 0 \\ u\big|_{\partial\Omega} = g_1 \\ \frac{\partial u}{\partial x_1}\big|_{\partial\Omega} = g_2. \end{array} \right.$$

In fact, take $g_1(x_1, x_2) = x_1, g_2(x_1, x_2) \equiv 1$. Notice that any function of the form $v(x_1, x_2) = x_1 + Cx_2$ satisfies

$$\left\{ \begin{array}{l} \Delta v = 0 \\ v\big|_{\partial\Omega} = x_1 \\ \frac{\partial v}{\partial x_1}\big|_{\partial\Omega} = 1. \end{array} \right.$$

Hence the problem is sensible, but it has infinitely many solutions.   ☐

Our goal is to be able to recognize problems that have one, and only one, solution.

Necessary Conditions on the Operators B_j

First, the degree of each B_j must be smaller than the degree of P. The reason is that P is elliptic. According to elliptic regularity theory, all derivatives of u of order m and above are controlled by f (the forcing term in the partial differential equation). Thus these derivatives are not free to be specified.

Now we develop the Lopatinski condition. We shall assume for simplicity (and in the end see that this entails no loss of generality) that our partial differential operator P is homogeneous of degree m. Thus it has the form

$$\sum_{|\alpha|=m} b_\alpha \left(\frac{\partial}{\partial x} \right)^\alpha.$$

For each fixed $\xi' = (\xi_1, \ldots, \xi_{N-1})$, we consider $P(\xi', D_N)$. We will solve

$$P(\xi', D_N)v(x_N) = 0$$

as an ordinary differential equation. Thus our problem has the form

$$\sum_{\ell=0}^{m} a_\ell(\xi') \left(\frac{d}{dx_N} \right)^{\ell} v = 0.$$

The solution v of such an equation will have the form

$$v(x_N) = \sum_{j=1}^{r(\xi')} \sum_{\ell=0}^{\nu_j-1} c_{j\ell}(\xi')(x_N)^\ell e^{ix_N \lambda_j(\xi')},$$

where $\lambda_1(\xi'), \lambda_2(\xi'), \ldots, \lambda_r(\xi')$ are the roots of $P(\xi', \cdot) = 0$ with multiplicities $\nu_1(\xi'), \nu_2(\xi'), \ldots, \nu_r(\xi')$.

We restrict attention to those λ_j with positive real part. Since these roots alone will be enough to enable us to carry out our program, this choice is justified in the end. However, an *a priori* theoretical justification for restricting attention to these λ's may be found in [HOR1].

After renumbering, let us say that we have retained the roots

$$\lambda_1(\xi'), \ldots, \lambda_{r_0}(\xi'), \qquad r_0 \leq r.$$

Then

$$v = \sum_{j=1}^{r_0} \sum_{\ell=0}^{\nu_j-1} c_{j\ell}(\xi')(x_N)^\ell e^{ix_N \lambda_j(\xi')}.$$

Fact: A fundamental observation for us will be that the number r_0 of roots (counting multiplicities) with positive imaginary part is independent of $\xi' \in \mathbb{R}^{N-1} \setminus \{0\}$. This is proved by way of the following two observations:

(*i*) Each $\lambda_j(\xi'), j = 1, \ldots, r_0$, depends continuously on ξ'.

(*ii*) There are no λ_j with zero imaginary part (except possibly $\lambda = 0$).

We leave it to the reader to supply the details verifying that these two observations imply that the number of roots is independent of ξ' (use the ellipticity for (ii)). Now we summarize the situation: our problem is to solve, for fixed ξ', the system

$$P(\xi', D_N)v = 0 \quad \text{on} \ \mathbb{R}^N_+$$

$$B_j(\xi', D_N)v = g_j(\xi') \quad \text{on} \ \partial\mathbb{R}^N_+, j = 1, \ldots, k. \tag{5.1.1}$$

From now on we take k to equal $\nu_0 + \cdots + \nu_{r_0}$.

For fixed ξ' we will formulate a condition that guarantees that our system has a unique solution:

The Lopatinski Condition: For each fixed $\xi' \neq 0$ and for each set of functions $\{g_j\}$, the system (5.1.1) has a unique solution.

Let us clarify what we are about to show: If, for each fixed ξ', the ordinary differential equation with boundary conditions (5.1.1) has a unique solution (no matter what the data $\{g_j\}$) then we will show that the full system described at the beginning of the section has a unique solution. The condition amounts, after some calculation, to demanding the invertibility of a certain matrix.

Let $p(\xi) = \sum_{j,k} a_{jk}\xi_j\xi_k$ with the matrix (a_{jk}) positive definite. Then

$$P(D) = \sum a_{jk}\frac{\partial}{\partial x_j}\frac{\partial}{\partial x_k} .$$

Observe that

$$C_1|\xi|^2 \leq |P(\xi)| \leq C_2|\xi|^2.$$

In this example, for fixed ξ', the polynomial $P(\xi',\eta)$ is quadratic in η. The positive definiteness implies that it has just one root with positive imaginary part. So, if Lopatinski's condition is to be satisfied, we can have just one boundary condition of degree 0 or 1. Let us consider two cases:

(i) B is of degree zero (that is, B consists of multiplication by a function). Thus, by Lopatinski, we must be able to solve

$$B(\xi', D_N)v = b(\xi')v = g(\xi')$$

for every g. This is the same as being able to find, for every $g(\xi')$, a function $c(\xi')$ such that

$$b(\xi')c(\xi')e^{ix_N\cdot\lambda_1(\xi')} = g(\xi')$$

when $x_N = 0$; in other words, we must be able to solve $b(\xi')c(\xi') = g(\xi')$. We can find such a c provided only that $b(\xi')$ does not vanish.

(ii) B is of degree one. It is convenient to write B as

$$B(D) = \sum_j b_j\frac{\partial}{\partial x_j}$$

where $D_j = i\partial/\partial x_j$. Assume for simplicity that the b_j's are real. According to the Lopatinksi condition, we must be able to solve

$$B(\xi', D_N)v = g$$

on $\mathbf{R}^{N-1}$ for any g. Recall that we have already ascertained that

$$v = c(\xi')e^{ix_N\lambda_1(\xi')}.$$

Because B has degree one we have

$$B(\xi', D_N) = b_1 i\xi_1 + b_2 i\xi_2 + \cdots + b_{N-1} i\xi_{N-1} + b_N i\frac{\partial}{\partial x_N}$$

$$B(\xi', D_N)v = ib_1\xi_1 c(\xi')e^{ix_N\cdot\lambda_1(\xi')} + ib_2\xi_2 c(\xi')e^{ix_N\cdot\lambda_1(\xi')}$$

$$+ \cdots + ib_{N-1}\xi_{N-1} c(\xi')e^{ix_N\cdot\lambda_1(\xi')} + ib_N i\lambda_1(\xi')c(\xi')e^{ix_N\cdot\lambda_1(\xi')}$$

$$= g(\xi').$$

Setting $x_N = 0$ gives

$$i\left[b_1\xi_1 c(\xi') + \cdots + b_{N-1}\xi_{N-1} c(\xi')\right] - b_N\lambda_1(\xi')c(\xi') = g(\xi').$$

Since the coefficients b_j are real, the hypothesis that $b_N \neq 0$ will guarantee that we can always solve this equation.

5.2   Well-Posedness

Let the system

$$P(\xi', D_N)v = f \text{ on } \mathbb{R}_+^N$$

$$B_j(\xi', D_N)v = g_j(\xi') \text{ on } \partial\mathbb{R}_+^N, \qquad j = 1,\ldots,k, \qquad (5.2.1)$$

have the property that the operators P, B_j have constant coefficients. The operator P is homogeneous of degree m. Assume that each operator B_j is homogeneous of order m_j and that $m > m_j$. The system is called *well-posed* if Conditions (A), (B), (C) below are met.

(A) *Regularity*. The space of solutions of

$$P(D)u = 0, \quad x_N > 0$$

$$B_j(D)u = 0, \quad x_N = 0, \qquad 1 \le j \le k,$$

in $H^m(\mathbb{R}_+^N)$ has finite dimension and there is a $C > 0$ such that

$$\|v\|_{H^m} \le C\left(\|P(D)v\|_{H^0} + \sum_{j=1}^{k}\|B_j(D)v\|_{H^{m-m_j-1/2}} + \|v\|_{H^0}\right)$$

$$\forall v \in H^m(\mathbb{R}_+^N).$$

(B) Existence. In the space

$$C_c^\infty(\mathbb{R}_+^N) \times \prod_{p=1}^k C_c^\infty(\mathbb{R}^{N-1})$$

there is a subspace $\mathcal{L}$ having finite codimension such that if $(f, g_1, g_2, \ldots, g_k)$ $\in \mathcal{L}$, then the boundary value problem (5.2.1) has a solution u in $H^m(\mathbb{R}_+^N)$.

(C) Let γ be the operator of restriction to the hyperplane

$$\{(x_1, \ldots, x_N) : x_N = 0\}.$$

Then the set

$$\{(P(D)u, \gamma B_1(D)u, \ldots \gamma B_k(D)u) : u \in H^m(\mathbb{R}_+^N)\}$$

is closed in

$$H^0(\mathbb{R}_+^N) \times H^{m-m_1-1/2}(\mathbb{R}^{N-1}) \times \cdots \times H^{m-m_k-1/2}(\mathbb{R}^{N-1}).$$

Now our theorem is

THEOREM 5.2.2
The system (5.2.1) is well posed (that is, conditions (A), (B), and (C) are satisfied) if and only if the system satisfies Lopatinski's condition.

The proof of this theorem will occupy the rest of the section. It will be broken up into several parts.

PART 1 OF THE PROOF   It is plain that the failure of the existence part of the Lopatinski condition implies that either B or C of well-posedness fails.

We will thus show in this part that if the uniqueness portion of the Lopatinski condition fails then Condition A of well-posedness fails. The failure of Lopatinski uniqueness for some $\xi_0' \neq 0$ means that the system

$$P(\xi_0', D_N)v = 0$$
$$B_j(\xi_0', D_N)v = 0$$

has a nonzero solution. Call it v. Let $\phi_1 \in C_c^\infty(\mathbb{R}^{N-1})$, $\phi_2 \in C_c^\infty(\mathbb{R})$ both be identically equal to 1 near 0. For $T > 0$ we define

$$u_T(x) = \phi_1(x')\phi_2(x_N)e^{iTx'\cdot\xi_0'}v(Tx_N).$$

We will substitute u_T into condition (A) of well-posedness and let $T \to +\infty$ to obtain a contradiction.

First observe that $\|u_T\|_{\sup} < \infty$. Now let $\alpha = (\alpha_1, \ldots, \alpha_{N-1}, 0)$ be a multiindex of order m. We calculate that

$$\|u_T\|_{H^m} \geq \|D^\alpha u_T\|_{H^0}^2$$

$$= \int \left| D_{x'}^\alpha \left[\phi_1(x')\phi_2(x_N)e^{iT(x'\cdot\xi_0')}v(Tx_N) \right] \right|^2 dx' dx_N$$

$$\geq C \int \left| \phi_1(x')\phi_2(x_N)(\xi_0')^\alpha T^{|\alpha|}v(Tx_N) \right|^2 dx' dx_N$$

$$- C \int \left| \sum_{\substack{\beta+\gamma=\alpha \\ |\beta|>0}} C_{\beta\gamma} D_{x'}^\beta [\phi_1(x')]\phi_2(x_N)(\xi_0')^\gamma T^{|\gamma|}v(Tx_N) \right|^2 dx' dx_N$$

$$= \frac{C}{T} \int \left| \phi_1(x')\phi_2(x_N/T)(\xi_0')^\alpha T^{|\alpha|}v(x_N) \right|^2 dx' dx_N$$

$$- \frac{C}{T} \int \left| \sum_{\substack{\beta+\gamma=\alpha \\ |\beta|>0}} C_{\beta\gamma} D_{x'}^\beta [\phi_1(x')]\phi_2(x_N/T)(\xi_0')^\gamma T^{|\gamma|}v(x_N) \right|^2 dx' dx_N$$

$$\geq cT^{2m}T^{-1} - C'T^{2m-2}T^{-1}$$

$$\geq CT^{2m-1}$$

for T large. Therefore

$$\|u_T\|_{H^m} \geq CT^{2m-1} \tag{5.2.2.1}$$

for T large. On the other hand,

$$P(D)u_T = P(D) \left[\phi_1(x')\phi_2(x_N)e^{iTx'\cdot\xi_0'}v(Tx_N) \right]$$

$$= \phi_1(x')\phi_2(x_N)P(D) \left[e^{iTx'\cdot\xi_0'}v(Tx_N) \right]$$

$$+ \text{(terms in which a derivative falls on a cutoff function)}$$

$$= 0 + \text{(terms in which fewer than } m \text{ derivatives}$$

$$\text{fall on } e^{iTx'\cdot\xi_0'}v(Tx_N)). \tag{5.2.2.2}$$

We have used here the fact that

$$P(D) \left[e^{iT(x'\cdot\xi_0')}v(Tx_N) \right] = P(T\xi_0', TD_N)v\big|_{Tx_N}$$

$$= T^m P(\xi_0', D_N)v\big|_{Tx_N}$$

$$= 0.$$

As a result, from (5.2.2.2), we obtain

$$\|P(D)u_T\|_0^2 \le CT^{2m-2}. \tag{5.2.2.3}$$

Similarly,

$$B_j(D)\left[e^{iT x' \cdot \xi_0'} v(T x_N)\right] = B_j(T\xi_0', TD_N)v\big|_{T x_N}$$

$$= T^{m_j} B_j(\xi_0', D_N)v\big|_{T x_N}.$$

Therefore

$$B_j(D)[u_T] = B_j(D)\left[\phi_1(x')\phi_2(x_N)e^{iT x' \cdot \xi_0'} v(T x_N)\right]$$

$$= 0 + \text{ terms in which derivatives of total order}$$

$$\text{not exceeding } m_j - 1 \text{ land on } e^{iT x' \cdot \xi_0'} v(T x_N).$$

As above,

$$\|B_j(D)u_T\|_{m-m_j-1/2}^2 \le C\left[T^{m_j-1+m-m_j-1/2}\right]^2 = CT^{2m-3}. \tag{5.2.2.4}$$

Therefore, substituting u_T into condition (A) of the definition of well-posedness, we obtain (from (5.2.2.1), (5.2.2.3), and (5.2.2.4)) that

$$CT^{2m-1} \le C'T^{2m-2} + C''.$$

This inequality leads to a contradiction if we let $T \to +\infty$.

PART 2 OF THE PROOF   Assume that the Lopatinski condition holds at all $\xi' \ne 0$. We will prove that the system is well posed. This argument will proceed in several stages and will take the remainder of the section.

Let $u \in H^m(\mathbf{R}_+^N)$ be a solution of

$$\begin{cases} P(D)u = 0 & \text{on } \mathbf{R}_+^N \\ B_j(D)u = 0 & \text{on } \partial\mathbf{R}_+^N, \quad j = 1, \dots, k. \end{cases}$$

Let

$$v(\xi', x_N) = \int e^{i\xi' \cdot x'} u(x', x_N)\, dx'.$$

Then

$$\begin{cases} P(\xi', D_N)v = 0 & \text{on } \Omega \\ B_j(\xi', D_N)v = 0 & \text{on } \partial\Omega. \end{cases}$$

The general solution of $P(\xi', D_N)v = 0$ looks like

$$v(\xi', x_N) = \sum_{j=1}^{r_0} \sum_{\ell=0}^{\nu_j-1} c_{j\ell}(\xi')(x_N)^\ell e^{ix_N \lambda_j(\xi')}.$$

But the Lopatinski hypothesis then guarantees that such a $v \equiv 0$ so that $u \equiv 0$.

To prove part (A) of well-posedness, it is therefore enough to show that whenever $f \in L^2(\mathbf{R}_+^N)$ and $g_j \in H^{m-m_j-1/2}(\mathbf{R}^{N-1})$, then the boundary value problem

$$P(D)u = f$$

$$B_j(D)u = g_j, \quad j = 1, \dots, k$$

has a solution that satisfies the desired inequality. We shall need the following lemma.

LEMMA 5.2.3
If P is a constant coefficient partial differential operator, then P always has a fundamental solution. That is, if P is of order m, then there is a bounded operator

$$\mathcal{E} : H^s \to \mathcal{D}'$$

for every $s \in \mathbf{R}$ such that $P\mathcal{E} = \mathcal{E}P = \delta_0$. If P is elliptic then $\mathcal{E}$ is of order $-m$.

PROOF   First consider the case $N = 1$. Select a number $T \in \mathbf{R}$ such that $P(\xi + iT)$ never vanishes. Then the fundamental solution operator is

$$\mathcal{E}(\phi) = \int \frac{\hat{\phi}(-\xi - iT)}{P(\xi + iT)} \, d\xi$$

for $\phi \in \mathcal{D}$. We check that

$$P(D)\mathcal{E}\phi = \mathcal{E}P(-D)\phi = \int \frac{(P(-D)\phi)\hat{\,}(-\xi - iT)}{P(\xi + iT)} \, d\xi$$

$$= \int \frac{P(\xi + iT)\hat{\phi}(-\xi - iT)}{P(\xi + iT)} \, d\xi = \int \hat{\phi}(-\xi - iT) \, d\xi$$

$$= \int \hat{\phi}(\xi - iT) \, d\xi = \int \hat{\phi}(\xi) \, d\xi$$

$$= \phi(0).$$

This proves the result in dimension 1.

If $N > 1$, then we can reduce the problem to the one-dimensional case as follows: By rotating coordinates, we may assume that the coefficient of ξ_1^m in P is not zero. Fix $\xi' = (\xi_2, \dots, \xi_N)$. We can find a $T, |T| \leq 1$, such that

$$|P(\xi_1 + iT, \xi')| \geq C_0 > 0 \qquad \text{for all } \xi_1.$$

Moreover, if we multiply P by a constant, we can assume that

$$|P(\xi_1 + iT, \xi')| > 1$$

for all ξ_1. This inequality—that is, the choice of the constant to normalize the inequality—will depend on ξ'. But the choice is uniform in a neighborhood of the fixed ξ'. Thus to each fixed ξ' we associate a neighborhood $W_{\xi'}$ and a real number $T, |T| \le 1$, such that

$$|P(\xi_1 + iT, \xi')| > 1$$

on $W_{\xi'}$.

Observe that $\mathbf{R}^{N-1}$ is covered by these neighborhoods $W_{\xi'}$. We may refine this covering to a locally finite one $W_1, W_2, \ldots$, with a T_j associated to each W_j. Now we replace the W_j with their disjoint counterparts: define

$$W_1' = W_1$$
$$W_2' = \overline{W}_2 \setminus W_1'$$
$$W_3' = \overline{W}_3 \setminus W_2' \setminus W_1'$$

$$\cdots$$

These sets still cover $\mathbf{R}^{N-1}$ and they are disjoint. Now we define the "Hörmander ladder"

$$H = \cup_j \cup_{\xi' \in W_j'} \{(\xi_1 + iT_j, \xi')\}.$$

Given $\phi \in \mathcal{D}$, we define

$$\mathcal{E}(\phi) = \int_H \frac{\hat{\phi}(-\xi_1 - iT, \xi')}{P(\xi_1 + iT, \xi')} \, d\xi.$$

Notice that, on H, we know that $|P| > 1$ and

$$(P(D)\mathcal{E})\,\phi = \mathcal{E}(P(-D)\phi) = \int_H \frac{(P(-D)\phi)\,\hat{}\,(-\xi_1 - iT, \xi')}{P(\xi_1 + iT, \xi')} \, d\xi$$

$$= \int_H \hat{\phi}(-\xi_1 - iT, \xi') \, d\xi$$

$$= \sum_j \int_{\xi' \in W_j'} \int_{\mathbf{R}} \hat{\phi}(-\xi_1 - iT_j, \xi') \, d\xi_1 \, d\xi'$$

$$= \sum_j \int_{\xi' \in W_j} \int_{\mathbf{R}} \hat{\phi}(-\xi_1, \xi') \, d\xi_1 \, d\xi'$$

$$= \int_{\mathbf{R}^{N-1}} \int_{\mathbf{R}} \hat{\phi}(\xi_1, \xi') \, d\xi_1 \, d\xi'$$

$$= \phi(0).$$

It is elementary to check that in case P is elliptic, then $\mathcal{E}$ is an operator of order $-m$.

That completes the proof of the lemma. ∎

We conclude this section by proving the inequality in part (A) of well-posedness. Notice that part (C) of well-posedness follows immediately from this inequality. Along the way, we prove the sufficiency of the Lopatinski condition for the existence of solutions to our system:

PROPOSITION 5.2.4
Let $m > m_j, j = 1, \ldots, k$. Assume that our system satisfies the Lopatinski condition. Then whenever $f \in L^2(\mathbb{R}^N_+)$ and $g_j \in H^{m-m_j-1/2}(\mathbb{R}^{N-1})$, we may conclude that the boundary value problem

$$P(D)u = f$$
$$B_j(D)u = g_j, \quad j = 1, \ldots, k$$

has a unique solution $u \in H^m(\mathbb{R}^N_+)$. The solution satisfies the inequality in part (A) of well-posedness.

PROOF First we notice that the Lopatinski condition guarantees that the kernel of the system is zero. Since the system is linear, we conclude that solutions are unique once they exist.

For existence, we begin by extending f to be L^2 on all of $\mathbb{R}^N$. We denote the extended function by f as well. Let $\mathcal{E}$ be the fundamental solution for the operator $P(D)$. Set $u_1 = \mathcal{E} * f$. Now define

$$v = u - u_1$$
$$h_j = g_j - \gamma B_j(D)u_1,$$

where γ is the operation of restriction to the boundary of $\mathbb{R}^N_+$. Thus our system becomes

$$P(D)v = 0 \qquad \text{on } \mathbb{R}^N_+$$
$$B_j(D)v = h_j \qquad \text{on } \partial\mathbb{R}^N_+.$$

Observe that

$$\|h_j\|_{m-m_j-1/2} \leq \|g_j\|_{H^{m-m_j-1/2}(\mathbb{R}^{N-1})} + \|\gamma B_j(D)u_1\|_{H^{m-m_j-1/2}(\mathbb{R}^{N-1})}$$
$$\leq \|g_j\|_{H^{m-m_j-1/2}(\mathbb{R}^{N-1})} + C\|u_1\|_{H^m(\mathbb{R}^N)}.$$

Here, of course, we have used the standard restriction theorem for Sobolev spaces.

Now this last line is

$$\leq \|g_j\|_{H^{m-m_j-1/2}(\mathbf{R}^{N-1})} + C\|f\|_{H^0(\mathbf{R}^N)}$$

because $\mathcal{E}$ is an operator of order $-m$. We apply the partial Fourier transform, denoted by $\tilde{\ }$, in the x' variable to transform our system to

$$P(\xi', D_N)\tilde{v} = 0 \qquad \text{on } \mathbf{R}_+^N$$
$$B_j(\xi', D_N)\tilde{v} = \tilde{h}_j \qquad \text{on } \partial\mathbf{R}_+^N, \; j = 1,\dots,k. \qquad (5.2.4.1)$$

By Lopatinksi's condition, the space of solutions of the first equation that decrease exponentially has finite dimension (indeed k dimensions) and the map

$$\tilde{v} \mapsto \{\gamma B_1(\xi', D_N)\tilde{v},\dots,\gamma B_k(\xi', D_N)\tilde{v}\}$$

is one-to-one and onto. Thus a solution to (5.2.4.1) exists. By the fact that all norms on a finite-dimensional vector space are comparable (alternatively, by the open mapping principle),

$$\sum_{j=0}^{m}\int_0^\infty |D_N^j\tilde{v}(\xi',x_N)|^2\,dx_N + \sum_{j=0}^{m-1}|D_N^j\tilde{v}(\xi',0)|^2 \leq C_2(\xi')\sum_{j=1}^{k}|\tilde{h}_j(\xi')|^2.$$

Here the right-hand side represents a norm on the space of k-tuples, while the left-hand side is a norm on the solution space of the boundary value problem.

Now by direct estimation, or using Theorem 10.2.1 of [HOR1], we obtain

$$\sum_j |\tilde{h}_j(\xi')|^2 \leq C \cdot \left(\sum_j |\tilde{g}_j(\xi')|^2 + \sum_j |B_j(\xi',D_N)\tilde{u}_1(\xi',0)|^2 \right)$$
$$\leq C_3(\xi')\left[\sum_j |\tilde{g}_j(\xi')|^2 + \int_0^\infty |\tilde{f}(\xi',x_N)|^2\,dx_N \right].$$

The constant $C(\xi')$ is a continuous function of ξ' (since it arises from the inversion of a matrix with continuous coefficients). Therefore it is bounded above and below on $\{\xi' : |\xi'| = 1\}$. We conclude that

$$\sum_{j=0}^{m}\int_0^\infty |D_N^j\tilde{v}(\xi',x_N)|^2\,dx_N + \sum_{j=0}^{m-1}|D_N^j\tilde{v}(\xi',0)|^2$$
$$\leq C_0\left[\sum_j |\tilde{g}_j(\xi')|^2 + \int_0^\infty |\tilde{f}(\xi',x_N)|^2\,dx_N \right]$$

provided that $|\xi'| = 1$. Let $r > 0$ be a constant. Then any inequality that holds for the original system

$$P(D)u = f$$
$$B_j(D)u = g_j, \qquad j = 1,\ldots,k$$

must also hold for the system

$$P(rD)u = r^m f$$
$$B_j(rD)u = r^{m_j} g_j, \qquad j = 1,\ldots,k$$

when $|\xi'| = 1/r$. Putting this information into our inequality, and substituting $|\xi'|$ for $1/r$, yields that

$$\sum_{j=0}^{m} |\xi'|^{2(m-j)} \int_0^{\infty} |D_N^j \tilde{v}(\xi', x_N)|^2 \, dx_N + \sum_{j=0}^{m-1} |\xi'|^{2(m-j)-1} |D_N^j \tilde{v}(\xi', 0)|^2$$

$$\leq C_0 \left\{ \int_0^{\infty} |\tilde{f}(\xi', x_N)|^2 \, dx_N + \sum_{j=1}^{k} |\xi'|^{2(m-m_j)-1} |\tilde{g}_j(\xi')|^2 \right\}$$

Adding $\int |\tilde{v}(\xi', x_N)|^2 \, dx_N$ to both sides of the inequality and integrating in the ξ' variable yields

$$\|v\|_{H^m(\mathbf{R}^N)}^2 + \|v(x', 0)\|_{H^{m-1/2}(\mathbf{R}^{N-1})}^2$$

$$\leq C_0 \left[\|f\|_{H^0(\mathbf{R}^N)}^2 + \sum_{j=1}^{k} \|g_j\|_{H^{m-m_j-1/2}(\mathbf{R}^{N-1})}^2 + \|v\|_{H^0(\mathbf{R}^N)} \right].$$

This is just the sort of estimate that we seek for the finite dimensional solution space of the system

$$P(D)v = 0$$
$$B_j(D)v = h_j.$$

Combining this with the obvious estimate

$$\|u_1\|_{H^m} \leq \|f\|_{H^0}$$

gives the estimate that we need for part (A) of well-posedness. Our proof is therefore complete. ∎

5.3   Remarks on the Solution of the Boundary Value Problem in the Constant Coefficient Case

We have considered the boundary value problem

$$
\text{B.V.P.} \qquad \begin{cases} P(u) = f \\ B_j u = g_j \quad \text{if } j = 1, \dots, k. \end{cases}
$$

Here the operator P is assumed to have constant coefficients, to be elliptic, and to be homogeneous of degree m. We assume that the B_j's satisfy the Lopatinski nondegeneracy condition.

For ξ' fixed, we examined

$$
\begin{cases} P(\xi', D_N)u = f & \text{if } x_N > 0 \\ B_p(\xi', D_N)u = g_j(\xi') & \text{if } x_N = 0, \qquad j = 1, \dots, k. \end{cases}
$$

Lopatinski's condition tells us that this is a well-posed linear system of ordinary differential equations. Thus the standard classical theory of ordinary differential equations (see [HOR1], [INCE], or [COL]) guarantees that there is a unique solution $u_1(\xi', x_N)$.

The solution that we seek for the B.V.P is

$$
u(x', x_N) = \int e^{-ix' \cdot \xi'} u_1(\xi', x_N) \, d\xi'.
$$

For if $\tilde{\ }$ is the partial Fourier transform, then

$$
(\widetilde{Pu}) = P(\xi', D_N)\tilde{u} = P(\xi', D_N)u_1 = \tilde{f}.
$$

Properties of this solution u are

1. u is unique.
2. The map $(f, g_1, \dots, g_k) \mapsto u$ is linear.
3. Define

$$
\mathcal{H} = L^2(\mathbf{R}_+^N) \times \prod_{j=1}^{k} H^{m-m_j-1/2}(\mathbf{R}^{N-1}).
$$

If $(f, g_1, \dots, g_k) \in \mathcal{H}$ then $u \in H^m$.

There are two basic ingredients to seeing why property (3) holds:

(i) Solving an ordinary differential equation in the x_N variable entails m integrations, so u should be m degrees smoother in the x_N direction than f.

(ii) The function u is obtained from f by division on the Fourier transform side by coefficients of $P(\xi', D_N)$. Therefore the Fourier transform of u decays at ∞ at a rate m degrees faster than f.

5.4 Solution of the Boundary Value Problem in the Variable Coefficient Case

Now the boundary value problem is

$$\text{B.V.P.} \quad \begin{cases} P(u) = f \\ B_j u = g_j \quad \text{if } j = 1, \ldots, k. \end{cases}$$

We assume that P is of order m, is elliptic, and has variable coefficients. Each B_j is of degree m_j and has variable coefficients. We shall not assume that P is homogeneous; however, we will continue to assume that each B_j *is* homogeneous. This last hypothesis is not necessary, but it is convenient. The index k is the number of roots with positive imaginary part for the polynomial

$$P_m(x, \xi' + T\xi_N) = 0.$$

Also the coefficients of P and of the B_j are smooth *functions*, not constants.

Definitions

In the present context, the Lopatinski condition takes the following form: For each fixed $x^0 \in \mathbb{R}^{N-1}$, we have that the constant coefficient system

$$\begin{cases} P(x^0, D)u &= f(x^0) \\ B_j(x^0, D)u &= g_j \end{cases}$$

satisfies the Lopatinski condition for constant coefficient systems.

A *parametrix* for the boundary problem is an operator

$$E : \mathcal{H} \longrightarrow H^m(\mathbb{R}^N_+)$$

such that

$$AEF = F + TF, \qquad F \in \mathcal{H}$$

$$EAu = u + T_1 u, \qquad u \in H^m(\mathbb{R}^N_+),$$

where $F = (f, g_1, \ldots, g_k)$. Here $\mathcal{H}$ is the usual product Hilbert space and the operator

$$A : H^m(\mathbb{R}_+^N) \to \mathcal{H}$$

is defined by

$$Au = (Pu, \gamma B_1 u, \ldots, \gamma B_k u).$$

The error operators T, T_1 are compact operators on $H^m(\mathbb{R}_+^N)$ and $\mathcal{H}$, respectively. The operator γ denotes restriction.

Main Results

THEOREM 5.4.1
Assume that our boundary value problem satisfies the Lopatinski condition as defined above. Assume that $f \in L^2(\mathbb{R}_+^N), g_j \in H^{m-m_j-1/2}(\mathbb{R}^{N-1})$. Then there is a linear operator

$$E : \mathcal{H} \to H^m(\mathbb{R}_+^N)$$

such that if $F \equiv (f, g_1, \ldots, g_k)$ and $u \equiv E(F)$, then

$$Au = F + TF,$$

where T is of order -1 and

$$\|u\|_{H^m} \leq C \left(\|f\|_{H^0(\mathbb{R}^N)} + \sum_j \|g_j\|_{H^{m-m_j-1/2}(\mathbb{R}^{N-1})} + \|u\|_{H^0} \right).$$

PROOF   We assume for convenience that the coefficients of P and of the B_j's have compact support. We will make decisive use of the hypothesis that the coefficients are smooth. [Much modern research concentrates in part on studying elliptic and other problems with rough coefficients. From the point of view of applications, such a study is rather natural (see [MOS] for some of the pioneering work and [FKP] for more recent work along these lines). But the necessary techniques are extremely complicated and we cannot explore them here.]

Now let $\epsilon > 0$. Then there is a $\delta > 0$ such that if $|x - y| < \delta$ and if a is any coefficient of P or of one of the B_j's then

$$|a(x) - a(y)| < \epsilon.$$

Let $\phi_j \in C_c^\infty(\mathbb{R}^N), \phi_j \geq 0, \sum \phi_j \equiv 1$ on $\mathbb{R}^N$. Assume that diam supp $\phi_j < \delta/2$. We can assume that no point x is in more than $M(N)$ of the supports of the ϕ_j's. (If we take the supports to be balls, then in fact we may take $M(N)$ to be $N + 1$.) Now choose functions $\psi_j \in C_c^\infty(\mathbb{R}^N)$ such that $\psi \equiv 1$ on the support of ϕ_j. We may also assume that diam supp $\psi_j < \delta/2$. (The trick of choosing cutoff functions ψ that are identically equal to 1 on the support of some smaller cutoff functions ϕ is a device of wide utility in this subject. We already saw it enter into our study of interior estimates for elliptic operators. We will see it put to particularly good use when we study the $\bar{\partial}$-Neumann problem in a later chapter.)

Define W_j to be the support of ψ_j. If, for some ℓ, $W_\ell \cap \partial \mathbb{R}^N_+ = \emptyset$, then we choose a point $x^\ell \in W_\ell$ and construct a parametrix E_ℓ for the constant coefficient problem

$$P_m(x^\ell, D)u = f,$$

where P_m is the principal symbol of P. Next we treat the other $\ell's$.

If $W_\ell \cap \partial \mathbb{R}^N_+ \neq \emptyset$ then select $x^\ell \in W_\ell \cap \partial \mathbb{R}^N_+$. By our work on the boundary value problem for the constant coefficient case, we may find for each ℓ an operator

$$E_\ell : \mathcal{H} \to H^m$$

such that the function

$$u_\ell = E_\ell(\psi_\ell f, \psi_\ell g_1, \ldots, \psi_\ell g_k)$$

satisfies

$$P_m(x^\ell, D)u_\ell = \psi_\ell f$$

and

$$B_j(x^\ell, D)u_\ell = \psi_\ell g_j.$$

Now we set

$$E_0 = \sum_\ell \phi_\ell E_\ell \psi_\ell.$$

Define an approximate solution of our boundary value problem by $u = E_0(F)$, where $F = (f, g_1, \ldots, g_k)$. Then

$$P(x, D)E_0(F) = \sum_\ell P(x, D) [\phi_\ell E_\ell \psi_\ell F]$$

$$= \sum_\ell \phi_\ell P(x, D)(E_\ell \psi_\ell F)$$

$$+ \sum_\ell \left(\begin{array}{c}\text{some} \\ \text{derivative} \\ \text{of } \phi_\ell\end{array}\right)\left(\begin{array}{c}\text{At most} \\ (m-1) \text{ derivatives} \\ \text{of } E_\ell \psi_\ell F\end{array}\right)$$

$$= \sum_\ell \phi_\ell P(x, D)(E_\ell \psi_\ell F) + \mathrm{OP}_{m-1}(E_\ell \psi_\ell F)$$

$$= \sum_\ell \phi_\ell P_m(x^\ell, D)(E_\ell \psi_\ell F)$$

$$+ \sum_\ell \phi_\ell \left[P_m(x, D) - P_m(x^\ell, D) \right] (E_\ell \psi_\ell F)$$

$$+ \sum_\ell \left[P(x, D) - P_m(x, D) \right] (E_\ell \psi_\ell F)$$

$$+ \mathrm{OP}_{m-1}(E_\ell \psi_\ell F)$$

$$= I + II + III + \mathrm{OP}_{m-1}(E_\ell \psi_\ell F).$$

By the definition of E_ℓ, we have

$$I = \sum_\ell \phi_\ell \psi_\ell f + \text{lower order errors}$$

$$= f + S^1 f,$$

where $S^1 f \in H^1$ when $f \in H^0$. Set $II = TF$ and $III = T^1 F$, where T is the sum of operators from $\mathcal{H}$ into H^0 with arbitrarily small norms (depending on ϵ) and T^1 is an operator mapping $\mathcal{H}$ to H^1. (Observe that the assertion about the size of the norm of T follows immediately from the presence of the coefficients $[P_m(x, D) - P_m(x^\ell, D)]$, which are uniformly small.)

In summary, we have

$$P(x, D)E_0 F = f + TF + T^1 F + S^1 F.$$

Similarly, we can calculate $B_j(x, D)E_0 F$ and obtain that

$$B_j(x, D)E_0 F = \sum_\ell \left[\phi_\ell \psi_\ell g_j \right] + \tilde{T}_j F + \tilde{T}_j^1 F + \tilde{S}_j^1 F.$$

Here T_j and T_j^1 are operators with properties analogous to those of T and T^1.

Let

$$A = \left(P(x, D), \gamma B_1(x, D), \ldots, \gamma B_k(x, D) \right).$$

Then we can summarize our findings with the equation

$$AE_0 F = F + UF + U^1 F, \tag{5.4.1.1}$$

where U maps $\mathcal{H}$ to $\mathcal{H}$ with arbitrarily small norm and U^1 is smoothing. Now we define $E = E_0(I + U)^{-1}$ (notice that this inverse exists because we may take the norm of U to be smaller than, say, $1/2$). Then

$$AEF = AE_0(I + U)^{-1}F$$
$$= (I + U)^{-1}F + U(I + U)^{-1}F + U^1(I + U)^{-1}F.$$

In the last equality we have applied equation (5.4.1.1) to $(I + U)^{-1}F$. Now the last line equals

$$(I + U)(I + U)^{-1}F + U^1(I + U)^{-1}F = F + U^1(I + U)^{-1}F.$$

Since the operator $U^1(I + U)^{-1}$ is a smoothing operator, we now see that E is a parametrix for our boundary value problem. ∎

REMARK   Crucial to the existence of E in this proof was the Lopatinski condition at each point. ∎

COROLLARY 5.4.2
If the B.V.P. satisfies the Lopatinski condition, then the solutions of the homogeneous problem

$$Pu = 0 \text{ in } \mathbb{R}^N_+$$
$$B_j u = 0 \text{ on } \partial\mathbb{R}^N_+, \qquad j = 1, \ldots, k$$

form a finite-dimensional subspace that consists of functions that are in $C^\infty(\overline{\mathbb{R}^N_+})$.

PROOF   One checks that $EA = I + T$, where T is in fact an operator of order -1.

Let $u \in H^m$ be in the null space of A. Then $u = -Tu$. Therefore, since T is of order -1, we conclude that $u \in H^{m+1}$. Iterating, we see that $u \in C^\infty$.

Note that, since T is of order -1, $T : H^m \to H^{m+1} \subseteq H^m$. By Rellich's lemma, this operator is compact. Let $\mathcal{M} \subseteq \text{Ker} A$ be the closed unit ball. Set $\mathcal{N} = T(\mathcal{M})$. Then, since T is compact, $\mathcal{N}$ is compact. From the equation $u = -Tu$ we then see that $\mathcal{M}$ itself is compact. Therefore the kernel of A is finite dimensional. ∎

Operations on Higher Sobolev Spaces

So far we have constructed a parametrix E for our B.V.P. that satisfies

$$E : \mathcal{H} = L^2 \times H^{m-m_1-1/2} \times \cdots \times H^{m-m_k-1/2} \longrightarrow H^m.$$

We would like now to show that if

$$K^s = H^s \times H^{s+m-m_1-1/2} \times \cdots \times H^{s+m-m_k-1/2},$$

then

$$E : K^s \longrightarrow H^{s+m}.$$

The experience that we have so far with pseudodifferential operators suggests that they will serve us well in this endeavor.

We begin by defining Λ_s to be the pseudodifferential operator with symbol $(1 + |\xi'|^2)^{s/2}$. (In other contexts this operator is known as a tangential Bessel potential of order s. See [STSI] and [FOK].) We will use these extensively in Chapter 7. Let $F \in K^s$. Then $\Lambda_s F \in \mathcal{H}$ and hence $E\Lambda_s F \in H^m$. But Λ_s commutes with A (up to a lower order error) so that $E\Lambda_s = \Lambda_s E + (\text{error})$ and we find that $E = \Lambda_s^{-1} E \Lambda_s$ modulo a smoothing error term. In conclusion, if $u = EF$ then $\Lambda_s u \in H^m$.

With these preliminaries out of the way, we begin by treating the case $0 < s \leq 1$. Write $P(x, D) = D_N^m + D_0$, where D_0 involves terms of the form $D_N^{m-j} D_t^\alpha, |\alpha| \leq j$, and $0 < j \leq m$. To see that $D_N^{m-j} D_t^\alpha u \in H^s$, it suffices for us to check that

$$\left| \widehat{D_t^\alpha u} \right| (1 + |\xi|^2)^{(m-j+s)/2} \in L^2.$$

But

$$\left| \widehat{D_t^\alpha u} \right| (1 + |\xi|^2)^{(m-j+s)/2} \leq |\hat{u}||\xi'|^j (1 + |\xi|^2)^{(m-j+s)/2}$$

$$= |\xi'|^s |\xi'|^{j-s} (1 + |\xi|^2)^{(m-j+s)/2} |\hat{u}|$$

$$\leq |\xi'|^s |\hat{u}| (1 + |\xi|^2)^{m/2}.$$

We have used here the fact that $0 < s \leq 1 \leq j$. The latter expression is clearly in L^2 since $\Lambda_s u \in H^m$. Finally, since $Pu = f$, we have that $D_N^m u \in H^s$ (from the partial differential equation itself). Hence $u \in H^{m+s}$.

Next, if $1 < s \leq 2$, then $\Lambda_{s-1} u \in H^{m+1}$ and with the same argument as above we may see that $u \in H^{m+s}$. The result for higher s follows inductively by similar arguments.

In summary, we have proved that

$$\|u\|_{s+m} \leq C \left(\|Au\|_s + \|u\|_s \right)$$

$$= C \left(\|Pu\|_{H^s(\mathbb{R}_+^N)} + \sum_{j=1}^k \|B_j u\|_{s+m-m_j-1/2} + \|u\|_s \right).$$

Now we vindicate the form that our program has taken (that is, concentrating on regularity estimates in the absence of existence results) by using our regularity theorems together with some functional analysis to prove an existence theorem.

COROLLARY 5.4.3   AN EXISTENCE THEOREM

If the B.V.P. satisfies the Lopatinski condition, then there are functions $v_1, \ldots, v_p \in C^\infty(\overline{\mathbb{R}^N_+})$ and $w_{j1}, \ldots, w_{jp} \in C^\infty(\partial\mathbb{R}^N_+), j = 1, \ldots, k$, such that the following is true. If $(f, g_1, \ldots, g_k) \in K^s$, $s > \max(m - m_j - 1/2)$, and $f \perp v_j \ \forall j = 1, \ldots, p$ and $g_j \perp w_{jn} \ \forall j = 1, \ldots, k$ and $n = 1, \ldots, p$, then there is a solution of the B.V.P.

$$\begin{cases} Pu = f \\ B_j u = g_j \qquad j = 1, \ldots, k. \end{cases}$$

PROOF   We have constructed a parametrix E such that

$$AE = I + T_{-1}.$$

If we take adjoints, we obtain

$$E^* A^* = I + (T_{-1})^*.$$

Thus E^* is a parametrix for A^*. By Corollary 5.4.2, we know that the dimension of the kernel of A^* is finite. Also $\operatorname{Ker} A^* \subseteq \cap K^s$. Hence the kernel consists of functions that are C^∞ on the closure of the half-space. Let $G_1, \ldots, G_p$ be a basis for the kernel of A^*. Then each G_n has the form $(v_n, w_{1n}, \ldots, w_{kn})$.

Because of the inequality

$$\|u\|_{s+m} \leq C \left(\|Au\|_s + \|u\|_s \right),$$

the range of A is closed. Then F is in the range of A if and only if it is perpendicular to the kernel of A^*. This is exactly what we want to prove.   ∎

5.5   Solution of the Boundary Value Problem Using Pseudodifferential Operators

In his seminal paper [HOR3], Hörmander used the new theory of pseudodifferential operators to study a class of noncoercive boundary value problems. However, the techniques that he introduced are already interesting when applied to the classical coercive problems that we have been studying. We will explain how Hörmander's techniques work in this section.

For simplicity we restrict attention to order-two partial differential equations. We adhere to the custom of letting $D_j = i\partial/\partial x_j, j = 1, \ldots, N$. This makes the formulas involving the Fourier transform come out more cleanly. Thus our boundary value problem takes the form

$$Pu = \sum_{j,\ell=1}^{N} a_{j\ell}(x) D_j D_\ell u + \sum_{j=1}^{N} b_j(x) D_j u + c(x) u = f(x) \quad \text{for } x \in \mathbb{R}_+^N$$

$$Bu = \sum_{j=1}^{N} \alpha_j(x) D_j u + \beta_0(x) u = g(x) \qquad\qquad \text{for } x \in \partial \mathbb{R}_+^N.$$

Here we have assumed that all the coefficients are real-valued and smooth. Notice that, because P is second order, there is only one boundary condition. The ellipticity condition on P is

$$\sum_{j,\ell=1}^{N} a_{j\ell}(x) \xi_j \xi_\ell \geq C_0 |\xi|^2.$$

We further assume that

$$\sum_{j=1}^{N} |\alpha_j|^2 + |\beta_0|^2 \neq 0.$$

We will use the power of the theory of distributions. To that end, we introduce a little notation. If $v \in C^\infty(\overline{\mathbb{R}_+^N})$, then

$$v^0 = \begin{cases} v & \text{on } \overline{\mathbb{R}_+^N} \\ 0 & \text{elsewhere.} \end{cases}$$

We let $\tilde{v}$ denote any smooth extension of v to $\mathbb{R}^N$. The symbol v_0 denotes the trace of v on $\partial \mathbb{R}_+^N$, and v_1 is the trace of $(\partial/\partial x_N)v$ on $\partial \mathbb{R}_+^N$. Now we require the following preliminary calculations. The reader is invited to sharpen his prowess at distribution theory by actually carrying out the details.

1. $D_N(v^0) = [D_N v]^0 + i\delta(x_N) \otimes v_0(x')$
2. $D_j(v^0) = [D_j(v)]^0, \quad j = 1, \ldots, N-1$
3. $D_N^2(v^0) = (D_N^2 v)^0 - \delta(x_N) \otimes v_1(x') - \delta'(x_N) \otimes v_0(x')$
4. $D_N D_j(v^0) = (D_N D_j v)^0 + i\delta(x_N) \otimes D_j v_0(x'), \quad j = 1, \ldots N-1$
5. $D_j D_\ell(v^0) = (D_j D_\ell v)^0, \quad \ell, j = 1, \ldots, N-1.$

What is lurking in the background here is the fact that the derivative, in the sense of distributions, of $-\chi_{[a,b]}$, where $\chi_{[a,b]}$ is the characteristic function of the interval $[a, b] \subseteq \mathbb{R}$, is $\delta_b - \delta_a$. This is the distribution-theoretic formulation of the fundamental theorem of calculus. The five properties listed above show us that the function v^0 is sensitive to normal derivatives but not to tangential derivatives at the boundary.

Perhaps a comment is in order here about the use of the tensor $\otimes$ notation. Technically speaking, it is not possible to take the product of $\delta(x_N)$ and

$v_0(x')$ because these two (generalized) functions have different domains. The $\otimes$ notation serves as a mediator to address this situation: we understand that $\delta(x_N) \otimes v_0(x')$ is a distribution that acts on a testing function $\phi(x_1, \ldots, x_N)$ according to the formula

$$\big(\delta(x_N) \otimes v_0(x')\big)(\phi) = \int \phi(x', 0) v_0(x') \, dx'.$$

For completeness, we now sketch the proofs of three of the five assertions:

PROOF OF (1)

$$
\begin{aligned}
D_N(v^0) &= D_N(\tilde{v}\chi_{\overline{\mathbf{R}^N_+}}) \\
&= D_N \tilde{v} \cdot \chi_{\overline{\mathbf{R}^N_+}} + \tilde{v} D_N(\chi_{\overline{\mathbf{R}^N_+}}) \\
&= (D_N \tilde{v})^0 + \tilde{v}(x') \otimes i\delta(x_N) \\
&= (D_N v)^0 + i v_0(x') \otimes \delta(x_N).
\end{aligned}
$$

PROOF OF (3)

$$
\begin{aligned}
D_N^2(v^0) &= D_N \left\{ (D_N v)^0 + i\delta(x_N) \otimes v_0(x') \right\} \\
&= D_N \left\{ (\widetilde{D_N v})\chi_{\overline{\mathbf{R}^N_+}} + i\delta(x_N) \otimes v_0(x') \right\} \\
&= [D_N(\widetilde{D_N v})]\chi_{\overline{\mathbf{R}^N_+}} + \widetilde{D_N v} \otimes i\delta(x_N) - \delta'(x_N) \otimes v_0(x') \\
&= (D_N^2 v)^0 - \delta(x_N) \otimes v_1(x') - \delta'(x_N) \otimes v_0(x').
\end{aligned}
$$

PROOF OF (4)

$$
\begin{aligned}
D_N D_j(v^0) &= D_N \big((D_j v)^0\big) \\
&= D_N \left(\widetilde{D_j v} \cdot \chi_{\overline{\mathbf{R}^N_+}} \right) \\
&= (D_N \widetilde{D_j v}) \cdot \chi_{\overline{\mathbf{R}^N_+}} + \widetilde{D_j v} \left(D_N \chi_{\overline{\mathbf{R}^N_+}} \right) \\
&= (D_N D_j v)^0 + \widetilde{D_j v}(x') \otimes i\delta(x_N) \\
&= (D_N D_j v)^0 + i\delta(x_N) \otimes D_j v_0(x').
\end{aligned}
$$

We want to calculate $P(v^0)$. To do so, we will use a parametrix E for P. Since E is the parametrix for an elliptic operator, it follows that E has the form

$$E(\phi) = (2\pi)^{-N} \int e(x, \xi) \hat{\phi}(\xi) e^{-ix\cdot\xi} \, d\xi.$$

Note that, in this section only, we shall keep track of the unfortunate constants that are part of the theory of Fourier integrals. This is necessary in order to make the calculations come out properly.

We will compute the following:

I. For w a smooth function on $\partial\mathbb{R}_+^N = \mathbb{R}^{N-1}$, we need to understand $E(w(x')\otimes \delta(x_N))$. Now

$$E(w(x') \otimes \delta(x_N)) = (2\pi)^{-N} \int e(x, \xi) \left(w(x') \otimes \delta(x_N)\right)^\wedge \cdot e^{-ix\cdot\xi} \, d\xi.$$

But

$$\left(w(x') \otimes \delta(x_N)\right)^\wedge = \int e^{ix\cdot\xi} w(x') \otimes \delta(x_N) \, dx$$

$$= \int e^{ix'\cdot\xi'} w(x') \, dx' \int e^{ix_N\xi_N} \delta(x_N) \, dx_N$$

$$= \hat{w}(\xi').$$

Therefore

$$E(w(x') \otimes \delta(x_N)) = (2\pi)^{-N} \int_{\xi'} \int_{\xi_N} e(x, \xi) e^{-ix_N\xi_N} \, d\xi_N \hat{w}(\xi') e^{-ix'\cdot\xi'} \, d\xi'$$

$$\equiv (2\pi)^{-(N-1)} \int_{\xi'} k(x, \xi') \hat{w}(\xi') e^{-ix'\cdot\xi'} \, d\xi'$$

$$\equiv K(w)(x).$$

Check that K has order -1.

II. For w a smooth function on $\partial\mathbb{R}_+^N = \mathbb{R}^{N-1}$ we also need to understand $E(w(x') \otimes \delta'(x_N))$. Now

$$E\left(w(x') \otimes \delta'(x_N)\right) = (2\pi)^{-N} \int e(x, \xi) \left(w(x') \otimes \delta'(x_N)\right)^\wedge e^{-ix\cdot\xi} \, d\xi.$$

But

$$\left(w(x') \otimes \delta'(x_N)\right)^\wedge = \widehat{w(\xi')} \cdot i\xi_N \hat{\delta} = \hat{w}(\xi') i\xi_N.$$

Therefore

$$E\left(w(x') \otimes \delta'(x_N)\right) = (2\pi)^{-N} \int_{\xi'} \int_{\xi_N} e(x,\xi) i\xi_N e^{-ix_N\xi_N} d\xi_N \widehat{w}(\xi') e^{-ix'\cdot\xi'} d\xi'$$

$$\equiv (2\pi)^{-(N-1)} \int_{\xi'} k_1(x,\xi') \widehat{w}(\xi') e^{-ix'\cdot\xi'} d\xi'$$

$$\equiv K_1(w)(x).$$

Check that K_1 has order 0.

Let us assume that the operator P has coefficient a_{NN} equal to 1. (It is easy to arrange for the coefficient to be nonzero just by a rotation of coordinates. Then the coefficient is easily normalized by division.) Then the principal symbol of P is

$$\sigma_{\text{prin}}(P) = \xi_N^2 + 2\xi_N \sum_{j=1}^{N-1} a_{jN}(x)\xi_j + \sum_{j,\ell=1}^{N-1} a_{j\ell}(x)\xi_j\xi_\ell$$

$$= (\xi_N - \alpha)(\xi_N - \beta).$$

Here $\alpha = \bar{\beta}$ since we assume that P is a partial differential operator with real coefficients. Also notice that α, β are expressions in $\xi_1, \ldots, \xi_{N-1}$ and x.
The principal symbol of E, for ξ large, is then

$$\sigma_{\text{prin}}(E) = \frac{1}{\sigma_{\text{prin}}(P)} = \frac{1}{(\xi_N - \alpha)(\xi_N - \beta)}.$$

Now we will compute the principal symbols of K and K_1. We know that

$$k(x,\xi') = \frac{1}{2\pi} \int_{\xi_N} e(x,\xi) e^{-ix_N\xi_N} d\xi_N.$$

Thus the principal part of $k(x,\xi')$ is

$$\frac{1}{2\pi} \int_{\xi_N} \frac{1}{(\xi_N - \alpha)(\xi_N - \beta)} e^{-ix_N\xi_N} d\xi_N \qquad (5.5.1)$$

for ξ' large. We want to think of ξ_N as the real part of a complex parameter z_N. Thus line (5.5.1) equals the limit, as the radius of the curve γ tends to ∞, of

$$\frac{1}{2\pi} \int_{\gamma} \frac{1}{(z_N - \alpha)(z_N - \beta)} e^{-ix_N z_N} dz_N = \frac{2\pi i}{2\pi} \text{Res}_\alpha \left(\frac{1}{(z_N - \alpha)(z_N - \beta)}\right)$$

$$= \frac{i}{\alpha - \beta}.$$

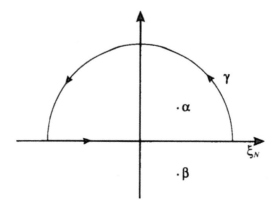

FIGURE 5.1

See Figure 5.1. Be sure to let $x_N \to 0^+$ when evaluating the residue. Similarly, the principal symbol of K_1 is

$$k_1(x, \xi') = \operatorname{Res}_\alpha \left(\frac{-z_N}{(z_N - \alpha)(z_N - \beta)} \right) = \frac{\alpha}{\beta - \alpha} \ .$$

Finally, we are in a position to calculate $P(v^0)$:

$$P(v^0) = D_N^2(v^0) + 2 \sum_{j=1}^{N-1} a_{jN} D_N D_j v^0 + \sum_{j,\ell=1}^{N-1} D_j D_\ell v^0$$

$$+ \sum_{j=1}^{N} b_j(x) D_j v^0 + c(x) v^0.$$

From our preliminary calculations (1)–(5), we may now see that

$$P(v^0) = \left[(D_N^2 v)^0 - \delta(x_N) \otimes v_1(x') - \delta'(x_N) \otimes v_0(x') \right]$$

$$+ 2 \sum_{j=1}^{N-1} a_{jN} \left[(D_N D_j v)^0 + i D_j v_0(x') \otimes \delta(x_N) \right]$$

$$+ \sum_{j,\ell=1}^{N-1} a_{j\ell}(x)(D_j D_\ell v)^0 + b_N(x) \left[(D_N v)^0 + i\delta(x_N) \otimes v_0(x') \right]$$

$$+ \sum_{j=1}^{N-1} b_j(x)(D_j v)^0 + c(x) v^0(x)$$

$$= (Pv)^0 - \delta(x_N) \otimes v_1(x') - \delta'(x_N) \otimes v_0(x')$$

$$+ 2i \sum_{j=1}^{N-1} a_{jN}(x)\delta(x_N) \otimes D_j v_0(x') + i\delta(x_N) \otimes v_0(x') \cdot b_N(x).$$

If we apply E to both sides of this equation, we obtain

$$v^0 + T_{-2}v^0 = E\left((Pv)^0\right) - K(v_1) - K_1(v_0)$$

$$+ 2iE\left(\sum_{j=1}^{N-1} a_{jN}(x)D_j v_0(x') \otimes \delta(x_N)\right)$$

$$+ iE\left(b_N(x)\delta(x_N) \otimes v_0(x')\right).$$

We assume, as we may, that the error is of order -2. Now

$$Kv_1 = -v^0 - T_{-2}v^0 + E\left((Pv)^0\right) - K_1(v_0)$$

$$+ 2i \sum_{j=1}^{N-1} K\left(a_{jN}(x)D_j v_0(x')\right) + iK\left(b_N(x')v_0\right). \qquad (5.5.2)$$

Now we restrict the x variable to the boundary. Recall that the quantities α and β are defined by

$$\alpha, \beta = \frac{-2\sum_j a_{jN}(x)\xi_j \pm \sqrt{(-2\sum_j a_{jN}(x)\xi_j)^2 - 4\sum_j a_{j\ell}\xi_j\xi_\ell}}{2}.$$

Then $\alpha - \beta \in S^1$ (here S^1 is the *symbol class*), $\sigma_{\mathrm{prin}}(K_1) \in S^0$, and $\sigma_{\mathrm{prin}}(K) \in S^{-1}$.

The operator K is elliptic of order -1, hence it has an inverse, up to a smoothing term. Call that inverse M. Applying M to both sides of (5.5.2), we find that

$$v_1 = -M(v_0) - MT_{-2}v^0 + ME(Pv)^0 - MK_1 v_0$$

$$+ 2i \sum_{j=1}^{N-1} a_{jN}(x)D_j v_0(x') + ib_N(x')v_0.$$

In short, we obtain

$$v_1 = A_1 v_0 + A_{-1}v^0 + A_0 v_0, \qquad (5.5.3)$$

where A_1 is of order 1, A_{-1} is of order -1, and A_0 is of order 0. Moreover,

$$A_1 = -M - MK_1 + 2i \sum_j a_{jN}(x)D_j.$$

As a result,

$$\sigma_1(A_1) = -\frac{1}{\sigma_{\text{prin}}(K)} - \frac{1}{\sigma_{\text{prin}}(K)}\sigma_0(K_1) + 2i \sum_j a_{jN}(x)\xi_j$$

$$= \frac{1}{i}(\alpha - \beta) + \frac{1}{i}(\alpha - \beta)\frac{\alpha}{\beta - \alpha} + 2i \sum_{j=1}^{N-1} a_{jN}(x)\xi_j$$

$$= \frac{1}{i}(\alpha - \beta) - \frac{1}{i}\alpha + 2i \sum_{j=1}^{N} a_{jN}(x)\xi_j$$

$$= i\left[\beta + 2\sum_{j=1}^{N} a_{jN}(x)\xi_j\right]$$

$$= i\left[\beta - (\alpha + \beta)\right]$$

$$= -i\alpha.$$

Now we examine the boundary condition: on $\partial\mathbb{R}_+^N$ we have

$$Bu = \sum_{j=1}^{N} \alpha_j(x)D_jv + \beta_0(x)v = g(x)$$

or

$$-\frac{1}{i}\alpha_N(x')\frac{\partial}{\partial x_N}v(x') + \sum_{j=1}^{N-1} \alpha_j(x')D_jv(x') + \beta_0(x')v = g(x')$$

or

$$i\alpha_N(x')v_1(x') + \sum_{j=1}^{N-1} \alpha_j(x')D_jv_0(x') + \beta_0(x')v_0 = g(x').$$

We substitute equation (5.5.3) for v_1 into this last equation to obtain

$$i\alpha_N(x)\left[A_1v_0 + A_{-1}v^0 + A_0v_0\right](x') + \sum_{j=1}^{N-1} \alpha_j(x')D_jv_0(x') + \beta_0(x')v_0 = g(x').$$

We want to be able to solve this equation for v_0.
    Our problem amounts to this: If

$$\Phi(v_0) = F(f, g),$$

where $F(f,g)$ is given data and Φ is some operator, then can we invert Φ? We know that, in our situation,

$$\sigma_1(\Phi) = i\alpha_N(x')\sigma_1(A_1) + \sigma_1\left(\sum_{j=1}^{N-1}\alpha_j(x)D_j\right)$$

$$= \alpha_N(x')\alpha(x,\xi') + \sum_{j=1}^{N-1}\alpha_j(x)\xi_j.$$

Recall that all the coefficients of P are real. It follows that $\alpha_N(x)$ is real, $\alpha(x,\xi')$ is complex, and $\sum_{j=1}^{N-1}\alpha_j(x)\xi_j$ is real. Thus we can invert the operation Φ provided that $\sigma_1(\Phi)$ is not zero. The condition $\alpha_N(x') \neq 0$ guarantees this nonvanishing. But $\alpha_N(x') \neq 0$ is just the Lopatinski condition for our second-order boundary value problem.

We have succeeded in solving the boundary value problem, using the calculus of pseudodifferential operators, under the hypothesis that the Lopatinski condition is satisfied.

5.6 Remarks on the Dirichlet Problem on an Arbitrary Domain, and a Return to Conformal Mapping

Now we investigate boundary value problems in their most natural setting: on a smoothly bounded domain in $\mathbb{R}^N$. Because we had the foresight to develop machinery to study a large and flexible class of operators—namely the elliptic ones—our task will be surprisingly simple.

Thus we begin by fixing a smoothly bounded domain $\Omega \subseteq \mathbb{R}^N$. Let us fix a well-posed elliptic boundary value problem

$$(B.V.P.) \qquad \begin{cases} Pu = f & \text{on } \Omega \\ B_j u\big|_{\partial\Omega} = g_j & \text{on } \partial\Omega, \ j = 1,\ldots,k. \end{cases}$$

Let $\{U_\ell\}$ be a finite open cover of $\bar{\Omega}$ with the property that each U_ℓ is topologically trivial and also such that each $U_\ell \cap \Omega$ is diffeomorphic to a ball. We fix one of the U_ℓ and consider two cases:

1. $U_\ell \cap \partial\Omega \neq \emptyset$;

2. $U_\ell \cap \partial\Omega = \emptyset$.

In case (1), we let $\phi_\ell : U_\ell \to W_\ell \subseteq \mathbb{R}^N$ be a diffeomorphism onto an open subset of $\mathbb{R}^N$ such that $\phi_\ell(U_\ell \cap \Omega) = W_\ell \cap \mathbb{R}^N_+$ and $\phi_\ell(U_\ell \cap \partial\Omega) = W_\ell \cap \partial\mathbb{R}^N_+$.

This map, under the standard push forward by ϕ_ℓ, induces a boundary value problem

$$(\widetilde{B.V.P.}_\ell) \qquad \begin{cases} \tilde{P}^\ell u = \tilde{f} & \text{on} \quad W_\ell \cap \mathbb{R}^N_+ \\ \tilde{B}^\ell_j \tilde{u}\big|_{\partial\Omega} = \tilde{g}^\ell_j & \text{on} \quad \partial\mathbb{R}^N_+ \cap W_\ell, \quad j = 1, \ldots, k. \end{cases}$$

Notice that this is still an elliptic boundary value problem, for we established long ago that the property of ellipticity is invariant under diffeomorphisms. We will say that our boundary value problem satisfies the Lopatinski condition if $(\widetilde{B.V.P.})$ does. By our previous work, we can obtain (assuming the Lopatinski condition) a parametrix $\tilde{E}_\ell$ for $(\widetilde{B.V.P.}_\ell)$. Then we use ϕ_ℓ^{-1} to pull $\tilde{E}_\ell$ back to the original U_ℓ and we thus obtain an operator E_ℓ on U_ℓ.

In case (2), let E_ℓ be any parametrix for P.

Now let $\{\mu_\ell\}$ be a partition of unity subordinate to the covering $\{U_\ell\}$. For each ℓ, let ψ_ℓ be such that $\operatorname{supp}\psi_\ell \subseteq U_\ell$ and $\psi_\ell \equiv 1$ on $\operatorname{supp}\mu_\ell$. Finally define

$$E = \sum_\ell \mu_\ell E_\ell \psi_\ell.$$

Exercise: Verify for yourself that E is a parametrix for the original boundary value problem (B.V.P.).

Recall that our motivation for studying elliptic boundary value problems was a consideration, in Chapter 1, of boundary regularity of the Riemann mapping function from the disc to a smoothly bounded, simply connected domain in the complex plane. We reduced that problem to the problem of proving boundary regularity for solutions of the Dirichlet problem for the Laplacian. That regularity is now established via the parametrix just constructed. We have

THEOREM 5.6.1
Let $\Omega \subseteq \mathbb{R}^N$ be a smoothly bounded domain. Then the unique solution to the Dirichlet problem

$$\begin{cases} \Delta u & = \quad 0 \\ u\big|_{\partial\Omega} & = \quad \phi, \end{cases}$$

with smooth data ϕ, is smooth on $\bar{\Omega}$.

Now we have established this regularity, and much more. Of course there are other, more direct approaches to the boundary regularity problem for conformal mappings. For examples, see both [BEK] and [KEL].

In the function theory of several complex variables, the boundary regularity of biholomorphic mappings is a much deeper problem—inaccessible by way of elliptic boundary value problems. In fact, the correct partial differential equation to study is the $\bar{\partial}$-Neumann problem (see Chapter 7 of the present book). The work of Bell in particular (for instance [BE1], [BE2]) makes the connection quite explicit. Although it is known for a large class of domains that biholomorphic mappings extend smoothly to diffeomorphisms of the closures of the respective domains, the problem in general remains open. The paper [BED] is a nice survey of what was known until 1984, beginning with the breakthrough paper of Fefferman [FEF].

Boundary estimates for solutions of elliptic boundary value problems of the sort we have been studying in this section and the last are commonly referred to as the "Schauder estimates" after Julius Schauder. Our setup using pseudo-differential operators makes it easy to derive estimates in the Sobolev topology *because our pseudodifferential operators have sharp bounds in the Sobolev topology.*

However analogous estimates hold in many other classical function spaces, including Lipschitz spaces (see [KR2] for definitions and a detailed study of these spaces) and, more generally, Triebel–Lizorkin spaces. Let us briefly discuss the first of these (which are a special case of the second). In order to obtain Lipschitz regularity for our boundary value problem, all that is required is to see that if σ lies in the symbol class S^m then the associated operator $\mathrm{Op}(\sigma)$ maps Λ_α to $\Lambda_{\alpha-m}$. This is a complicated business and we we will say just a few words about the proof at this time (a good reference for this and related matters is [KR2]). But we wish to emphasize that the problem is entirely harmonic analysis: the partial differential equation has been solved.

In order to study the mapping properties of a translation invariant operator on a Lipschitz space, it is useful to have a new description of these spaces. Fix $\phi \in C_c^\infty(\mathbb{R}^N)$. For $\epsilon > 0$ we set $\phi_\epsilon(x) = \epsilon^{-N}\phi(x/\epsilon)$. The function ϕ_ϵ is called a *function of "thickness"* ϵ because it has the property that

$$\left|\nabla^k \phi_\epsilon\right| \leq C_k \cdot \epsilon^{-k}$$

with C_k independent of ϵ. Then a bounded function f on $\mathbb{R}^N$ lies in $\Lambda_\alpha, \alpha > 0$, if and only if the functions $f_\epsilon \equiv f * \phi_\epsilon$ satisfy

$$|f - f_\epsilon| \leq C \cdot \epsilon^\alpha.$$

If T_σ is a pseudodifferential operator, then one studies $T_\sigma f$, for $f \in \Lambda_\alpha \cap L^p$, by considering

$$T_\sigma f = T_\sigma(f_\epsilon) + T_\sigma(f - f_\epsilon).$$

We can say no more about the matter here. Again we refer the reader to [KR2] and references therein.

Because restriction theorems for Lipschitz spaces are trivial (namely the restriction of a Λ_α function to a smooth submanifold is still in Λ_α), the regularity statement for elliptic boundary value problems is rather simple in the Lipschitz topology. For the record, we record here one small part of the Schauder theory that is of particular interest for this monograph.

THEOREM 5.6.2

Let $\Omega \subseteq \mathbb{R}^N$ be a smoothly bounded domain. Let P be a uniformly elliptic operator of order two (in the sense that we have been studying) with smooth coefficients on $\bar{\Omega}$. The unique solution to the boundary value problem

$$Pu = 0 \quad on \ \Omega$$
$$u = f \quad on \ \partial\Omega$$

has the following regularity property: If $f \in \Lambda_\alpha(\partial\Omega)$, then the solution u of the problem satisfies $u \in \Lambda_\alpha(\bar{\Omega})$.

This result bears a moment's discussion. It is common in partial differential equations books for the regularity to be formulated thus:

$$\text{If } f \in C^k(\partial\Omega), \text{ then } u \in C^{k-\epsilon}(\bar{\Omega}).$$

This is essentially the sharpest result that can be proved when using the C^k norms. That is because, from the point of view of integral operators, C^k norms are flawed. On the other hand, Lipschitz spaces (where at integer values of α we use Zygmund's definition with higher order differences—see [KR2]) are well behaved under pseudodifferential operators. Thus one obtains sharp regularity in the Lipschitz topology.

Similar comments apply to the interior regularity. The correct regularity statement, in the Lipschitz topology, for the equation

$$Pu = g$$

with $g \in \Lambda_\alpha^{loc}$ is that $u \in \Lambda_{\alpha+m}^{loc}$ (where m is the degree of P elliptic). *This is true even when α is an integer, provided that we use the correct definition of Lipschitz space as in* [KR2]. Again this is at variance with the more commonly cited regularity statement that $g \in C^k$ implies that $u \in C^{k+m-\epsilon}$. It is important to use function spaces that are well suited to the problem in question.

5.7   A Coda on the Neumann Problem

Besides the Dirichlet problem, the other fundamental classical elliptic boundary value problem is the *Neumann problem*. It may be formulated as follows:

$$\begin{cases} \Delta u & = \ 0 \\ \frac{\partial}{\partial\nu}u\big|_{\partial\Omega} & = \ \phi \end{cases} \qquad (*)$$

for smooth data ϕ. Here $\partial/\partial\nu$ denotes the unit outward normal vector field to $\partial\Omega$. Unlike the Dirichlet problem, the data for the Neumann problem is not completely arbitrary. For we may apply Green's theorem (see [KR1]) as follows: if u is a solution to $(*)$, then

$$\int_{\partial\Omega} \frac{\partial u}{\partial \nu}\, d\sigma = \int_\Omega \triangle u\, dV = 0.$$

Subject to this caveat, the theory that we have developed certainly applies to the Neumann problem. One must check that the problem is well-posed (we leave this as an exercise). We may conclude that a solution of the Neumann problem must satisfy the expected interior and boundary regularity.

Because of the noted compatibility condition, existence is more delicate. Because we are working on a bounded domain, there are algebraic-topological conditions at play (recall the earlier discussion of the maximum principle in this light). Thus other considerations would apply if we were to treat existence. Note in passing that the existence theorem that we have established for elliptic boundary value problems does not apply directly to the Dirichlet problem either, and for a philosophically similar reason.

6

A Degenerate Elliptic Boundary Value Problem

6.1 Introductory Remarks

Let us take a new look at the Laplacian on the disc in $D \subseteq \mathbb{C}$. Recall that, for $|a| < 1$, $a \in \mathbb{C}$, the function

$$\phi_a(\zeta) = \frac{\zeta - a}{1 - \bar{a}\zeta}$$

defines a holomorphic, one-to-one, and surjective mapping from the disc to itself. These mappings are known as *Möbius transformations*. In fact if $|\zeta| = 1$ then

$$|\phi_a(\zeta)| = \left| \frac{\zeta - a}{1 - \bar{a}\zeta} \right|$$

$$= \left| \frac{\zeta - a}{\bar{\zeta}(1 - \bar{a}\zeta)} \right|$$

$$= \left| \frac{\zeta - a}{\bar{\zeta} - \bar{a}} \right|$$

$$= 1.$$

All of our assertions about ϕ_a follow easily from this.

Next observe that $\phi_{-a}(0) = a, \phi_a(a) = 0$, and $(\phi_a)^{-1} = \phi_{-a}$. The group $G = \mathrm{Aut}\,(D)$ of one-to-one, surjective, holomorphic transformations of the disc is generated by $\{\phi_a\}$ together with the collection of all rotations. Indeed, any such transformation ψ of the disc may be written as

$$\psi(\zeta) = e^{i\theta_0} \cdot \phi_a(\zeta)$$

for some real θ_0 and $a \in \mathbb{C}$ of modulus less than 1. Observe that the group G acts transitively on the disc: if $\alpha, \beta \in D$ then $(\phi_{-\beta} \circ \phi_\alpha)(\alpha) = \beta$.

Next we introduce the differential operators

$$\frac{\partial}{\partial z} = \frac{1}{2}\left(\frac{\partial}{\partial x} - i\frac{\partial}{\partial y}\right) \quad \text{and} \quad \frac{\partial}{\partial \bar{z}} = \frac{1}{2}\left(\frac{\partial}{\partial x} + i\frac{\partial}{\partial y}\right).$$

It is straightforward to check that the Laplacian $\triangle$ satisfies

$$\triangle = 4\frac{\partial}{\partial z}\frac{\partial}{\partial \bar{z}} = 4\frac{\partial}{\partial \bar{z}}\frac{\partial}{\partial z}.$$

Also,

$$\frac{\partial}{\partial z}z = 1 = \frac{\partial}{\partial \bar{z}}\bar{z}$$

and

$$\frac{\partial}{\partial z}\bar{z} = 0 = \frac{\partial}{\partial \bar{z}}z.$$

If f is a C^1 complex-valued function, then we may write $f = u + iv$ with u and v real-valued. Then

$$\frac{\partial f}{\partial \bar{z}} = \frac{1}{2}\left[\frac{\partial u}{\partial x} - \frac{\partial v}{\partial y}\right] + \frac{i}{2}\left[\frac{\partial u}{\partial y} + \frac{\partial v}{\partial x}\right].$$

We see that $\partial f/\partial \bar{z} = 0$ if and only if f satisfies the Cauchy–Riemann equations, that is, if and only if f is holomorphic.

Now consider an arbitrary second-order partial differential operator in $\mathbb{C}$:

$$L = a(z)\frac{\partial^2}{\partial z\partial \bar{z}} + b(z)\frac{\partial^2}{\partial z^2} + c(z)\frac{\partial^2}{\partial \bar{z}^2} + d(z)\frac{\partial}{\partial z} + e(z)\frac{\partial}{\partial \bar{z}} + f(z).$$

We want to study those L's that satisfy the following properties:

1. L at the origin equals the Laplacian.
2. If $f \in C^\infty(D)$ and $\phi \in \text{Aut}(D)$ then

$$L(f \circ \phi)(z)\big|_{z=0} = \left[(Lf) \circ \phi(z)\right]\big|_{z=0}.$$

These two properties uniquely determine L up to a constant multiple. Suitably normalized, the operator becomes unique. We leave these assertions as an exercise. In fact, it turns out that

$$L = 2(1 - |\zeta|^2)^2\,\triangle. \qquad (6.1.1)$$

This operator is called the *invariant Laplacian* (or sometimes the *Laplace–Beltrami operator* for the Poincaré–Bergman metric). We note that the analogous second-order operator on $\mathbb{R}^N$ that commutes with the group of rigid motions of the plane (translations, rotations, and reflections) is the Laplacian (or constant multiples thereof). For many purposes the pre-factor of $(1 - |\zeta|^2)^2$ in the invariant Laplacian is of no interest. So we end up with just the familiar Laplacian, and it seems that we have discovered nothing new.

On the unit ball in $\mathbb{C}^n$, matters are more complicated. The uniquely determined second-order partial differential operator that commutes with biholomorphic transformations of the ball is not the standard Laplacian. It will turn out to be the Laplace–Beltrami operator for the Bergman metric of the ball, and will have a different form. (We say more about Bergman geometry in the next section.) We provide the details in the next few sections.

For simplicity of notation, we restrict attention to the unit ball B in complex dimension two. A function on B is called *holomorphic* if it is holomorphic, in the usual one-variable sense, in each variable separately. A mapping $F = (f_1, f_2) : B \to B$ is called holomorphic if each of f_1, f_2 is holomorphic. It is *biholomorphic* if F is one-to-one, surjective, and has a holomorphic inverse. (Background in these matters may be found in [KR1].) We begin by determining the biholomorphic self-maps of B.

The unitary maps of $\mathbb{C}^2$ are those 2×2 complex matrices satisfying $U^{-1} = U^*$. Here $*$ is the conjugate transpose, or Hermitian adjoint. A geometrically more appealing description of the unitary group is that it is the collection of those mappings that preserve the Hermitian inner product of vectors $z = \langle z_1, z_2 \rangle$ and $w = \langle w_1, w_2 \rangle$:

$$z \cdot w = z_1 \bar{w}_1 + z_2 \bar{w}_2.$$

Obviously any unitary map is a biholomorphic self-mapping of the ball.

The other important biholomorphic self-mappings of the ball are the Möbius transformations. A thorough consideration of these mappings appears in [RU2]. Here we shall be brief. For $a \in \mathbb{C}, |a| < 1$, we define

$$\phi_a(z_1, z_2) = \left(\frac{z_1 - a}{1 - \bar{a}z_1}, \frac{\sqrt{1 - |a|^2}\, z_2}{1 - \bar{a}z_1} \right). \tag{6.1.2}$$

We first check that $\phi_a : B \to B$. Now

$$\left| \frac{z_1 - a}{1 - \bar{a}z_1} \right|^2 + \left| \frac{\sqrt{1 - |a|^2}\, z_2}{1 - \bar{a}z_1} \right|^2 < 1$$

if and only if

$$|z_1 - a|^2 + (1 - |a|^2)|z_2|^2 < |1 - \bar{a}z_1|^2,$$

that is,

$$|z_1|^2 + |a|^2 - 2\mathrm{Re}\,\bar{a}z_1 + (1 - |a|^2)|z_2|^2 < 1 + |a|^2|z_1|^2 - 2\mathrm{Re}\,\bar{a}z_1.$$

This is equivalent to

$$\left(|z_1|^2 + |z_2|^2\right)\left(1 - |a|^2\right) < \left(1 - |a|^2\right),$$

which is the defining inequality for the ball. Notice that $\phi_{-a}(0,0) = (a,0)$ and $\phi_a(a,0) = 0$. If a, b are complex numbers each with modulus less than one, then $\phi_{-b} \circ \phi_a(a,0) = \phi_{-b}(0,0) = (b,0)$. Thus we see that the group generated by the Möbius transformations (6.1.2) together with the unitary transformations acts transitively on B. In fact, one may prove using the one-variable Schwarz lemma that this group coincides with the group of all biholomorphic self-maps of B (exercise, or see [RUD2]).

6.2 The Bergman Kernel

Let Ω be a bounded domain in $\mathbb{C}^n$. Set

$$A^2(\Omega) = \left\{ f \text{ holomorphic on } \Omega : \int_\Omega |f|^2 \, dV < \infty. \right\}$$

Here dV represents the standard Euclidean volume element. As on the ball, a function of two complex variables is holomorphic precisely when it is holomorphic in each variable separately.

For $f, g \in A^2(\Omega)$ we define

$$\langle f, g \rangle = \int_\Omega f \bar{g} \, dV.$$

With this inner product, $A^2(\Omega)$ becomes an inner product space. The corresponding norm on A^2 is

$$\|f\| = \sqrt{\langle f, f \rangle}.$$

In a moment we shall see that it is in fact complete.

LEMMA 6.2.1
If K is a compact subset of Ω then there exists a constant $C = C(K, \Omega)$ such that for all $f \in A^2(\Omega)$ we have

$$\sup_K |f(z)| \le C \cdot \left(\int_\Omega |f|^2 \, dV \right)^{1/2}.$$

PROOF If f is holomorphic, then in particular f is harmonic in each variable separately and hence harmonic as a function of several real variables. Thus f

satisfies the mean value property with respect to balls. Since K is compact, there exists an $r > 0$ such that $B(z,r) \subseteq \Omega$ for all $z \in K$. Now

$$|f(z)| \leq \frac{1}{V(B(z,r))} \int_{B(z,r)} |f(\zeta)| \, dV(\zeta)$$

$$\leq \frac{1}{\sqrt{V(B(z,r))}} \left(\int_{B(z,r)} |f(\zeta)|^2 \, dV(\zeta) \right)^{1/2}$$

$$= \frac{1}{\sqrt{V(B(z,r))}} \|f\|_{A^2(\Omega)},$$

where in the penultimate line we have used the Schwarz inequality.   ∎

PROPOSITION 6.2.2
The space $A^2(\Omega)$ is a Hilbert space when equipped with the inner product

$$\langle f, g \rangle = \int_\Omega f \bar{g} \, dV.$$

PROOF   It is only necessary to show that the space is complete. If $\{f_j\}$ is a Cauchy sequence in $A^2(\Omega)$ (i.e., in the topology of $L^2(\Omega)$), then it converges uniformly on compact subsets of Ω by the lemma. Then there exists a function f to which $\{f_j\}$ converges normally. Hence f is holomorphic. Since f is also the L^2 limit of the sequence, it follows that $f \in A^2(\Omega)$.   ∎

Now fix a point $z \in \Omega$. The functional

$$e_z : A^2(\Omega) \ni f \longmapsto f(z) \in \mathbb{C}$$

is continuous by the lemma. Then, by the Riesz representation theorem, there exists a $k_z \in A^2(\Omega)$ such that

$$f(z) = e_z(f) = \langle f, k_z \rangle$$

$$= \int f(\zeta) \overline{k_z(\zeta)} \, dV(\zeta). \tag{6.2.3}$$

Thus we have

DEFINITION 6.2.4   *The function*

$$K(z, \zeta) \equiv \overline{k_z(\zeta)}$$

*is called the **Bergman kernel** for the domain Ω.*

PROPOSITION 6.2.5
If $f \in A^2(\Omega)$ then for all $z \in \Omega$ we have

$$f(z) = \int_\Omega f(\varsigma) K(z, \varsigma) \, dV(\varsigma).$$

PROOF This is just a restatement of line (6.2.3) above. ∎

PROPOSITION 6.2.6
The Bergman kernel satisfies

$$K(z, \varsigma) = \overline{K(\varsigma, z)}.$$

PROOF Since, by its very definition, $\overline{K(z, \cdot)}$ is an element of A^2, we have for $w \in \Omega$ that

$$\overline{K(z, w)} = \int \overline{K(z, \varsigma)} K(w, \varsigma) \, d\varsigma$$

$$= \overline{\int \overline{K(w, \varsigma)} K(z, \varsigma) \, d\varsigma}$$

$$= \overline{\overline{K(w, z)}}$$

$$= K(w, z). \qquad \blacksquare$$

Our existence argument for the Bergman kernel is abstract; we now present a constructive method for understanding the kernel. There are in fact two constructive approaches: (1) the use of conformal invariance and (2) using a basis for $A^2(\Omega)$. We begin by discussing (2).

PROPOSITION 6.2.7
Let $H(z, \varsigma)$ be a function on $\Omega \times \Omega$ satisfying

1. $H(\cdot, \varsigma) \in A^2(\Omega)$;
2. $H(z, \varsigma) = \overline{H(\varsigma, z)}$;
3. $f(z) = \int_\Omega f(\varsigma) H(z, \varsigma) \, dV(\varsigma)$ for all $f \in A^2(\Omega)$.

Then H is the Bergman kernel: $H = K$.

PROOF This is similar to the proof of the last proposition. We leave the details as an exercise. ∎

Now $A^2(\Omega)$ is a subspace of L^2 so it is separable. Let $\{\phi_j\}$ be a complete orthonormal basis for $A^2(\Omega)$. We consider the formal sum

$$\sum_j \phi_j(z)\overline{\phi_j(\zeta)} \equiv \tilde{K}(z,\zeta).$$

Fix a compact set $L \subseteq \Omega$. If $w \in L$ then

$$\left(\sum_j |\phi_j(w)|^2\right)^{1/2} = \sup_{\substack{\{a_j\}\in\ell^2 \\ \|\{a_j\}\|_{\ell^2}=1}} \left|\sum_j a_j\phi_j(w)\right|$$

$$= \sup_{\substack{f\in A^2(\Omega) \\ \|f\|=1}} |f(w)|$$

$$\leq C_L$$

by the lemma. By the Schwarz lemma we conclude that

$$\sum_j \phi_j(z)\overline{\phi_j(\zeta)}$$

converges uniformly on $L \times L$. Now it is easy to see that $\tilde{K}$ satisfies the first two properties that characterize the Bergman kernel. Finally, by the Riesz–Fisher theory,

$$\int f(\zeta)\tilde{K}(z,\zeta)\,dV(\zeta) = \sum_j \left(\int f(\zeta)\overline{\phi_j(\zeta)}\,dV(\zeta)\right)\phi_j(z)$$

$$= f(z).$$

Hence $\tilde{K}$ *is the Bergman kernel.*

Now we can explicitly calculate the Bergman kernel on some particular domains. First consider the disc $D \subseteq \mathbb{C}$. We take $\phi_j(\zeta) = \zeta^{j-1}/\gamma_j, j = 1, 2, \ldots,$ where the constants γ_j will be specified in a moment. These functions are pairwise orthogonal, as one easily sees by introducing polar coordinates. They span $A^2(D)$, for if $f \in A^2$ has power series expansion $f(z) = \sum a_j z^j$ and if $\langle f, \phi_j \rangle = 0$ for all j then all a_j are zero. Finally, we select the constants γ_j to make each ϕ_j have norm one: We calculate that

$$\iint_D |z^j|^2\,dx\,dy = \int_0^1 \int_0^{2\pi} r^{2j+1}\,d\theta\,dr$$

$$= 2\pi \cdot \frac{1}{2j+2}$$

$$= \frac{\pi}{j+1}.$$

Therefore we take $\gamma_j = \sqrt{\pi}/\sqrt{j}$ and set

$$\phi_j(z) = \frac{\sqrt{j}}{\sqrt{\pi}} z^{j-1}.$$

Then

$$K(z,\zeta) = \sum_{j=1}^{\infty} \phi_j(z)\overline{\phi_j(\zeta)}$$

$$= \frac{1}{\pi} \sum_{j=0}^{\infty} (j+1)(z\bar{\zeta})^j.$$

Recall that

$$\sum_{j=0}^{\infty} (j+1)\lambda^j = \frac{d}{d\lambda} \sum_{j=0}^{\infty} \lambda^j = \frac{d}{d\lambda} \frac{1}{1-\lambda} = \frac{1}{(1-\lambda)^2}.$$

We conclude that

$$K(z,\zeta) = \frac{1}{\pi} \frac{1}{(1-z\bar{\zeta})^2}.$$

Exercise: Give an alternate derivation of the Bergman kernel for the disc by using the Cauchy integral formula together with Stokes's theorem.

Now we return our attention to the ball in $\mathbb{C}^2$. In analogy with what we did on the disc, we could set

$$\phi_{jk}(z) = \frac{z_1^j z_2^k}{\gamma_{jk}}$$

and proceed to calculate suitable values for the γ_{jk} (this procedure is carried out in [KR1]. However we shall instead use approach (1) for calculating the Bergman kernel. In order to carry out this procedure, we shall need the following proposition:

PROPOSITION 6.2.8
Let $\Phi : \Omega_1 \to \Omega_2$ be biholomorphic. Then the Bergman kernels K_{Ω_1} and K_{Ω_2} of these two domains are related by the formula

$$K_{\Omega_1}(z,\zeta) = \det \operatorname{Jac}_{\mathbb{C}} \Phi(z) K_{\Omega_2}(\Phi(z), \Phi(\zeta)) \overline{\det \operatorname{Jac}_{\mathbb{C}} \Phi(\zeta)}. \qquad (6.2.8.1)$$

Before we prove this proposition, we make some remarks. First, recall that in real multivariable calculus when we do a change of variables in an integral we use the *real* Jacobian determinant $\det \operatorname{Jac}_{\mathbb{R}}$. But now we are doing complex calculus and we use the *complex* Jacobian determinant $\det \operatorname{Jac}_{\mathbb{C}}$. In the context of $\mathbb{C}^n$, the real Jacobian is a real $2n \times 2n$ matrix. On the other hand, the complex Jacobian is a complex $n \times n$ matrix. How are these two concepts of the Jacobian related? It turns out that, when Φ is holomorphic, then

$$\det \mathrm{Jac}_{\mathbf{R}} \Phi = |\det \mathrm{Jac}_{\mathbf{C}} \Phi|^2 .$$

The reader may wish to prove this as an exercise, or consult [KR1] for details.

LEMMA 6.2.9
If $g \in A^2(\Omega_2)$ then

$$(g \circ \Phi) \cdot \det \mathrm{Jac}_{\mathbf{C}} \Phi \in A^2(\Omega_1).$$

PROOF We calculate that

$$\int_{\Omega_1} \left| (g \circ \Phi(z)) \det \mathrm{Jac}_{\mathbf{C}} \Phi(z) \right|^2 dV(z)$$

$$= \int_{\Omega_2} |g(w)|^2 \left| \det \mathrm{Jac}_{\mathbf{C}} \Phi(\Phi^{-1}(w)) \right|^2 \left| \det \mathrm{Jac}_{\mathbf{R}} \Phi^{-1}(w) \right| dV(w)$$

$$= \int_{\Omega_2} |g(w)|^2 \, dV(w) < \infty. \qquad\qquad \blacksquare$$

COROLLARY 6.2.10
The right-hand side of equation (6.2.8.1) is square integrable and holomorphic in the z variable and square integrable and conjugate holomorphic in the ζ variable.

PROOF Obvious. ∎

Now, in order to prove formula (6.2.8.1), it remains to verify property (3) of the three properties characterizing the Bergman kernel. Let $f \in A^2(\Omega_1)$. Let us write $\mathcal{J}$ for $\det \mathrm{Jac}_{\mathbf{C}}$. Then we have

$$\int_{\Omega_1} f(\zeta) \mathcal{J}\Phi(z) K_{\Omega_2}(\Phi(z), \Phi(\zeta)) \overline{\mathcal{J}\Phi(\zeta)} \, dV(\zeta)$$

$$= \int_{\Omega_2} f(\Phi^{-1}(\xi)) \mathcal{J}\Phi(z) K_{\Omega_2}(\Phi(z), \xi) \overline{\mathcal{J}(\Phi(\Phi^{-1}(\xi)))} \cdot \left| \mathcal{J}\Phi^{-1}(\xi) \right|^2 dV(\xi)$$

$$\equiv \mathcal{J}\Phi(z) \int_{\Omega_2} g(\xi) K_{\Omega_2}(\Phi(z), \xi) \, dV(\xi),$$

where

$$g(\xi) = f(\Phi^{-1}(\xi)) \mathcal{J}\Phi^{-1}(\xi) \in A^2(\Omega_2)$$

by 6.2.9. Thus the term on the right side of our chain of equalities is equal to

$$J\Phi(z)g(\Phi(z)) = J\Phi(z)f\left(\Phi^{-1}(\Phi(z))\right)J\Phi^{-1}(\Phi(z))$$
$$= f(z).$$

That verifies the reproducing property for the right side of (6.2.8.1) and the proposition is proved. ∎

Armed with our preliminary calculations, we now turn to our calculation of the Bergman kernel for the ball $B \subseteq \mathbb{C}^2$. If $f \in A^2(B)$ then, by the mean value property,

$$f(0) = \frac{1}{V(B)} \int_B f(\zeta)\, dV(\zeta).$$

This equality leads us to surmise that $K_B(0, \zeta) = 1/V(B)$ hence, in particular, $K_B(0, 0) = 1/V(B)$. Assuming this, we use the result of our proposition to calculate $K_B((a, 0), (a, 0))$ when $|a| < 1$. Define

$$\Phi(z_1, z_2) = \left(\frac{z_1 + a}{1 + \bar{a}z_1}, \frac{\sqrt{1 - |a|^2}z_2}{1 + \bar{a}z_1}\right).$$

Set $\alpha = \Phi(0, 0) = (a, 0)$. Using formula (6.2.8.1) we see that

$$K_B(0, 0) = \det \operatorname{Jac}_{\mathbb{C}}\Phi(0)K_B\left(\Phi(0), \Phi(0)\right)\overline{\det \operatorname{Jac}_{\mathbb{C}}\Phi(0)}.$$

In other words,

$$\frac{1}{V(B)} = \det \operatorname{Jac}_{\mathbb{C}}\Phi(0)K_B(\alpha, \alpha)\overline{\det \operatorname{Jac}_{\mathbb{C}}\Phi(0)}.$$

Observe that

$$\operatorname{Jac}_{\mathbb{C}}\Phi(0, 0) = \begin{pmatrix} 1 - |a|^2 & 0 \\ 0 & \sqrt{1 - |a|^2} \end{pmatrix}.$$

Therefore

$$\det \operatorname{Jac}_{\mathbb{C}}\Phi(0, 0) = (1 - |a|^2)^{3/2}.$$

Hence

$$K_B(\alpha, \alpha) = \frac{1}{V(B)}\frac{1}{(1 - |a|^2)^3}.$$

Now for every $\beta \in B$ there exists a point $\alpha = (a, 0) \in B$ and a unitary transformation U such that $U\beta = \alpha$. Moreover notice that if U is unitary then $\operatorname{Jac}_{\mathbb{C}}U = U$. Therefore the proposition implies that

$$K_B(z, \zeta) = K_B(Uz, U\zeta).$$

Thus

$$K_B(\beta,\beta) = K_B(U\beta, U\beta)$$

$$= \frac{1}{V(B)} \cdot \frac{1}{(1 - U\beta \cdot \overline{U\beta})^3}$$

$$= \frac{1}{V(B)} \cdot \frac{1}{(1 - \beta \cdot \bar{\beta})^3} \cdot \tag{6.2.11}$$

Now we need the following observation: If $F(z,w)$ is a function holomorphic in $z \in \Omega$ and conjugate holomorphic in $w \in \Omega$ and if $F(z,w) = 0$ when $z = w$ then $F \equiv 0$. Assume this claim for the moment. Then we may conclude that

$$K_B(z,w) = \frac{1}{V(B)} \frac{1}{(1 - z \cdot \bar{w})^3} \cdot \tag{6.2.12}$$

Let us review the logic: We have demonstrated that this function K_B satisfies the three properties that characterize the Bergman kernel so it must be the Bergman kernel.

We conclude by proving the observation. Define $G(z,w) = F(z,\bar{w})$. Then G is holomorphic in both z and w. Moreover, $G \equiv 0$ on $S = \{(z,w) : z = \bar{w}\}$. Consider the mapping

$$\mathcal{H} : (z,w) \longmapsto (z + w, i(z - w)).$$

Then $G \circ \mathcal{H}$ is holomorphic and equals 0 when $z + w = \overline{i(z - w)}$, that is on

$$T \equiv \{(z,w) : \operatorname{Re} z = -\operatorname{Im} z \text{ and } \operatorname{Re} w = \operatorname{Im} w\}.$$

It now follows from elementary one-variable considerations that $G \circ \mathcal{H} \equiv 0$, hence $F \equiv 0$.

REMARK   Let Ω be any bounded domain in $\mathbb{C}^n$. It is a corollary of the representation $K_\Omega(z,z) = \sum_{j=1}^\infty |\phi_j(z)|^2$ that $K(z,z) \neq 0$ for all z. This observation will prove useful later.   ∎

6.3   The Szegö and Poisson–Szegö Kernels

Let $\Omega \subset\subset \mathbb{C}^n$ be a smoothly bounded domain, and define

$$A(\Omega) \equiv C(\bar{\Omega}) \cap H(\Omega),$$

where $H(\Omega)$ is the space of holomorphic functions on Ω. Define

$$\|f\|_{H^2(\Omega)} = \left(\int_{\partial\Omega} |f|^2 \, d\sigma \right)^{1/2},$$

where $d\sigma$ is area measure on the boundary of Ω (see [KR1], Appendix II). and let $H^2(\Omega)$ be the closure of $A(\Omega)$ with respect to this norm.

REMARK It is natural to wonder how this definition of $H^2(\Omega)$ relates to other, perhaps more familiar, definitions of the space. On the unit disc $H^2(D)$ is usually defined to be the space of those functions f holomorphic on D such that

$$\sup_{0<r<1} \int_0^{2\pi} |f(re^{i\theta})|^2 \, d\theta < \infty.$$

There is an analogous definition on more general domains (see [KR1]). Yet a third approach to Hardy spaces is due to Lumer [LUM]. According to Lumer, a holomorphic function f on a domain Ω is in H^2 if there is a harmonic function h such that $|f|^2 \le h$. On domains in $\mathbb{C}^n$ with C^2 boundary, all three definitions are equivalent (see [KR1]). The equivalence of the first definition with the other two on an arbitrary pseudoconvex domain is difficult—see [BEA]. ∎

LEMMA 6.3.1
*If K is a compact set contained in Ω, then there exists a constant $C = C(\Omega, K)$
such that for every $f \in H^2(\Omega)$ it holds that*

$$\sup_{z \in K} |f(z)| \le C\|f\|_{H^2}.$$

PROOF Exercise: The shortest proof uses the Bochner–Martinelli formula, for which see [KR1]. ∎

Equipped with this lemma, one can imitate the Bergman theory to prove the existence of a reproducing kernel $S(z, \zeta)$ for the space H^2. This kernel is known as the Szegö or Cauchy–Szegö kernel. Further, if $\{\phi_j\}$ is an orthonormal basis for H^2 then

$$S(z, \zeta) = \sum_j \phi_j(z)\overline{\phi_j(\zeta)}.$$

For the unit ball in $\mathbb{C}^n$, we may calculate (as we did for the Bergman kernel in the last section) that the Szegö kernel is given by

$$S(z, \zeta) = \frac{1}{\sigma(\partial B)} \cdot \frac{1}{(1 - z \cdot \bar{\zeta})^n} \cdot$$

In the case $n = 1$,

$$S(z, \zeta) = \frac{1}{2\pi} \cdot \frac{1}{1 - z\bar{\zeta}} \cdot$$

If $f \in A(D)$ and $z = re^{i\theta} \in D$, then the Szegö reproducing formula becomes

$$f(z) = \frac{1}{2\pi} \int_{\partial D} \frac{1}{1 - z\bar{\zeta}} f(\zeta) \, d\sigma(\zeta)$$

$$= \frac{1}{2\pi} \int_0^{2\pi} \frac{1}{1 - re^{i\theta}e^{-i\phi}} f(e^{i\phi}) \, d\phi$$

$$= \frac{1}{2\pi i} \int_0^{2\pi} \frac{1}{e^{i\phi} - re^{i\theta}} f(e^{i\phi}) i e^{i\phi} \, d\phi$$

$$= \frac{1}{2\pi i} \oint_{\partial D} \frac{1}{\zeta - z} f(\zeta) \, d\zeta.$$

So we see that, on the disc, the Szegö reproducing formula is just the standard Cauchy integral formula.

Associated with the Szegö kernel on a domain Ω is the *Poisson–Szegö kernel* defined by

$$\mathcal{P}(z, \zeta) = \frac{|S(z, \zeta)|^2}{S(z, z)} .$$

PROPOSITION 6.3.2
The Poisson–Szegö kernel has the following properties:

1. $\mathcal{P}(z, \zeta) \geq 0$ *for* $z \in \Omega, \zeta \in \partial\Omega$.
2. *For* $f \in A(\Omega), z \in \Omega$, *we have*

$$f(z) = \int_{\partial\Omega} \mathcal{P}(z, \zeta) f(\zeta) \, d\zeta.$$

PROOF   Recall that $S(z, \zeta) = \sum_j \phi_j(z)\overline{\phi_j(\zeta)}$ and that $S(z, z) = \sum_j |\phi_j(z)|^2$. If there were a $z_0 \in \Omega$ such that $S(z_0, z_0) = 0$, then $\phi_j(z_0) = 0$ for every j hence $S(z_0, \zeta) = 0$ for all ζ. But then it would follow from the reproducing property that $f(z_0) = 0$ for every $f \in H^2$. That is plainly false. Thus $S(z, z)$ is never zero so that $\mathcal{P}$ is well defined. Also, $\mathcal{P}(z, \zeta) \geq 0$ by its very definition.

Finally, if $f \in A(\Omega)$ and $z \in \Omega$ is fixed then set

$$g(\zeta) = \frac{f(\zeta)\overline{S(z, \zeta)}}{S(z, z)} .$$

Because f is bounded, $\overline{S(z, \cdot)} \in H^2$, and $S(z, z)$ is a positive constant, it follows that $g \in H^2$. By the Szegö reproducing property, we then have that

$$g(z) = \int_{\partial\Omega} g(\zeta) S(z, \zeta) \, d\sigma(\zeta).$$

In other words,

$$\frac{f(z)\overline{S(z,z)}}{S(z,z)} = \int_{\partial\Omega} f(\zeta)\frac{\overline{S(z,\zeta)}}{S(z,z)}S(z,\zeta)\,d\sigma(\zeta)$$

or

$$f(z) = \int_{\partial\Omega} f(\zeta)\mathcal{P}(z,\zeta)\,d\sigma(\zeta). \qquad \blacksquare$$

Let us consider our new kernels on the disc and the ball. Let ω_{2n-1} denote the surface area measure of ∂B in $\mathbb{C}^n$. Then

$$S(z,\zeta) = \frac{1}{\omega_{2n-1}} \cdot \frac{1}{(1 - z \cdot \overline{\zeta})^n} ,$$

thus

$$\mathcal{P}(z,\zeta) = \frac{1}{\omega_{2n-1}} \cdot \frac{(1 - |z|^2)^n}{|1 - z \cdot \overline{\zeta}|^{2n}} .$$

In case $n = 1$, we may calculate for $z = re^{i\theta} \in D$ and $\zeta = e^{i\phi} \in \partial D$ that

$$\begin{aligned}
\mathcal{P}(z,\zeta) &= \frac{1}{2\pi}\frac{1 - |z|^2}{|1 - z \cdot \overline{\zeta}|^2} \\
&= \frac{1}{2\pi}\frac{1 - r^2}{|1 - re^{i(\theta-\phi)}|^2} \\
&= \frac{1}{2\pi}\frac{1 - r^2}{1 - 2r\cos(\theta - \phi) + r^2} .
\end{aligned}$$

Thus in one complex variable the Poisson–Szegö kernel is the classical Poisson kernel. In dimensions two and above this is not the case. In fact, we should like to stress a fundamental difference between the two kernels in dimension two. The singularity (denominator) of the classical Poisson kernel for the Laplacian in $\mathbb{C}^2$ is $|z - \zeta|^4$. However, the singularity (denominator) of the Poisson–Szegö kernel is $|1 - z \cdot \overline{\zeta}|^4$. Whereas in one complex variable the expressions $|z - \zeta|$ and $|1 - z \cdot \overline{\zeta}|$ are equal (for $z \in D, \zeta \in \partial D$), in several complex variables they are not. Formally, this is because $z \cdot \overline{\zeta} = z_1\overline{\zeta_1} + \cdots + z_n\overline{\zeta_n}$; whereas we can "undo" the multiplication $z \cdot \overline{\zeta}$ in $\mathbb{C}^1$ by multiplication by ζ^{-1}, which has modulus 1, there is no analogous operation in several variables.

The difference between the two singularities in $\mathbb{C}^2$, for instance, has a profound geometric aspect. The natural geometry in ∂B associated with the singularity $s = |\zeta - z|$ of the Poisson kernel is suggested by the balls

$$B(z,r) = \{\zeta : |s| < r\}.$$

These balls are *isotropic*: they measure the same in all directions.

However, the natural geometry in ∂B associated with the singularity $p = |1 - z \cdot \bar{\zeta}|$ of the Poisson–Szegö kernel is suggested by the balls

$$\beta(z,r) = \{\zeta : |p| < r\}.$$

To understand the shape of one of these balls, let us take, in dimension 2, $z = (1,0) \in \partial B$. Then

$$\beta(z,r) = \{\zeta : |1 - \bar{\zeta}_1| < r\}.$$

If $(\zeta_1, \zeta_2) \in \beta(z,r)$ then, because this "ball" is a subset of the boundary, we may calculate that

$$\begin{aligned}
|\zeta_2|^2 &= 1 - |\zeta_1|^2 \\
&= (1 - |\zeta_1|)(1 + |\zeta_1|) \\
&\leq 2(1 - |\zeta_1|) \\
&\leq 2|1 - \zeta_1| \\
&< 2r.
\end{aligned}$$

Thus we see that whereas the ball has size r in the ζ_1 direction, it has size $\sqrt{r}$ in the z_2 direction. Therefore these balls are *nonisotropic*.

The contrast of the isotropic geometry of real analysis and the nonisotropic geometry of several complex variables will be a prevailing theme throughout this chapter and for much of the remainder of the book.

The Szegö kernel reproduces all of H^2 while the Poisson–Szegö kernel reproduces (on a formal level) only the subspace $A(\Omega)$. But $\mathcal{P}$ is real—indeed, nonnegative. In particular, if $f \in A(\Omega)$ then we may take the real part of both sides of the formula

$$f(z) = \int_{\partial\Omega} f(\zeta)\mathcal{P}(z,\zeta)\,d\sigma(\zeta)$$

to obtain

$$\text{Re } f(z) = \int \text{Re } f(\zeta)\mathcal{P}(z,\zeta)\,d\sigma(\zeta).$$

Thus the Poisson–Szegö kernel reproduces pluriharmonic functions (real parts of holomorphic functions) that are continuous on $\bar{\Omega}$. (See [KR1] for more on pluriharmonic functions.)

The interesting fact for us will be that, on the ball in $\mathbb{C}^n$, the kernel $\mathcal{P}(z,\zeta)$ solves the Dirichlet problem for a certain invariant second-order elliptic partial differential operator. This operator, which we shall study in detail below, is the Laplace–Beltrami operator for the Bergman kernel. In order to determine this operator, we shall require some knowledge of the Bergman metric.

6.4 The Bergman Metric

Let Ω be a domain in $\mathbb{R}^N$. A *Riemannian metric* on Ω is a matrix $g = (g_{ij}(x))_{i,j=1}^N$ of C^2 functions on Ω such that if $0 \neq \xi \in \mathbb{R}^N$ then

$$\xi \left(g_{ij}(x)\right)_{i,j=1}^N {}^t\xi > 0. \tag{6.4.1}$$

In this way we assign to each x a positive definite quadratic form. The associated notion of vector length (for each x) is

$$\|\xi\|_x = \sqrt{\xi(g_{ij}(x))^t\xi}. \tag{6.4.2}$$

Recall that in calculus we define the length of a curve $\gamma : [0,1] \to \mathbb{R}$ to be

$$\ell(\gamma) = \int_0^1 |\gamma'(t)| \, dt.$$

The philosophy of Riemannian geometry is to allow the method of calculating the length of $\gamma'(t)$ to vary with t—that is, to let the norm used to calculate the length of $\gamma'(t)$ depend on the base point $\gamma(t)$. Thus, in the Riemannian metric g explicated in $(6.4.1), (6.4.2)$ the length of a curve γ is given by

$$\ell_g(\gamma) = \int_0^1 \|\gamma'(t)\|_{\gamma(t)} \, dt.$$

Example 1
Let $\Omega = D$, the unit disc in $\mathbb{R}^2$. Fix $0 < \lambda < 1$. We let $\gamma(t) = (\lambda t, 0), 0 \leq t \leq 1$. Then, with the metric $g(x) = \{\delta_{ij}\}$, we calculate that

$$\gamma'(t) = (\lambda, 0)$$

and

$$\|\gamma'(t)\|_{\gamma(t)} = \sqrt{(\lambda,0) \begin{pmatrix} 1 & 0 \\ 0 & 1 \end{pmatrix} \begin{pmatrix} \lambda \\ 0 \end{pmatrix}} = \sqrt{\lambda^2} = \lambda.$$

Therefore

$$\lambda_g(\gamma) = \int_0^1 \lambda \, dt = \lambda.$$

In this example we see that we have calculated the Euclidean length of a segment in the usual fashion. ▯

Example 2
Let

$$g_{ij}(z) = \begin{pmatrix} \frac{1}{(1-|z|^2)^2} & 0 \\ 0 & \frac{1}{(1-|z|^2)^2} \end{pmatrix}$$

and let γ be as in Example 1. Then

$$\ell_g(\gamma) = \int_0^1 \sqrt{(\lambda,0)\begin{pmatrix} \frac{1}{(1-|\lambda t|^2)^2} & 0 \\ 0 & \frac{1}{(1-|\lambda t|^2)^2} \end{pmatrix}\begin{pmatrix} \lambda \\ 0 \end{pmatrix}}\, dt$$

$$= \lambda \int_0^1 \frac{1}{1-(\lambda t)^2}\, dt$$

$$= \frac{1}{2}\log\left(\frac{1+\lambda}{1-\lambda}\right).$$

We see that in the metric in this example, a version of the Poincaré metric on the disc, our segment connecting 0 to λ has length which is unbounded as $\lambda \to 1^-$. □

DEFINITION 6.4.1 *Let Ω be a bounded domain in $\mathbb{C}^n$ and let $K = K_\Omega$ be the Bergman kernel for Ω. For $i, j = 1, \ldots, n$ we define*

$$g_{ij}(z) = \frac{\partial^2}{\partial z_i \partial \bar{z}_j}\log K(z,z).$$

The matrix $g = \left(g_{ij}(z)\right)_{i,j=1}^n$ is called the Bergman metric on Ω. It acts as a Hermitian quadratic form on the complex tangent space.

REMARK   From today's perspective—especially in view of the theory of Kähler manifolds, in which all metrics arise locally as the complex Hessians of potential functions—the definition of the Bergman metric is quite natural. The use of logarithmic potential functions is particularly well established. However, half a century ago, when Bergman gave this definition, it was quite an original idea.

We shall do some explicit calculations of the Bergman metric in what follows. ∎

Let us discuss the definition of the Bergman metric. First, it makes good sense since (as already noted) $K(z,z) > 0$ for all $z \in \Omega$. We shall now check that the Bergman metric is invariant under biholomorphic mappings. To this end, let

$$\Phi : \Omega_1 \to \Omega_2$$

be a biholomorphic mapping. Let $\gamma : [0, 1] \to \Omega_1$ be a smooth curve. Then

$$\Phi_* \gamma \equiv \Phi \circ \gamma : [0, 1] \to \Omega_2.$$

LEMMA 6.4.2
With the above notation,

$$\ell_{\text{Berg},\Omega_2}(\Phi_* \gamma) = \ell_{\text{Berg},\Omega_1}(\gamma).$$

PROOF   We want to show that

$$\int_0^1 \|(\Phi_* \gamma)'(t)\|_{\Omega_2, \Phi_* \gamma(t)} \, dt = \int_0^1 \|\gamma'(t)\|_{\Omega_1, \gamma(t)} \, dt.$$

Of course it suffices (and, as it turns out, it is also necessary) to show that the integrands are equal. By the transformation formula for the Bergman kernel, we know that

$$\mathcal{J}\Phi(z) K_{\Omega_2}(\Phi(z), \Phi(z)) \overline{\mathcal{J}\Phi(z)} = K_{\Omega_1}(z, z).$$

Taking logarithms, we find that

$$\log \mathcal{J}\Phi(z) + \log \left(K_{\Omega_2}(\Phi(z), \Phi(z)) \right) + \log \overline{\mathcal{J}\Phi(z)} = \log \left(K_{\Omega_1}(z, z) \right).$$

Let us now apply $\partial^2 / \partial z_i \partial \bar{z}_j$ on both sides. This second derivative applied to the right-hand side is $g_{ij}^{\Omega_1}(z)$. The second derivative applied to the left-hand side equals

$$0 + \sum_{\ell=1}^n \sum_{m=1}^n \frac{\partial^2 \log K_{\Omega_2}}{\partial s_\ell \partial \bar{w}_m}(\Phi(z), \Phi(z)) \cdot \frac{\partial \Phi_\ell}{\partial z_i} \frac{\partial \bar{\Phi}_m}{\partial \bar{z}_j} + 0.$$

Therefore

$$g_{ij}^{\Omega_1}(z) = \sum_{\ell,m=1}^n g_{\ell m}^{\Omega_2}(\Phi(z)) \frac{\partial \Phi_\ell}{\partial z_i} \frac{\partial \bar{\Phi}_m}{\partial \bar{z}_j} \ .$$

Finally,

$$\|\xi\|_{\Omega_1, z} = \sqrt{\sum_{i,j} g_{ij}^{\Omega_1}(z) \xi_i \bar{\xi}_j}$$

$$= \sqrt{\sum_{i,j} \sum_{\ell,m} g_{\ell m}^{\Omega_2}(\Phi(z)) \frac{\Phi_\ell}{\partial z_i}(z) \frac{\bar{\Phi}_m}{\partial \bar{z}_j}(z) \xi_i \bar{\xi}_j}$$

$$= \sqrt{\sum_{\ell,m} g_{\ell m}^{\Omega_2}(\Phi(z)) (\Phi_* \xi)_\ell \overline{(\Phi_* \xi)}_m}$$

$$= \|\Phi_* \xi\|_{\Omega_2, \Phi(z)}.$$

Example: Let us calculate the Bergman metric on the ball. Let

$$K(z, \zeta) = K_B(z, \zeta) = \frac{1}{V(B)} \frac{1}{(1 - z \cdot \bar{\zeta})^{n+1}}$$

be the Bergman kernel. Then

$$\log K(z, z) = -\log V(B) - (n + 1) \log(1 - |z|^2).$$

Further,

$$\frac{\partial}{\partial z_i} \left(-(n + 1) \log(1 - |z|^2) \right) = (n + 1) \frac{\bar{z}_i}{1 - |z|^2}$$

and

$$\frac{\partial^2}{\partial z_i \partial \bar{z}_j} \left(-(n + 1) \log(1 - |z|^2) \right) = (n + 1) \left[\frac{\delta_{ij}}{1 - |z|^2} + \frac{\bar{z}_i z_j}{(1 - |z|^2)^2} \right]$$

$$= \frac{(n + 1)}{(1 - |z|^2)^2} \left[\delta_{ij}(1 - |z|^2) + \bar{z}_i z_j \right]$$

$$\equiv g_{ij}(z).$$

If $n = 1$, then

$$g_{ij}(z) = g = \frac{2}{(1 - |z|^2)^2} ;$$

this is just the Poincaré metric on the disc.

When $n = 2$ we have

$$g_{ij}(z) = \frac{3}{(1 - |z|^2)^2} \left[\delta_{ij}(1 - |z|^2) + \bar{z}_i z_j \right].$$

Thus

$$(g_{ij}(z)) = \frac{3}{(1 - |z|^2)^2} \begin{pmatrix} 1 - |z_2|^2 & \bar{z}_1 z_2 \\ \bar{z}_2 z_1 & 1 - |z_1|^2 \end{pmatrix}.$$

Let

$$\left(g^{ij}(z) \right)_{i,j=1}^{n}$$

represent the inverse of the matrix

$$\left(g_{ij}(z) \right)_{i,j=1}^{n} \quad .$$

Then an elementary computation shows that

$$\left(g^{ij}(z) \right)_{i,j=1}^{n} = \frac{1 - |z|^2}{3} \begin{pmatrix} 1 - |z_1|^2 & -z_2 \bar{z}_1 \\ -z_1 \bar{z}_2 & 1 - |z_2|^2 \end{pmatrix} = \frac{1 - |z|^2}{3} \left(\delta_{ij} - \bar{z}_i z_j \right)_{i,j}.$$

Let

$$g \equiv \det \left(g_{ij}(z) \right).$$

Then

$$g = \frac{9}{(1 - |z|^2)^3} \cdot$$

☐

DEFINITION 6.4.3 *If Ω is a domain in $\mathbb{C}^n$ and $g(z) = (g_{ij}(z))_{ij}$ is a Hermitian metric on Ω, then the **Laplace–Beltrami operator** associated to g is defined to be the differential operator*

$$\Delta_B = \frac{2}{g} \sum_{i,j} \left\{ \frac{\partial}{\partial \bar{z}_i} \left(gg^{ij} \frac{\partial}{\partial z_j} \right) + \frac{\partial}{\partial z_j} \left(gg^{ij} \frac{\partial}{\partial \bar{z}_i} \right) \right\}.$$

Now let us calculate. If $(g_{ij})_{i,j=1}^n$ is the Bergman metric on the ball in $\mathbb{C}^n$ then we have

$$\sum_{i,j} \frac{\partial}{\partial \bar{z}_i} \left(gg^{ij} \right) = 0$$

and

$$\sum_{i,j} \frac{\partial}{\partial z_j} \left(gg^{ij} \right) = 0.$$

We verify these assertions in detail in dimension 2: Now

$$gg^{ij} = \frac{9}{(1 - |z|^2)^3} \cdot \frac{1 - |z|^2}{3} (\delta_{ij} - \bar{z}_i z_j)$$

$$= \frac{3}{(1 - |z|^2)^2} (\delta_{ij} - \bar{z}_i z_j).$$

It follows that

$$\frac{\partial}{\partial \bar{z}_i} gg^{ij} = \frac{6z_i}{(1 - |z|^2)^3} (\delta_{ij} - \bar{z}_i z_j) - \frac{3z_j}{(1 - |z|^2)^2} \cdot$$

Therefore

$$\sum_{i,j=1}^2 \frac{\partial}{\partial \bar{z}_i} gg^{ij} = \sum_{i,j=1}^2 \left[\frac{6z_i(\delta_{ij} - \bar{z}_i z_j)}{(1 - |z|^2)^3} - \frac{3z_j}{(1 - |z|^2)^2} \right]$$

$$= 6 \sum_i \frac{z_i}{(1 - |z|^2)^3} - 6 \sum_{i,j} \frac{|z_i|^2 z_j}{(1 - |z|^2)^3} - 6 \sum_j \frac{z_j}{(1 - |z|^2)^2}$$

$$= 6 \sum_j \frac{z_j}{(1 - |z|^2)^2} - 6 \sum_j \frac{z_j}{(1 - |z|^2)^2}$$

$$= 0.$$

The other derivative is calculated similarly.

Our calculations show that, on the ball in $\mathbb{C}^2$,

$$\Delta_B \equiv \frac{2}{g} \sum_{i,j} \left\{ \frac{\partial}{\partial \bar{z}_i} \left(g g^{ij} \frac{\partial}{\partial z_j} \right) + \frac{\partial}{\partial z_j} \left(g g^{ij} \frac{\partial}{\partial \bar{z}_i} \right) \right\}$$

$$= 4 \sum_{i,j} g^{ij} \frac{\partial}{\partial \bar{z}_i} \frac{\partial}{\partial z_j}$$

$$= 4 \sum_{i,j} \frac{1 - |z|^2}{3} \left(\delta_{ij} - \bar{z}_i z_j \right) \frac{\partial^2}{\partial z_j \partial \bar{z}_i} \ .$$

This formula, which we have derived in detail in dimension two, has an analog (with 3 replaced by $n + 1$) that is valid in any dimension.

Exercise: Verify that if $\Phi : B \rightarrow B$ is biholomorphic and if $\Delta_B u = 0$ then $\Delta(u \circ \Phi) = 0$. More generally, check that if v on B is any smooth function and Φ biholomorphic, then $\Delta_B(v \circ \Phi) = (\Delta_B v) \circ \Phi$.

Notice that, on the disc,

$$\Delta_D = 4 \frac{(1 - |z|^2)}{2} (1 - |z|^2) \frac{\partial^2}{\partial z \partial \bar{z}}$$

$$= 2(1 - |z|^2)^2 \Delta,$$

where Δ (without the subscript) is just the ordinary Laplacian. Thus the Laplace–Beltrami operator for the Poincaré–Bergman metric on the disc is just the Laplacian followed by a smooth function. It exhibits no features that are essentially different from those of the Laplacian. We see, however, that in two or more variables the Laplace–Beltrami operator is a genuinely new object of study. We shall learn more about it in later sections.

6.5   The Dirichlet Problem for the Invariant Laplacian on the Ball

We will study the following Dirichlet problem on $B \subseteq \mathbb{C}^2$:

$$\begin{cases} \Delta_B u = 0 & \text{on } B \\ u\big|_{\partial B} = \phi, \end{cases} \tag{6.5.1}$$

where ϕ is a given continuous function on ∂B.

Exercise: Is this a well-posed boundary value problem (in the sense of Lopatinski)?

The remarkable fact about this relatively innocent-looking boundary value problem is that there exist data functions $\phi \in C^\infty(\partial B)$ with the property that the (unique) solution to the boundary value problem is not even C^2 on $\bar{B}$. This result appears in [RGR1] and was also discovered independently by Garnett and Krantz. It is in striking contrast to the situation that obtains for the Dirichlet problem on a uniformly elliptic operator such as we studied in Chapter 5.

Observe that for $n = 1$ our Dirichlet problem becomes

$$\begin{cases} (1 - |z|^2)^2 \, \triangle u = 0 & \text{on } D \subseteq \mathbb{C} \\ u|_{\partial D} = \phi, \end{cases}$$

which is just the same as

$$\begin{cases} \triangle u = 0 & \text{on } D \subseteq \mathbb{C} \\ u|_{\partial D} = \phi. \end{cases}$$

This is the standard Dirichlet problem for the Laplacian—a uniformly strongly elliptic operator. Thus there is a complete existence and regularity theory: the solution u will be as smooth on the closure as is the data ϕ (provided that we measure this smoothness in the correct norms). Our problem in dimensions $n > 1$ yields some surprises. We begin by developing some elementary geometric ideas.

Let $\zeta, \xi \in \partial B$. Define

$$\rho(\zeta, \xi) = |1 - \zeta \cdot \bar{\xi}|^{1/2},$$

where $\zeta \cdot \bar{\xi} \equiv \zeta_1 \bar{\xi}_1 + \zeta_2 \bar{\xi}_2$. Then we have

PROPOSITION 6.5.1
The binary operator ρ is a metric on ∂B.

PROOF   Let $z, w, \zeta \in \partial B$. We shall show that

$$\rho(z, \zeta) \le \rho(z, w) + \rho(w, \zeta).$$

Assume for simplicity that the dimension $n = 2$. We learned the following argument from R. R. Coifman.

After applying a rotation, we may assume that $w = (0, i)$. For $z = (z_1, z_2) \in \partial B$, set $\mu(z) = \rho(z, w) = \sqrt{|1 - i\bar{z}_2|}$. Then, for any $z, \zeta \in \partial B$, we have

$$|1 - z \cdot \bar{\zeta}| = |-z_1\bar{\zeta}_1 + (1 + iz_2)(i\bar{\zeta}_2) + (1 - i\bar{\zeta}_2)|$$

$$\le |z_1\bar{\zeta}_1| + |1 + iz_2||\bar{\zeta}_2| + |1 - i\bar{\zeta}_2|$$

$$\le |\bar{\zeta}_1||z_1| + \mu^2(z) + \mu^2(\zeta). \tag{*}$$

However

$$|\zeta_1|^2 = (1 - |\zeta_2|)(1 + |\zeta_2|) \le 2(1 - |\zeta_2|) \le 2|i - \zeta_2| = 2\mu^2(\zeta).$$

Of course the same estimate applies to z_1. Therefore

$$|\zeta_1 z_1| \le 2\mu(\zeta)\mu(z).$$

Substituting this into $(*)$ gives

$$|1 - z \cdot \bar{\zeta}| \le 2\mu(\zeta)\mu(z) + \mu^2(z) + \mu^2(\zeta) = (\rho(z,w) + \rho(w,\zeta))^2. \quad \blacksquare$$

Now we define balls using ρ: for $P \in \partial B$ and $r > 0$ we define $\beta(P,r) = \{\zeta \in \partial B : \rho(P,\zeta) < r\}$. [These skew balls (see the discussion in Section 6.3) play a decisive role in the complex geometry of several variables. We shall get just a glimpse of their use here.] Let $0 \ne z \in B$ be fixed and let P be its orthogonal projection on the boundary: $\tilde{z} = z/|z|$. If we fix $r > 0$ then we may verify directly that

$$\mathcal{P}(z,\zeta) \to 0 \quad \text{uniformly in} \quad \zeta \in \partial B \setminus \beta(\tilde{z}, r) \quad \text{as} \quad z \to \tilde{z}.$$

PROPOSITION 6.5.2
Let $B \subseteq \mathbb{C}^n$ be the unit ball and $g \in C(\partial B)$. Then the function

$$G(z) = \begin{cases} \int_{\partial B} \mathcal{P}(z,\zeta) g(\zeta) \, d\sigma(\zeta) & \text{if } z \in B \\ g(z) & \text{if } z \in B \end{cases}$$

solves the Dirichlet problem (6.5.1) for the Laplace–Beltrami operator $\triangle_B$. Here $\mathcal{P}$ is the Poisson–Szegö kernel.

PROOF It is straightforward to calculate that

$$\triangle_B G(z) = \int_{\partial B} [\triangle_B \mathcal{P}(z,\zeta)] g(\zeta) \, d\sigma(\zeta)$$
$$= 0$$

because $\triangle_B \mathcal{P}(\cdot, \zeta) = 0$.

For simplicity, let us now restrict attention once again to dimension $n = 2$. We wish to show that G is continuous on $\bar{B}$. First recall that

$$\mathcal{P}(z,\zeta) = \frac{1}{\sigma(\partial B)} \frac{(1 - |z|^2)^2}{|1 - z \cdot \bar{\zeta}|^4}.$$

Notice that

$$\int_{\partial B} |\mathcal{P}(z,\zeta)| \, d\sigma(\zeta) = \int_{\partial B} \mathcal{P}(z,\zeta) \, d\sigma(\zeta) = \int_{\partial B} \mathcal{P}(z,\zeta) \cdot 1 \, d\sigma(\zeta) = 1$$

since the identically 1 function is holomorphic on Ω and is therefore reproduced by integration against $\mathcal{P}$. We also have used the fact that $\mathcal{P} \ge 0$.

Now we enter the proof proper of the proposition. Fix $\epsilon > 0$. By the uniform continuity of g we may select a $\delta > 0$ such that if $P \in \partial B$ and $\zeta \in \beta(P, \delta)$ then $|g(P) - g(\zeta)| < \epsilon$. Then, for any $0 \neq z \in B$ and P its projection to the boundary, we have

$$
\begin{aligned}
|G(z) - g(P)| &= \left| \int_{\partial B} \mathcal{P}(z, \zeta) g(\zeta) \, d\sigma(\zeta) - g(P) \right| \\
&= \left| \int_{\partial B} \mathcal{P}(z, \zeta) g(\zeta) \, d\sigma(\zeta) - \int_{\partial B} \mathcal{P}(z, \zeta) g(P) \, d\sigma(\zeta) \right| \\
&\leq \int_{\partial B} \mathcal{P}(z, \zeta) |g(\zeta) - g(P)| \, d\sigma \\
&= \int_{\beta(P, \delta)} \mathcal{P}(z, \zeta) |g(\zeta) - g(P)| \, d\sigma(\zeta) \\
&\quad \int_{\partial B \backslash \beta(P, \delta)} \mathcal{P}(z, \zeta) |g(\zeta) - g(P)| \, d\sigma(\zeta) \\
&\leq \epsilon + 2\|g\|_{L^\infty} \int_{\partial B \backslash \beta(P, \delta)} \mathcal{P}(z, \zeta) \, d\sigma(\zeta).
\end{aligned}
$$

By the remarks preceding this argument, we may choose r sufficiently close to 1 that $\mathcal{P}(z, \zeta) < \epsilon$ for $|z| > r$ and $\zeta \in \partial B \setminus \beta(P, \delta)$. Thus, with these choices, the last line does not exceed $C \cdot \epsilon$.

We conclude the proof with an application of the triangle inequality: Fix $P \in \partial B$ and suppose that $0 \neq z \in B$ satisfies both $|P - z| < \delta$ and $|z| > r$. If $\tilde{z} = z/|z|$ is the projection of z to ∂B then we have

$$
|G(z) - g(P)| \leq |G(z) - g(\tilde{z})| + |g(\tilde{z}) - g(P)|.
$$

The first term is majorized by ϵ by the argument that we just concluded. The second is less than ϵ by the uniform continuity of g on $\partial \Omega$.

That concludes the proof. ∎

Now we know how to solve the Dirichlet problem for Δ_B; next we want to consider regularity for this operator. The striking fact, in contrast with the uniformly elliptic case, is that for g even in $C^\infty(\partial B)$ we may not conclude that the solution G of the Dirichlet problem is C^∞ on $\bar{B}$. In fact, in dimension n, the function G is not generally in $C^n(\bar{B})$. Consider the following example:

Example: Let $n = 2$. Define

$$
g(z_1, z_2) = |z_1|^2.
$$

Of course $g \in C^{\infty}(\partial B)$. We now calculate $\mathcal{P}g(z)$ rather explicitly. We have

$$\mathcal{P}g(z) = \frac{1}{\sigma(\partial B)} \int_{\partial B} \frac{(1 - |z|^2)^2}{|1 - z \cdot \bar{\zeta}|^4} |\zeta_1|^2 \, d\sigma(\zeta).$$

Let us restrict our attention to points z in the ball of the form $z = (r + i0, 0)$. We set

$$\mathcal{P}g(r + i0) = \phi(r).$$

We shall show that ϕ fails to be C^2 on the interval $[0, 1]$ at the point 1. We have

$$\phi(r) = \frac{1}{\sigma(\partial B)} \int_{\partial B} \frac{(1 - r^2)^2}{|1 - r\zeta_1|^4} |\zeta_1|^2 \, d\sigma(\zeta)$$

$$= \frac{(1 - r^2)^2}{\sigma(\partial B)} \int_{|\zeta_1| < 1} \int_{|\zeta_2| = \sqrt{1 - |\zeta_1|^2}} \frac{|\zeta_1|^2}{|1 - r\zeta_1|^4} \cdot \frac{1}{\sqrt{1 - |\zeta_1|^2}} \, ds(\zeta_2) \, dA(\zeta_1)$$

$$= \frac{(1 - r^2)^2}{\sigma(\partial B)} \int_{|\zeta_1| < 1} \frac{|\zeta_1|^2}{|1 - r\zeta_1|^4} \frac{2\pi\sqrt{1 - |\zeta_1|^2}}{\sqrt{1 - |\zeta_1|^2}} \, dA(\zeta_1)$$

$$= \frac{2\pi}{\sigma(\partial B)} (1 - r^2)^2 \int_{|\zeta_1| < 1} \frac{|\zeta_1|^2}{|1 - r\zeta_1|^4} \, dA(\zeta_1).$$

Now we set $\zeta_1 = \tau e^{i\psi}, 0 \leq \tau < 1, 0 \leq \psi \leq 2\pi$. The integral is then

$$\frac{2\pi}{\sigma(\partial B)} (1 - r^2)^2 \int_0^{2\pi} \int_0^1 \frac{\tau^2}{|1 - r\tau e^{i\psi}|^4} \tau \, d\tau \, d\psi.$$

We perform the change of variables $r\tau = s$ and set $C = 2\pi/\sigma(\partial B)$. The integral becomes

$$C \frac{(1 - r^2)^2}{r^4} \int_0^{2\pi} \int_0^r \frac{s^3}{|1 - se^{i\psi}|^4} \, ds \, d\psi$$

$$C \frac{(1 - r^2)^2}{r^4} \int_0^r s^3 \int_0^{2\pi} \frac{1}{|1 - se^{i\psi}|^4} \, d\psi \, ds.$$

Now let us examine the inner integral. It equals

$$\int_0^{2\pi} \frac{1}{(1 - se^{i\psi})^2 (1 - se^{-i\psi})^2} \, d\psi = \int_0^{2\pi} \frac{e^{2i\psi}}{(1 - se^{i\psi})^2 (e^{i\psi} - s)^2} \, d\psi$$

$$= \int_0^{2\pi} \frac{\frac{e^{i\psi}}{(1 - se^{i\psi})^2}}{(e^{i\psi} - s)^2} e^{i\psi} \, d\psi$$

$$= 2\pi \cdot \frac{1}{2\pi i} \oint_{|\eta| = 1} \frac{\frac{\eta}{(1 - s\eta)^2}}{(\eta - s)^2} \, d\eta.$$

Applying the theory of residues to this Cauchy integral, we find that the last line equals

$$2\pi \frac{d}{d\eta}\left(\frac{\eta}{(1-s\eta)^2}\right)\bigg|_{\eta=s} = 2\pi \frac{1+s^2}{(1-s^2)^3}\,.$$

Thus

$$
\begin{aligned}
\phi(r) &= 2\pi C\frac{(1-r^2)^2}{r^4}\int_0^r \frac{s^3(1+s^2)}{(1-s^2)^3}\,ds \\[2mm]
&\overset{(s\mapsto\sqrt{s})}{=} \pi C\frac{(1-r^2)^2}{r^4}\int_0^{r^2} \frac{s(1+s)}{(1-s)^3}\,ds \\[2mm]
&= \pi C\frac{(1-r^2)^2}{r^4}\int_0^{r^2} \frac{2}{(1-s)^3} - \frac{3}{(1-s)^2} + \frac{1}{(1-s)}\,ds \\[2mm]
&= \pi C\frac{(1-r^2)^2}{r^4}\left\{\left[\frac{1}{(1-s)^2} - \frac{3}{(1-s)} - \log(1-s)\right]_0^{r^2}\right\} \\[2mm]
&= \pi C\frac{(1-r^2)^2}{r^4}\left\{\frac{1}{(1-r^2)^2} - \frac{3}{1-r^2} - \log(1-r^2) + 2\right\} \\[2mm]
&= \frac{\pi C}{r^4}\left\{1 - 3(1-r^2) - (1-r^2)^2\log(1-r^2) + 2(1-r^2)^2\right\} \\[2mm]
&= \pi C\left\{\frac{1-3(1-r^2)+2(1-r^2)^2}{r^4} - \frac{(1-r^2)^2\log(1-r^2)}{r^4}\right\}.
\end{aligned}
$$

Thus we see that $\phi(r)$ is the sum of two terms. The first of these is manifestly smooth at the point 1. However, the second is not C^2 (from the left) at 1. Therefore the function ϕ is not C^2 at 1 and we conclude that $\mathcal{P}g$ is not smooth at the point $(1,0) \in \partial B$, even though g itself is. ☐

The phenomenon described in this example was discovered by Garnett and Krantz in 1977 (unpublished) and independently by C. R. Graham. Graham [RGR1] subsequently developed a regularity theory for Δ_B using weighted function spaces. He also used Fourier analysis to explain the failure of boundary regularity in the usual function space topologies.

It turns out that these matters were anticipated by G. B. Folland in 1975 (see [FOL]). Using spherical harmonics, one can see clearly that the Poisson–Szegö integral of a function $g \in C^\infty(\partial B)$ will be smooth on $\bar{B}$ if and only if g is the boundary function of a pluriharmonic function (these arise naturally as the real parts of holomorphic functions—see [KR1]). We shall explicate these matters by beginning, in the next section, with a discussion of spherical harmonics.

6.6 Spherical Harmonics

Spherical harmonics are for many purposes the natural generalization of the Fourier analysis of the circle to higher dimensions. Spherical harmonics are also intimately connected to the representation theory of the orthogonal group. As a result, analogs of the spherical harmonics play an important role in general representation theory.

Our presentation of spherical harmonics owes a debt to [STW]. We begin with a consideration of spherical harmonics in real Euclidean N-space. For $k = 0, 1, 2, \ldots$ we let $\mathcal{P}_k$ denote the linear space over $\mathbb{C}$ of all homogeneous polynomials of degree k. Then $\{x^\alpha\}_{|\alpha|=k}$ is a basis for $\mathcal{P}_k$. Let d_k denote the dimension, over the field $\mathbb{C}$, of $\mathcal{P}_k$. We need to calculate d_k. This will require a counting argument.

We need to determine the number of N-tuples $\alpha = (\alpha_1, \ldots, \alpha_N)$ such that $\alpha_1 + \cdots \alpha_N = k$. Imagine $N + k - 1$ boxes as shown in Figure 6.1. We mark any $N - 1$ of these boxes. Let $\alpha_1 \geq 0$ be the number of boxes preceding the first one marked, $\alpha_2 \geq 0$ the number of boxes between the first and second that were marked, and so on. This defines N nonnegative integers $\alpha_1, \ldots, \alpha_N$ such that $\alpha_1 + \cdots + \alpha_N = k$. Also, every such N-tuple $(\alpha_1, \ldots, \alpha_N)$ arises in this way. Thus we see that

$$d_k = \binom{N+k-1}{N-1} = \binom{N+k-1}{k} = \frac{(N+k-1)!}{(N-1)!k!} \, .$$

Now we want to define a Hermitian inner product on $\mathcal{P}_k$. In this chapter, if $P = \sum_\alpha c_\alpha x^\alpha$ is a polynomial then the differential operator $P(D)$ is defined to be

$$P(D) = \sum_\alpha c_\alpha \frac{\partial^\alpha}{\partial x^\alpha} \, .$$

Here α is a multiindex. For $P, Q \in \mathcal{P}_k$ we then define

$$\langle P, Q \rangle \equiv P(D) \left(\overline{Q} \right) .$$

FIGURE 6.1

If $P = \sum_{|\alpha|=k} p_\alpha x^\alpha$ and $Q = \sum_{|\alpha|=k} q_\alpha x^\alpha$ then we have

$$\langle P, Q \rangle = P(D)\,(\overline{Q}) = \sum_{|\alpha|=k} p_\alpha \partial^\alpha \left(\sum_{|\beta|=k} \bar{q}_\beta x^\beta \right)$$

$$= \sum_{|\alpha|,|\beta|=k} p_\alpha \bar{q}_\beta \partial^\alpha x^\beta$$

$$= \sum_{|\alpha|,|\beta|=k} p_\alpha \bar{q}_\beta \delta_{\alpha\beta} \alpha!,$$

where $\delta_{\alpha\beta} = 0$ if $\alpha \neq \beta$ and $= 1$ if $\alpha = \beta$. Also $\alpha! \equiv \alpha_1! \cdots \alpha_N!$. Therefore $\langle P, Q \rangle$ is scalar-valued. It is linear in each entry and Hermitian symmetric. Moreover, we see that

$$\langle P, P \rangle = \sum_\alpha |p_\alpha|^2 \alpha!$$

so that

$$\langle P, P \rangle \geq 0 \quad \text{and} \quad = 0 \quad \text{iff} \quad P = 0.$$

Thus $\langle \, \cdot \, , \, \cdot \, \rangle$ is a Hermitian, nondegenerate inner product on $\mathcal{P}_k$.

PROPOSITION 6.6.1
Let $P \in \mathcal{P}_k$. Then we can write

$$P(x) = P_0(x) + |x|^2 P_1(x) + \cdots + |x|^{2\ell} P_\ell(x),$$

where each polynomial P_j is homogeneous and harmonic with degree $k-2j, 0 \leq j \leq \ell$, and $\ell = [k/2]$.

PROOF Any polynomial of degree less than 2 is harmonic, so there is nothing to prove in this case. We therefore assume that $k \geq 2$. Define the map

$$\phi_k : \mathcal{P}_k \to \mathcal{P}_{k-2}$$
$$P \mapsto \Delta P,$$

where Δ is the (classical) Laplacian. Now consider the adjoint operator

$$\phi_k^* : \mathcal{P}_{k-2} \to \mathcal{P}_k.$$

This adjoint is determined by the equalities

$$\langle Q, \Delta P \rangle = Q(D)\overline{\Delta P} = \Delta Q(D)\bar{P} = \langle R, P \rangle,$$

where $R(x) = |x|^2 Q(x)$. Therefore

$$\phi_k^*(Q)(x) = |x|^2 Q(x).$$

Notice that ϕ_k^* is one-to-one. Recall also that ϕ_k is surjective if and only if ϕ_k^* is one-to-one (for $Q \perp$ range ϕ_k if and only if $\phi_k^*(Q) = 0$—we are in a finite-dimensional vector space). Moreover, the kernel of ϕ_k is perpendicular to the image of ϕ_{k-2}^*. In symbols,

$$\mathcal{P}_k \cong \ker \phi_k \oplus \operatorname{im} \phi_{k-2}^*.$$

That is,

$$\mathcal{P}_k = \mathcal{A}_k \oplus \mathcal{B}_k,$$

where

$$\mathcal{A}_k = \ker\phi_k = \{P \in \mathcal{P}_k : \Delta P = 0\}$$

and

$$\mathcal{B}_k = \operatorname{im} \phi_{k-2}^* = \{P \in \mathcal{P}_k : P(x) = |x|^2 Q(x), \text{some } Q \in \mathcal{P}_{k-2}\}.$$

Hence, for $P \in \mathcal{P}_k$,

$$P(x) = P_0 + |x|^2 Q(x)$$

where P_0 is harmonic and $Q \in \mathcal{P}_{k-2}$.

The result now follows immediately by induction.   ∎

COROLLARY 6.6.2
The restriction to the surface of the unit sphere Σ_{N-1} of any polynomial of N variables is a sum of restrictions to Σ_{N-1} of harmonic polynomials.

PROOF Use the preceding proposition. The expressions $|x|^{2j}$ become 1 when restricted to the sphere.   ∎

DEFINITION 6.6.3 *The **spherical harmonics** of degree k, denoted $\mathcal{H}_k$, are the restrictions to the unit sphere of the elements of $\mathcal{A}_k$, that is, the restrictions to the unit sphere of the harmonic polynomials of degree k.*

If $Y = P\big|_{\Sigma_{N-1}}$ for some $P \in \mathcal{A}_k$, then

$$P(x) = Y(x/|x|) \cdot |x|^k$$

so that the restriction is an isomorphism of $\mathcal{A}_k$ onto $\mathcal{H}_k$. In particular,

$$
\begin{aligned}
\dim \mathcal{H}_k &= \dim \mathcal{A}_k \\
&= \dim \mathcal{P}_k - \dim \mathcal{P}_{k-2} \\
&= d_k - d_{k-2} \\
&= \binom{N+k-1}{k} - \binom{N+k-3}{k-2}
\end{aligned}
$$

for $k \geq 2$. Notice that $\dim \mathcal{H}_0 = 1$ and $\dim \mathcal{H}_1 = N$.

For $N = 2$, it is easy to see that

$$
\mathcal{H}_k = \text{span} \{\cos k\theta, \sin k\theta\}.
$$

Then $\dim \mathcal{H}_k = 2$ for all $k \geq 1$. This is, of course, consistent with the formula for the dimension of $\mathcal{H}_k$ that we just derived for all dimensions. For $N = 3$, one sees that $\dim \mathcal{H}_k = 2k + 1$ for all $k \geq 0$. We denote $\dim \mathcal{H}_k = \dim \mathcal{A}_k$ by a_k.

The space $\mathcal{A}_k$ is called the space of *solid spherical harmonics* and the space $\mathcal{H}_k$ is the space of *surface spherical harmonics*.

PROPOSITION 6.6.4
The finite linear combinations of elements of $\cup_k \mathcal{H}_k$ is uniformly dense in $C(\Sigma_{N-1})$ and L^2 dense in $L^2(\Sigma_{N_1}, d\sigma)$.

PROOF The first statement clearly implies the second. For the first we invoke the Stone–Weierstrass theorem. ∎

PROPOSITION 6.6.5
If $Y^{(k)} \in \mathcal{H}_k$ and $Y^{(\ell)} \in \mathcal{H}_\ell$ with $k \neq \ell$ then

$$
\int_{\Sigma_{N-1}} Y^{(k)}(x') Y^{(\ell)}(x') \, d\sigma(x') = 0.
$$

PROOF We will use Green's theorem (see [KR1] for a proof): if $u, v \in C^2(\bar{\Omega})$, where Ω is a bounded domain with C^2 boundary, then

$$
\int_{\partial \Omega} u \frac{\partial}{\partial \nu} v - v \frac{\partial}{\partial \nu} u \, d\sigma = \int_{\Omega} u \, \triangle v - v \, \triangle u \, dV.
$$

Here $\partial/\partial \nu$ is the unit outward normal derivative to $\partial \Omega$.

Now for $x \in \mathbb{R}^N$ we write $x = rx'$ with $r = |x|$ and $|x'| = 1$. Then

$$u(x) \equiv |x|^k Y^{(k)}(x') = r^k Y^{(k)}(x')$$

and

$$v(x) \equiv |x|^\ell Y^{(\ell)}(x') = r^\ell Y^{(\ell)}(x')$$

are harmonic polynomials.

If one of k or ℓ is zero, then one of u or v is constant and what we are about to do reduces to the well-known fact that for a harmonic function f on B, C^1 on $\bar{B}$, we have

$$\int_{\partial B} \frac{\partial}{\partial \nu} f \, d\sigma = 0$$

(see [KR1]). Details are left for the reader.

In case both k and ℓ are nonzero then

$$\frac{\partial}{\partial \nu} u(x') = \frac{\partial}{\partial r} \left(r^k Y^{(k)}(x') \right)$$
$$= k r^{k-1} Y^{(k)}(x')$$
$$= k Y^{(k)}(x')$$

(since $r = 1$) and, similarly,

$$\frac{\partial}{\partial \nu} v(x') = \ell Y^{(\ell)}(x').$$

By Green's theorem, then,

$$0 = \int_B u(x) \, \Delta \, v(x) - v(x) \, \Delta \, u(x) \, dV(x)$$
$$= \int_{\partial B} u(x')\ell Y^{(\ell)}(x') - v(x')k Y^{(k)}(x') \, d\sigma(x')$$
$$= \int_{\partial B} \ell Y^{(k)}(x')Y^{(\ell)}(x') - k Y^{(\ell)}(x')Y^{(k)}(x') \, d\sigma(x')$$
$$= (\ell - k) \int_{\partial B} Y^{(k)}(x')Y^{(\ell)}(x') \, d\sigma(x').$$

Since $\ell \neq k$, the assertion follows. ∎

We endow $L^2(\partial B, d\sigma)$ with the usual inner product. So of course each $\mathcal{H}^k$ inherits this inner product as well. For $k = 0, 1, 2, \ldots$ we let $\{Y_1^{(k)}, \ldots, Y_{a_k}^{(k)}\}$, $a_k = d_k - d_{k-2}$, be an orthonormal basis for $\mathcal{H}_k$. By Propositions 6.6.4 and 6.6.5 it follows that

$$\bigcup_{k=0}^{\infty} \left\{ Y_1^{(k)}, \ldots, Y_{a_k}^{(k)} \right\}$$

is an orthonormal basis for $L^2(\Sigma_{N-1}, d\sigma)$. If $f \in L^2(\Sigma_{N-1})$ then there is a unique representation

$$f = \sum_{k=0}^{\infty} Y^{(k)},$$

where the series converges in the L^2 topology and $Y^{(k)} \in \mathcal{H}_k$. Furthermore, by linear algebra,

$$Y^{(k)} = \sum_{j=1}^{a_k} b_j Y_j^{(k)},$$

where $b_j = \langle Y^{(k)}, Y_j^{(k)} \rangle, j = 1, \ldots, a_k$.

As an example of these ideas, we see for $N = 2$ that

$$k = 0: \quad Y_1^{(0)} = \frac{1}{\sqrt{2\pi}}$$

$$k \geq 1: \quad \begin{cases} Y_1^{(k)} = \frac{1}{\sqrt{\pi}} \cos k\theta \\ Y_2^{(k)} = \frac{1}{\sqrt{\pi}} \sin k\theta, \end{cases}$$

that is,

$$\left\{ \frac{1}{\sqrt{2\pi}}, \frac{1}{\sqrt{\pi}} \cos k\theta, \frac{1}{\sqrt{\pi}} \sin k\theta \right\}$$

is a complete orthonormal system in $L^2(\mathbf{T})$.

Claim: For $N = 2$ we can recover the Poisson kernel for the Laplacian from the spherical harmonics. If $f \in L^2(\partial D)$ then consider

$$F(re^{i\theta}) = \sum_{j,k} r^k \langle f, Y_j^{(k)} \rangle Y_j^{(k)}(e^{i\theta}).$$

Then we have

$$F(re^{i\theta}) = \int_0^{2\pi} f(e^{i\phi}) \frac{1}{\sqrt{2\pi}} d\phi \cdot \frac{1}{\sqrt{2\pi}}$$

$$+ \sum_{k=1}^{\infty} r^k \int_0^{2\pi} f(e^{i\phi}) \frac{\cos k\phi}{\sqrt{\pi}} d\phi \frac{\cos k\theta}{\sqrt{\pi}}$$

$$+ \sum_{k=1}^{\infty} r^k \int_0^{2\pi} f(e^{i\phi}) \frac{\sin k\phi}{\sqrt{\pi}} d\phi \frac{\sin k\theta}{\sqrt{\pi}}$$

$$= \frac{1}{2\pi} \int_0^{2\pi} f(e^{i\phi}) \, d\phi$$

$$+ \frac{1}{\pi} \sum_{k=1}^{\infty} r^k \int_0^{2\pi} f(e^{i\phi}) \cos k(\theta - \phi) \, d\phi$$

$$= \frac{1}{\pi} \int_0^{2\pi} f(e^{i\phi}) \left[\frac{1}{2} + \sum_{k=1}^{\infty} r^k \cos k(\theta - \phi) \right] d\phi.$$

But the expression in brackets equals

$$\frac{1}{2} + \mathrm{Re} \left\{ \sum_{k=1}^{\infty} r^k e^{ik(\theta - \phi)} \right\}$$

$$= \frac{1}{2} + \mathrm{Re} \left\{ r e^{i(\theta - \phi)} \cdot \sum_{k=0}^{\infty} r^k e^{ik(\theta - \phi)} \right\}$$

$$= \frac{1}{2} + \mathrm{Re} \left\{ r e^{i(\theta - \phi)} \cdot \frac{1}{1 - r e^{i(\theta - \phi)}} \right\}$$

$$= \frac{1}{2} \cdot \frac{1 - r^2}{1 - 2r \cos(\theta - \phi) + r^2} \cdot$$

Thus

$$F(re^{i\theta}) = \frac{1}{2\pi} \int_0^{2\pi} \frac{1 - r^2}{1 - 2r \cos(\theta - \phi) + r^2} f(e^{i\phi}) \, d\phi.$$

It follows from elementary Hilbert space considerations that $F(re^{i\theta}) \to f(e^{i\theta})$ (first check this claim on finite linear combinations of spherical harmonics, which are dense). Thus, at least formally, we have recovered the classical Poisson integral formula from spherical harmonic analysis.

6.7 Advanced Topics in the Theory of Spherical Harmonics: the Zonal Harmonics

Since the case $N \leq 2$ has now been treated in some detail, and has been seen to be familiar, let us assume from now on that $N > 2$.

Fix a point $x' \in \Sigma_{N-1}$ and consider the linear functional on $\mathcal{H}_k$ given by

$$e_{x'} : Y \mapsto Y(x').$$

Of course $\mathcal{H}_k$ is a Hilbert space so there exists a unique spherical harmonic $Z_{x'}^{(k)}$ such that

$$Y(x') = e_{x'}(Y) = \int_{\Sigma_{N-1}} Y(t') Z_{x'}^{(k)}(t') \, dt'$$

for all $Y \in \mathcal{H}_k$. (The reader will note here some formal parallels between the zonal harmonic theory and the Bergman kernel theory covered earlier. In fact, this parallel goes deeper. See, for instance, [ARO] for more on these matters.)

DEFINITION 6.7.1   The function $Z_{x'}^{(k)}$ is called the **zonal harmonic of degree** k **with pole at** x'.

LEMMA 6.7.2
If $\{Y_1, \ldots, Y_{a_k}\}$ is an orthonormal basis for $\mathcal{H}_k$, then

(a) $\sum_{m=1}^{a_k} \overline{Y_m(x')} Y_m(t') = Z_{x'}^{(k)}(t')$;

(b) $Z_{x'}^{(k)}$ is real-valued and $Z_{x'}^{(k)}(t') = Z_{t'}^{(k)}(x')$;

(c)   If ρ is a rotation then $Z_{\rho x'}^{(k)}(\rho t') = Z_{x'}^{(k)}(t')$.

PROOF   Let $Z_{x'}^{(k)} = \sum_{m=1}^{a_k} \langle Z_{x'}^{(k)}, Y_m \rangle Y_m$ be the standard representation of $Z_{x'}^{(k)}$ with respect to the orthonormal basis $\{Y_1, \ldots, Y_{a_k}\}$. Then

$$\langle Z_{x'}^{(k)}, Y_m \rangle = \int_{\Sigma_{N-1}} \overline{Y_m(t')} Z_{x'}^{(k)}(t') \, dt' = \overline{Y_m(x')};$$

we have used here the reproducing property of the zonal harmonic (note that since Y_m is harmonic then so is $\overline{Y_m}$). This proves (a), for we know that

$$Z_{x'}^{(k)}(t') = \sum_{m=1}^{a_k} \langle Z_{x'}^{(k)}, Y_m \rangle Y_m(t') = \overline{Y_m(x')} Y_m(t').$$

To prove (b), let $f \in \mathcal{H}_k$. Then

$$\bar{f}(x') = \int_{\Sigma_{N-1}} \bar{f}(t') Z_{x'}^{(k)}(t') \, dt'$$

$$= \overline{\int_{\Sigma_{N-1}} f(t') \overline{Z_{k'}^{(k)}(t')} \, dt'}.$$

That is,

$$f(x') = \int_{\Sigma_{N-1}} f(t') \overline{Z_{x'}^{(k)}(t')} \, dt'.$$

Thus we see that $\overline{Z_{x'}^{(k)}}$ reproduces $\mathcal{H}_k$ at the point x'. By the uniqueness of the zonal harmonic at x', we conclude that $Z_{x'}^{(k)} = \overline{Z_{x'}^{(k)}}$. Hence $Z_{x'}^{(k)}$ is real-valued. Now, using (a), we have

$$
Z_{x'}^{(k)}(t') = \sum_{m=1}^{a_k} \overline{Y_m(x')} Y_m(t')
$$

$$
= \overline{\sum_{m=1}^{a_k} Y_m(x') \overline{Y_m(t')}}
$$

$$
= \overline{Z_{t'}^{(k)}(x')}
$$

$$
= Z_{t'}^{(k)}(x').
$$

This establishes (b).

To check that (c) holds, it suffices by uniqueness to see that $Z_{\rho x'}^{(k)}(\rho t')$ reproduces $\mathcal{H}_k$ at x'. This is a formal exercise which we omit. ∎

LEMMA 6.7.3
Let $\{Y_1, \ldots, Y_{a_k}\}$ *be any orthonormal basis for* $\mathcal{H}^k$. *The following properties hold for the zonal harmonics:*

(a) $Z_{x'}^{(k)}(x') = \frac{a_k}{\sigma(\Sigma_{N-1})}$, *where* $a_k = \dim A_k = \dim \mathcal{H}_k$;

(b) $\sum_{m=1}^{a_k} |Y_m(x')|^2 = \frac{a_k}{\sigma(\Sigma_{N-1})}$;

(c) $|Z_{t'}^{(k)}(x')| \leq \frac{a_k}{\sigma(\Sigma_{N-1})}$.

PROOF   Let $x_1', x_2' \in \Sigma_{N-1}$ and let ρ be a rotation such that $\rho x_1' = x_2'$. Then by parts (a) and (c) of 6.7.2 we know that

$$
\sum_{m=1}^{a_k} |Y_m(x_1')|^2 = Z_{x_1'}^{(k)}(x_1') = Z_{x_2'}^{(k)}(x_2') = \sum_{m=1}^{a_k} |Y_m(x_2')|^2 \equiv c.
$$

Then

$$
a_k = \sum_{m=1}^{a_k} \int_{\Sigma_{N-1}} |Y_m(x')|^2 \, d\sigma(x')
$$

$$
= \int_{\Sigma_{N-1}} \sum_{m=1}^{a_k} |Y_m(x')|^2 \, dx'
$$

$$
= c\sigma(\Sigma_{N-1}).
$$

This proves parts (a) and (b).

For part (c), notice that

$$\|Z_{x'}^{(k)}\|_{L^2}^2 = \int_{\Sigma_{N-1}} |Z_{x'}^{(k)}(t')|^2 \, dt'$$

$$= \int_{\Sigma_{N-1}} \left(\sum_m \overline{Y_m(x')} Y_m(t') \right) \left(\overline{\sum_\ell \overline{Y_\ell(x')} Y_\ell(t')} \right) \, dt'$$

$$= \sum_m |Y_m(x')|^2$$

$$= \frac{a_k}{\sigma(\Sigma_{N-1})} \, .$$

Finally, we use the reproducing property of the zonal harmonics to see that

$$|Z_{t'}^{(k)}(x')| = \left| \int_{\Sigma_{N-1}} Z_{t'}^{(k)}(w') Z_{x'}^{(k)}(w') \, dw' \right|$$

$$\leq \|Z_{t'}^{(k)}\|_{L^2} \cdot \|Z_{x'}^{(k)}\|_{L^2}$$

$$= \frac{a_k}{\sigma(\Sigma_{N-1})} \, . \qquad\blacksquare$$

Now we wish to present a version of the expansion of the Poisson kernel in terms of spherical harmonics in higher dimensions. Recall that the Poisson kernel for the ball in $\mathbf{R}^N$ is

$$P(x, t') = \frac{1}{\sigma(\Sigma_{N-1})} \frac{1 - |x|^2}{|x - t'|^N}$$

for $0 \leq |x| < 1$ and $|t'| = 1$ (see [STW]). Now we have

THEOREM 6.7.4
If $x \in B$ then we write $x = rx'$ with $|x'| = 1$. It holds that

$$P(x, t') = \sum_{k=0}^{\infty} r^k Z_{x'}^{(k)}(t') = \sum_{k=0}^{\infty} r^k Z_{t'}^{(k)}(x')$$

is the Poisson kernel for the ball. That is, if $f \in C(\partial B)$ then

$$\int_{\partial B} P(x, t') f(t') \, d\sigma(t') \equiv u(x)$$

solves the Dirichlet problem on the ball with data f.

PROOF Observe that

$$a_k = d_k - d_{k-2} = \binom{N+k-1}{k} - \binom{N+k-3}{k-2}$$

$$= \frac{(N+k-3)!}{(k-1)!(N-2)!} \left\{ \frac{(N+k-1)(N+k-2)}{k(N-1)} - \frac{k-1}{N-1} \right\}$$

$$= \binom{N+k-3}{k-1} \left\{ \frac{N+2k-2}{k} \right\}$$

$$\le C \cdot \binom{N+k-3}{k-1}$$

$$\le C \cdot k^{N-2}.$$

Here $C = C(N)$ depends on the dimension, but not on k. With this estimate, and the estimate on the size of the zonal harmonics from the preceding lemma, we see that the series

$$\sum_{k=0}^{\infty} r^k Z_{t'}^{(k)}(x')$$

converges uniformly on compact subsets of B. Indeed, for $|x| \le s < 1, x = rx'$, we have that

$$\sum_{k=0}^{\infty} \left| r^k Z_{t'}^{(k)}(x') \right| \le \sum_{k=0}^{\infty} s^k \frac{a_k}{\sigma(\Sigma_{N-1})} \le \sum_{k=0}^{\infty} s^k \frac{C \cdot k^{N-2}}{\sigma(\Sigma_{N-1})} = C' \cdot \sum_{k=0}^{\infty} s^k k^{N-2} < \infty.$$

Now let $u(t') = \sum_{m=0}^{P} Y_m(t')$ be a finite linear combination of spherical harmonics with all $Y_m \in \mathcal{H}_k$. Then

$$\sum_{m=0}^{P} |x|^k Y_m \left(\frac{x}{|x|} \right) \equiv u(x) = \int_{\Sigma_{N-1}} u(t') P(x, t') \, dt'$$

is the solution to the classical Dirichlet problem with data Y. Here $P(x, t')$ is the classical Poisson kernel. On the other hand,

$$\int_{\Sigma_{N-1}} u(t') \sum_{k=0}^{\infty} |x|^k Z_{t'}^{(k)}(x') \, dt' = \sum_{m=0}^{P} \int_{\Sigma_{N-1}} Y_m(t') \sum_{k=0}^{\infty} |x|^k Z_{t'}^{(k)}(x') \, dt'$$

$$= \sum_{m=0}^{P} \sum_{k=0}^{\infty} |x|^k \int_{\Sigma_{N-1}} Y_m(t') Z_{x'}^{(k)}(t') \, dt'$$

$$= \sum_{m=0}^{P} |x|^m Y_m(x')$$

$$= u(x).$$

Thus

$$\int_{\Sigma_{N-1}} \left[P(x,t') - \sum_k |x|^k Z_x'^{(k)}(t') \right] u(t') \, dt' = 0$$

for all finite linear cominations of spherical harmonics. Since the latter are dense in $L^2(\Sigma_{N-1})$ the desired assertion follows. ∎

Our immediate goal now is to obtain an explicit formula for each zonal harmonic $Z_{x'}^{(k)}$. We begin this process with some generalities about polynomials.

LEMMA 6.7.5
Let P be a polynomial in $\mathbb{R}^N$ such that

$$P(\rho x) = P(x)$$

for all $\rho \in O(N)$ and $x \in \mathbb{R}^N$. Then there exist constants $c_0 \ldots, c_p$ such that

$$P(x) = \sum_{m=0}^{p} c_m \left(x_1^2 + \cdots + x_N^2 \right)^m.$$

PROOF We write P as a sum of homogeneous terms:

$$P(x) = \sum_{\ell=0}^{q} P_\ell(x),$$

where P_ℓ is homogeneous of degree ℓ. Now for any $\epsilon > 0$ and $\rho \in O(N)$ we have

$$\begin{aligned}
\sum_{\ell=0}^{q} \epsilon^\ell P_\ell(x) &= \sum_{\ell=0}^{q} P_\ell(\epsilon x) \\
&= P(\epsilon x) \\
&= P(\epsilon \rho x) \\
&= \sum_{\ell=0}^{q} P_\ell(\epsilon \rho x) \\
&= \sum_{\ell=0}^{q} \epsilon^\ell P_\ell(\rho x).
\end{aligned}$$

For fixed x, we think of the far left and far right of this last sequence of equalities as identities *of polynomials in* ϵ. It follows that $P_\ell(x) = P_\ell(\rho x)$ for every ℓ. The result of these calculations is that we may concentrate our attentions on P_ℓ.

Consider the function $|x|^{-\ell}P_\ell(x)$. It is homogeneous of degree 0 and still invariant under the action of $O(N)$. Then

$$|x|^{-\ell}P_\ell(x) = c_\ell,$$

for some constant c_ℓ. This forces ℓ to be even (since P_ℓ is a *polynomial* function); the result follows. $\blacksquare$

DEFINITION 6.7.6 *Let $e \in \Sigma_{N-1}$. A **parallel of** Σ_{N-1} **orthogonal to** e is the intersection of Σ_{N-1} with a hyperplane (not necessarily through the origin) orthogonal to the line determined by e and the origin.*

Notice that a parallel of Σ_{N-1} orthogonal to e is a set of the form

$$\{x' \in \Sigma : x' \cdot e = c\},$$

$-1 \le c \le 1$. Observe that a function F on Σ_{N-1} is constant on parallels orthogonal to $e \in \Sigma_{N-1}$ if and only if for all $\rho \in O(N)$ that fix e it holds that $F(\rho x') = F(x')$.

LEMMA 6.7.7
Let $e \in \Sigma_{N-1}$. An element $Y \in \mathcal{H}_k$ is constant on parallels of Σ orthogonal to e if and only if there exists a constant c such that

$$Y = cZ_e^{(k)}.$$

PROOF   Recall that we are assuming that $N \ge 3$. Let ρ be a rotation that fixes e. Then, for each $x' \in \Sigma$, we have

$$Z_e^{(k)}(x') = Z_{\rho e}^{(k)}(\rho x') = Z_e^{(k)}(\rho x').$$

Hence $Z_e^{(k)}$ is constant on the parallels of Σ orthogonal to e.

To prove the converse direction, assume that $Y \in \mathcal{H}_k$ is constant on the parallels of Σ orthogonal to e. Let $e_1 = (1, 0, \ldots, 0) \in \Sigma$ and let τ be a rotation such that $e = \tau e_1$. Define

$$W(x') = Y(\tau x').$$

Then $W \in \mathcal{H}_k$ is constant on the parallels of Σ orthogonal to e_1. Suppose we can show that $W = cZ_{e_1}^{(k)}(x')$ for some constant c. Then

$$Y(x') = W(\tau^{-1}x') = cZ_{e_1}^{(k)}(\tau^{-1}x')$$
$$= cZ_{\tau e_1}^{(k)}(x') = cZ_e^{(k)}(x').$$

So the lemma will follow. Thus we examine W and take $e = e_1$.

Define

$$P(x) = \begin{cases} |x|^k W(x/|x|) & \text{if } x \neq 0 \\ 0 & \text{if } x = 0. \end{cases}$$

Let ρ be a rotation that fixes e_1. We write

$$P(x) = \sum_{j=0}^{k} x_1^{k-j} P_j(x_2, \dots, x_N).$$

Since ρ fixes the powers of x_1 it follows that ρ leaves each P_j invariant. Then each P_j is a polynomial in $(x_2, \dots, x_N) \in \mathbb{R}^{N-1}$ that is invariant under the rotations of $\mathbb{R}^{N-1}$. We conclude that $P_j = 0$ for odd j and

$$P_j(x_2, \dots, x_N) = c_j \left(x_2^2 + \dots + x_N^2 \right)^{j/2} \equiv c_j R^j(x_2, \dots, x_N)$$

for j even. Therefore

$$P(x) = c_0 x_1^k + c_2 x_1^{k-2} R^2 + \cdots c_{2\ell} x_1^{2\ell} R^{k-2\ell},$$

for some $\ell \le k/2$. Of course P is harmonic, so $\triangle P \equiv 0$. A direct calculation then shows that

$$0 = \triangle P = \sum_p \left[c_{2p} \alpha_p + c_{2(p+1)} \beta_p \right] x_1^{k-2(p+1)} R^{2p},$$

where

$$\alpha_p \equiv (k - 2p)(k - 2p - 1)$$

and

$$\beta_p \equiv 2(p + 1)(N + 2p - 1).$$

Therefore we find the following recursion relation for the c's:

$$c_{2(p+1)} = -\frac{\alpha_p c_{2p}}{\beta_p}.$$

In particular, c_0 determines all the other c's.

From this it follows that all the elements of $\mathcal{H}_k$ that are constant on parallels of Σ orthogonal to e_1 are constant multiples of each other. Since $Z_{e_1}^{(k)}$ is one such element of $\mathcal{H}_k$, this proves our result. ∎

LEMMA 6.7.8
Fix k. Let $F_{y'}(x')$ be defined for all $x', y' \in \Sigma$. Assume that

(i) $F_{y'}(\,\cdot\,)$ *is a spherical harmonic of degree k for every $y' \in \Sigma$;*

(ii) *for every rotation ρ we have $F_{\rho y'}(\rho x') = F_{y'}(x')$, all $x', y' \in \Sigma$.*

Then there is a constant c such that

$$F_{y'}(x') = cZ_{y'}^{(k)}(x').$$

Exercise: Show that a function that is invariant under a Lie group action must be smooth (because the group is). Thus it follows immediately that the function F in the lemma is *a priori* smooth.

PROOF OF THE LEMMA   Fix $y' \in \Sigma$ and let $\rho \in O(N)$ be such that $\rho(y') = y'$. Then

$$F_{y'}(x') = F_{\rho y'}(\rho x') = F_{y'}(\rho x').$$

Therefore, by the preceding lemma,

$$F_{y'}(x') = c_{y'} Z_{y'}^{(k)}(x').$$

(Here the constant $c_{y'}$ may in principle depend on y'.) We need to see that for $y_1', y_2' \in \Sigma$ arbitrary, it in fact holds that $c_{y_1'} = c_{y_2'}$. Let $\sigma \in O(N)$ be such that $\sigma(y_1') = y_2'$. By hypothesis (ii),

$$
\begin{aligned}
c_{y_2'} Z_{y_2'}^{(k)}(\sigma x') &= F_{y_2'}(\sigma x') \\
&= F_{\sigma y_1'}(\sigma x') \\
&= F_{y_1'}(x') \\
&= c_{y_1'} Z_{y_1'}^{(k)}(x') \\
&= c_{y_1'} Z_{\sigma y_1'}^{(k)}(\sigma x') \\
&= c_{y_1'} Z_{y_2'}^{(k)}(\sigma x').
\end{aligned}
$$

Since these equalities hold for all $x' \in \Sigma$, we conclude that

$$c_{y_2'} = c_{y_1'}.$$

That is,

$$F_{y'}(x') = cZ_{y'}^{(k)}(x'). \qquad \blacksquare$$

DEFINITION 6.7.9 *Let $0 \le |z| < 1, |t| \le 1$, and fix $\lambda > 0$. Consider the equation $z^2 - 2tz + 1 = 0$. Then $z = t \pm \sqrt{t^2 - 1}$ so that $|z| = 1$. Hence*

$z^2 - 2tz + 1$ *is zero-free in the disc* $\{z : |z| < 1\}$ *and the function* $z \mapsto$ $(1 - 2tz + z^2)^{-\lambda}$ *is well defined and holomorphic in the disc. Set, for* $0 \leq r < 1$,

$$(1 - 2rt + r^2)^{-\lambda} = \sum_{k=0}^{\infty} P_k^\lambda(t) r^k.$$

Then $P_k^\lambda(t)$ *is defined to be the **Gegenbauer polynomial of degree** k associated to the parameter* λ.

PROPOSITION 6.7.10
The Gegenbauer polynomials satisfy the following properties:

1. $P_0^\lambda(t) \equiv 1$.
2. $\frac{d}{dt} P_k^\lambda(t) = 2\lambda P_{k-1}^{\lambda+1}(t)$ *for* $k \geq 1$.
3. $\frac{d}{dt} P_1^\lambda(t) = 2\lambda P_0^{\lambda+1}(t) = 2\lambda$.
4. P_k^λ *is actually a polynomial of degree* k *in* t.
5. *The monomials* $1, t, t^2, \ldots$ *can be obtained as finite linear combinations of* $P_0^\lambda, P_1^\lambda, P_2^\lambda, \ldots$.
6. *The linear space spanned by the* P_k^λ's *is uniformly dense in* $C[-1, 1]$.
7. $P_k^\lambda(-t) = (-1)^k P_k^\lambda(t)$ *for all* $k \geq 0$.

PROOF   We obtain (1) by simply setting $r = 0$ in the defining equation for the Gegenbauer polynomials.
    For (2), note that

$$2r\lambda \sum_{k=0}^{\infty} P_k^{\lambda+1}(t) r^k \equiv 2r\lambda \left(1 - 2rt + r^2\right)^{-(\lambda+1)}$$

$$= \frac{d}{dt} \left(1 - 2rt + r^2\right)^{-\lambda}$$

$$= \sum_{k=0}^{\infty} \frac{d}{dt} P_k^\lambda(t) r^k.$$

The result now follows by identifying coefficients of like powers of r.
    For (3), observe that (using (1) and (2))

$$\frac{d}{dt} P_1^\lambda(t) = 2\lambda P_0^{\lambda+1}(t) = 2\lambda.$$

It follows from integration that P_1^λ is a polynomial of degree 1 in t. Applying (2) and iterating yields (4).

Now (5) follows from (4) (inductively) and (6) is immediate from (5) and the Weierstrass approximation theorem.

Finally,

$$\sum_{k=0}^{\infty} P_k^\lambda(-t)r^k \equiv \left(1 - 2r(-t) + r^2\right)^{-\lambda}$$

$$= \left(1 - 2t(-r) + (-r)^2\right)^{-\lambda}$$

$$= \sum_{k=0}^{\infty} P_k^\lambda(t)(-r)^k$$

$$= \sum_{k=0}^{\infty} (-1)^k P_k^\lambda(t)r^k.$$

Now (7) follows from comparing coefficients of like powers of r. ∎

THEOREM 6.7.11
Let $N > 2, \lambda = (N-2)/2, k \in \{0,1,2,\ldots\}$. Then there exists a constant $c_{k,N}$ such that

$$Z_{y'}^{(k)}(x') = c_{k,N} P_k^\lambda(x' \cdot y').$$

Exercise: Compute by hand what the analogous statement is for $N = 2$. (Recall that the zonal harmonics in dimension 2 are just $\cos k\theta/\sqrt{\pi}$ and $\sin k\theta/\sqrt{\pi}$ for $k \geq 1$.)

PROOF   Let $y' \in \Sigma$ be fixed. For $x \in \mathbb{R}^N$ define

$$F_{y'}(x) = |x|^k P_k^\lambda\left(\frac{x}{|x|} \cdot y'\right).$$

By part (7) of 6.7.10, if k is even then

$$P_k^\lambda(t) = \sum_{j=0}^{m} d_{2j}t^{2j} \qquad \text{with } 2m = k;$$

also if k is odd then

$$P_k^\lambda(t) = \sum_{j=0}^{m} d_{2j+1}t^{2j+1} \qquad \text{with } 2m + 1 = k.$$

In both cases, $F_{y'}(x)$ is then a homogeneous polynomial of degree k. For instance, if k is even then

$$F_{y'}(x) = |x|^k P_k^\lambda \left(\frac{x \cdot y'}{|x|} \right)$$

$$= |x|^{2m} \sum_{j=0}^m d_{2j} \left(\frac{x \cdot y'}{|x|} \right)^{2j}$$

$$= \sum_{j=0}^m d_{2j} \left(|x|^2 \right)^{m-j} (x \cdot y')^{2j}.$$

We want to check that the hypotheses of Lemma 6.7.8 are satisfied when $F_{y'}(x')$ is so defined. Once this is done then the conclusion of our Proposition follows immediately. Thus we need to check that $F_{y'}(x')$ is rotationally invariant and that $F_{y'}(\cdot)$ is harmonic.

If $\rho \in O(N)$ and $x' \in \Sigma_{N-1}$ then

$$F_{\rho y'}(\rho x') = |x'|^k P_k^\lambda \left(\frac{\rho x' \cdot \rho y'}{|x|} \right)$$

$$= |x'|^k P_k^\lambda \left(\frac{x' \cdot y'}{|x|} \right)$$

$$= F_{y'}(x').$$

This establishes the rotational invariance.

To check harmonicity, recall that the map $x \mapsto |x - (y'/s)|^{2-N}$ is harmonic on $\mathbb{R}^N \setminus \{y'/s\}$ when $N \geq 3, s \neq 0$, and $y' \in \Sigma$. Then, with $\lambda = (N-2)/2$, we have

$$s^{2-N} \left| x - \frac{y'}{s} \right|^{2-N} = [(sx - y') \cdot (sx - y')]^{(2-N)/2}$$

$$= \left[|sx|^2 - 2(sx) \cdot y' + 1 \right]^{(2-N)/2}$$

$$= \left[1 - 2(s|x|) \left(\frac{x}{|x|} \cdot y' \right) + (s|x|)^2 \right]^{-\lambda}$$

$$\equiv \left[1 - 2rt + r^2 \right]^{-\lambda}$$

$$= \sum_{k=0}^\infty s^k |x|^k P_k^\lambda \left(\frac{x}{|x|} \cdot y' \right). \qquad (6.7.11.1)$$

Here we have taken $r = s|x|$ and $t = (x/|x|) \cdot y'$. Thus the sum at the end of this calculation is a harmonic function of x in $R_s = \{x \in \mathbb{R}^N : 0 < |x| < 1/s\}$ for $y' \in \Sigma$ fixed.

To see that each coefficient

$$|x|^k P_k^\lambda \left(\frac{x}{|x|} \cdot y' \right)$$

in the series is a harmonic function of $x \in \mathbb{R}^N$ we proceed as follows. Fix $0 \neq x_0 \in \mathbb{R}^N$. Then, for every s such that $0 < s < 1/|x_0|$, formula (6.7.11.1) tells us that the function

$$x \longmapsto \sum_{k=0}^{\infty} s^k |x|^k P_k^\lambda \left(\frac{x}{|x|} \cdot y' \right)$$

is harmonic. Therefore this function satisfies the mean value property. By uniform convergence we can switch the order of summation and integration in the mean value property to obtain

$$\sum_{k=0}^{\infty} s^k \frac{1}{\sigma(\partial B(x_0, r))} \int_{\partial B(x_0, r)} |x|^k P_k^\lambda \left(\frac{x}{|x|} \cdot y' \right) d\sigma(x)$$

$$= \frac{1}{\sigma(B(x_0, r))} \int_{\partial B(x_0, r)} \sum_{k=0}^{\infty} s^k |x|^k P_k^\lambda \left(\frac{x}{|x|} \cdot y' \right) d\sigma(x)$$

$$= \sum_{k=0}^{\infty} s^k |x_0|^k P_k^\lambda \left(\frac{x_0}{|x_0|} \cdot y' \right)$$

for $0 < r < |x_0|$. Since this equality holds for $0 < s < 1/|x_0|$, the identity principle for power series tells us that

$$\frac{1}{\sigma(B(x_0, r))} \int_{\partial B(x_0, r)} |x|^k P_k^\lambda \left(\frac{x}{|x|} \cdot y' \right) d\sigma(x) = |x_0|^k P_k^\lambda \left(\frac{x_0}{|x_0|} \cdot y' \right)$$

for every $0 < r < |x_0|$. It is a standard fact (see [KR1, Ch. 1]) that any function satisfying a mean value property of this sort—for any x_0 and all small r—must be harmonic. We conclude that

$$F_{y'}(x) = |x|^k P_k^\lambda \left(\frac{x}{|x|} \cdot y' \right)$$

is harmonic. The theorem follows. ∎

6.8   Spherical Harmonics in the Complex Domain and Applications

Now we give a rendition of "bigraded spherical harmonics" that is suitable for the study of functions of several complex variables. Our purpose is to return finally to the study of the regularity for the Laplace–Beltrami operator for the Bergman metric on the ball. Because of the detailed exposition that has gone on before, and because much of this new material is routine, we shall perform many calculations in $\mathbb{C}^2$ only and shall leave several others to the reader.

DEFINITION 6.8.1 *Let $\mathcal{H}^{p,q}$ be the space consisting of all restrictions to the unit sphere in $\mathbb{C}^n$ of harmonic (in the classical sense) polynomials that are homogeneous of degree p in z and homogeneous of degree q in $\bar{z}$.*

Observe that

$$\mathcal{H}_k = \bigcup_{p+q=k} \mathcal{H}^{p,q}.$$

PROPOSITION 6.8.2
The spaces $\mathcal{H}^{p,q}$ enjoy the following properties:

1. $D(p,q;n) \equiv \dim_{\mathbb{C}} \mathcal{H}^{p,q} = \dfrac{(p+q+n-1)(p+n-2)!(q+n-2)!}{p!q!(n-1)!(n-2)!}$.

2. *The space $\mathcal{H}^{p,q}$ is $U(n)$-irreducible. That is, $\mathcal{H}^{p,q}$ has no proper linear subspace L such that, for each $U \in U(n)$, U maps L into L.*

3. *If $f_1, \ldots, f_D$ is an orthonormal basis for $\mathcal{H}^{p,q}$, $D = D(p,q;n)$, then*

$$H_n^{p,q}(\zeta, \eta) \equiv \sum_{j=1}^{D} f_j(\zeta)\overline{f_j(\eta)}$$

reproduces $\mathcal{H}^{p,q}$. That is, if $\phi \in \mathcal{H}^{p,q}, \zeta \in \Sigma$, then

$$\phi(\zeta) = \int_\Sigma H_n^{p,q}(\zeta, \eta)\phi(\eta)\, d\sigma(\eta).$$

4. *The orthogonal projection $\pi_{p,q} : L^2(\Sigma) \to \mathcal{H}^{p,q}$ is given by*

$$\pi_{p,q}(f)(\zeta) = \int_\Sigma f(\eta)H_n^{p,q}(\zeta, \eta)\, d\sigma(\eta).$$

PROOF   We leave the proofs of parts (1) and (2) as exercises.
  To prove (3), notice that if $\phi \in \mathcal{H}^{p,q}$ then we may write $\phi = \sum_{j=1}^{D} a_j f_j$. Then

$$\int_\Sigma H_n^{p,q}(\zeta, \eta)\phi(\eta)\, d\sigma(\eta) = \int_\Sigma \left(\sum_{j=1}^{D} f_j(\zeta)\overline{f_j(\eta)} \right) \left(\sum_{j=1}^{D} a_k f_k(\eta) \right) d\sigma(\eta)$$

$$= \sum_{j,k=1}^{D} a_k f_j(\zeta) \int_\Sigma \overline{f_j(\eta)} f_k(\eta)\, d\sigma(\eta)$$

$$= \sum_{j=1}^{D} a_j f_j(\zeta)$$

$$= \phi(\zeta).$$

To prove (4), select $g \in L^2(\Sigma)$. Then

$$\pi_{p,q}g(\zeta) = \int_\Sigma g(\eta) \sum_{j=1}^D \overline{f_j(\eta)} f_j(\zeta) \, d\sigma(\eta)$$

$$= \sum_{j=1}^D \left(\int_\Sigma g(\eta) \overline{f_j(\eta)} \, d\sigma(\eta) \right) f_j(\zeta)$$

so that $\pi_{p,q}$ maps L^2 into $\mathcal{H}^{p,q}$. By (3) it follow that $\pi_{p,q} \circ \pi_{p,q} = \pi_{p,q}$. Finally, $\pi_{p,q}$ is plainly self-adjoint. Thus $\pi_{p,q}$ is the orthogonal projection onto $\mathcal{H}^{p,q}$. ∎

In order to present the solution of the Dirichlet problem for the Laplace–Beltrami operator Δ_B, we need to define another special function. This one is defined by means of an ordinary differential equation.

DEFINITION 6.8.3 *Let $a, b \in \mathbb{R}$ and $c > 0$. The linear differential equation*

$$x(1-x)y'' + \left[c - (a+b+1)x \right] y' - aby = 0 \qquad (6.8.3.1)$$

is called the **hypergeometric equation**.

If we divide the hypergeometric equation through by the leading factor $x(1 - x)$ we see that this is an ordinary differential equation with a regular singularity at 0. It follows (see [COL]) that (6.8.3.1) has a solution of the form

$$x^\lambda \sum_{j=0}^\infty a_j x^j, \qquad (6.8.4)$$

where $a_0 \neq 0$ and the series converges for $|x| < 1$. Let us now sketch what transpires when the expression (6.8.4) is substituted into the differential equation (6.8.3.1).

We find that

$$\sum_{j=0}^\infty a_j x^{\lambda+j-1}(\lambda+j)(\lambda+j-1+c) - \sum_{j=0}^\infty a_j x^{\lambda+j}(\lambda+j+a)(\lambda+j+b) = 0,$$

which gives the following system of equations for determining the exponent λ and the coefficients a_j:

$$a_0 \lambda(\lambda - 1 + c) = 0$$

$$a_j(\lambda+j)(\lambda+j-1+c) - a_{j-1}(\lambda+j-1+a)(\lambda+j-1+b) = 0, \quad j \geq 1.$$

The first of these equations yields that either $\lambda = 0$ or $\lambda = 1 - c$.

First consider the case $\lambda = 0$. We find that

$$a_j = \frac{(j-1+a)(j-1+b)}{j(j-1+c)} a_{j-1}, \qquad j = 1, 2, \ldots.$$

Setting $a_0 = 1$ we obtain

$$a_j = \frac{a(a+1)\cdots(a+j-1)b(b+1)\cdots(b+j-1)}{j!c(c+1)\cdots(c+j-1)}$$

$$= \frac{\Gamma(a+j)\Gamma(b+j)\Gamma(c)}{j!\Gamma(a)\Gamma(b)\Gamma(c+j)},$$

where Γ is the classical gamma function (see [CCP]). Thus for $\lambda = 0$ a particular solution to (6.8.3.1) is

$$y(x) = F(a, b, c; x) \equiv \sum_{j=0}^{\infty} \frac{\Gamma(a+j)\Gamma(b+j)\Gamma(c)}{\Gamma(a)\Gamma(b)\Gamma(c+j)} \cdot \frac{x^j}{j!}.$$

Now consider the case $\lambda = 1 - c$. Arguing in the same manner, if $c \neq 2, 3, 4, \ldots$ we find (setting $a_0 = 1$ again) that a particular solution of the differential equation is given by

$$y(x) = F(1 - c + a, 1 - c + b, 2 - c; x)$$

$$\equiv x^{1-c} \sum_{j=0}^{\infty} \frac{\Gamma(1-c+a+j)\Gamma(1-c+b+j)\Gamma(2-c)}{\Gamma(1-c+a)\Gamma(1-c+b)\Gamma(2-c+j)} \cdot \frac{x^j}{j!}.$$

We leave as an exercise the following statement: by checking the asymptotic behavior of $F(1 - c + a, 1 - c + b, 2 - c; x)$ at the origin, one may see that this function is linearly independent from that found when $\lambda = 0$. The functions F are known as the *hypergeometric functions*. See [ERD] for more on these matters.

In the case that $c = 2, 3, 4, \ldots$ then a modification of the above calculations (again see [COL, p. 165]) gives rise to a solution with a logarithmic singularity at 0.

Define

$$S_n^{p,q}(r) = r^{p+q} \frac{F(p, q, p+q+n; r^2)}{F(p, q, p+q+n; 1)}.$$

We want to show that $S_n^{p,q}$ is C^∞ on the interval $(-1, 1)$ and continuous on $[-1, 1]$. We will make use of the following classical summation tests for series. For more on these tests, see [STR].

LEMMA 6.8.5   DINI–KUMMER
For $j = 1, 2, \ldots$ let $a_j, b_j > 0$ and put

$$D_j = b_j - b_{j+1}\frac{a_{j+1}}{a_j}\ .$$

If $\liminf_{j\to\infty} D_j > 0$ then the series $\sum_j a_j$ converges.

REMARK   Notice that if $b_j = 1$ for all j then this test reduces to the ratio test.   ∎

PROOF   By hypothesis, we may find a $\beta > 0$ and an integer $j_0 > 0$ such that if $j \geq j_0$ then $D_j > \beta$. Thus

$$\beta < b_j - b_{j+1}\frac{a_{j+1}}{a_j}$$

so that

$$0 < a_j < \frac{a_j b_j - b_{j+1}a_{j+1}}{\beta} \tag{6.8.5.1}$$

for $j \geq j_0$.
   Now

$$\sum_{j=j_0}^{\infty} \frac{1}{\beta}\left(a_j b_j - a_{j+1}b_{j+1}\right) = \lim_{J\to\infty}\frac{1}{\beta}\sum_{j=j_0}^{J}\left(a_j b_j - a_{j+1}b_{j+1}\right).$$

By our hypothesis, $a_j b_j > a_{j+1}b_{j+1} > 0$ for all $j \geq j_0$. Therefore we may set $\gamma = \lim_{j\to\infty} a_j b_j$. The number γ is finite and nonnegative. Using (6.8.5.1) we have

$$\sum_{j=j_0}^{\infty} a_j < \frac{1}{\beta}\sum_{j=j_0}^{\infty}\left(a_j b_j - a_{j+1}b_{j+1}\right)$$

$$= \frac{1}{\beta}\left(a_{j_0}b_{j_0} - \gamma\right)$$

$$< \infty.\qquad ∎$$

COROLLARY 6.8.6   RAABE
If $a_j > 0$ for $j = 1, 2, \ldots$, then we set $Q_j = j(1 - a_{j+1}/a_j)$. If it holds that

$$\liminf Q_j > 1 \tag{6.8.6.1}$$

then $\sum_j a_j$ converges.

PROOF   Let $b_1 = 1$ and $b_j = j - 1$ for $j \geq 2$. Then

$$Q_j - 1 = j\left(1 - \frac{a_{j+1}}{a_j}\right) - 1$$

$$= (j - 1) - j\frac{a_{j+1}}{a_j}$$

$$= D_j,$$

where we are using the notation of the Lemma. Then $\liminf_{j \to \infty} Q_j > 1$ if and only if $\liminf_{j \to \infty} D_j > 0$.   ∎

PROPOSITION 6.8.7
Take

$$F(a, b, c; x) = \sum_{j=0}^{\infty} \frac{\Gamma(a + j)\Gamma(b + j)\Gamma(c)}{\Gamma(a)\Gamma(b)\Gamma(c + j)} \cdot \frac{x^j}{j!}$$

as usual. If $|x| = 1$ and $c > a + b$ then the series converges absolutely.

PROOF   We want to apply Raabe's test. Thus we need to calculate the terms Q_j. Denote the absolute value of the jth summand by α_j. Then, since $|x| = 1$, we have

$$\frac{\alpha_{j+1}}{\alpha_j} = \frac{(a + j)(b + j)}{(j + 1)(c + j)}.$$

Set $c = a + b + \delta$, where this equality defines $\delta > 0$. Then

$$\frac{\alpha_{j+1}}{\alpha_j} = \frac{ab + aj + bj + j^2}{(j + 1)(a + b + \delta + j)}$$

$$= 1 - \frac{\delta j + a + b + \delta + j - ab}{(j + 1)(a + b + \delta + j)}$$

$$= 1 - \frac{(1 + \delta)j}{(j + 1)(a + b + \delta + j)} + O(1/j^2).$$

As a result,

$$Q_j \equiv j\left(1 - \frac{\alpha_{j+1}}{\alpha_j}\right)$$

$$= j\left(\frac{(1 + \delta)j}{(j + \delta)(a + b + \delta + j)} + O(1/j^2)\right)$$

and $\liminf_{j \to \infty} Q_j = 1 + \delta > 1$. Thus Raabe's test implies our result.   ∎

It follows from the Proposition that $S_n^{p,q}$ is continuous on $[-1, 1]$ and C^∞ on $(-1, 1)$. We need to know when the function is in fact C^∞ up to the endpoints.

If either $p = 0$ or $q = 0$ then the order-zero term of the hypergeometric equation drops out. One may solve this hypergeometric equation for solutions of the form

$$\sum_{j=0}^{\infty} a_j (x - 1)^{j+\lambda}. \tag{6.8.8}$$

The solutions are *real analytic* near 1; in particular they are smooth. On the other hand, if both p and q are not zero, then the hypergeometric equation never has real analytic solutions near 1 as we may learn by substituting (6.8.8) into the differential equation. In fact the solutions are never C^n, where n is the dimension of the complex space that we are studying.

REMARKS   Gauss found that

$$\lim_{x \to 1^-} F(a, b, c; x) = \frac{\Gamma(c)\Gamma(c - a - b)}{\Gamma(c - a)\Gamma(c - b)} .$$

Also, one may substitute the function

$$y = \int_{[0,1]} (u - x)^{\xi - 1} \phi(u)\, du,$$

where ξ is a constant to be selected, into the hypergeometric equation. Some calculations, together with standard uniqueness theorems for ordinary differential equations, lead to the formula

$$F(a, b, c; x) = \frac{\Gamma(c)}{\Gamma(b)\Gamma(a)} \int_0^1 t^{b-1}(1 - t)^{a-b-1}(1 - xt)^{-a}\, dt$$

for $0 < x < 1$. It is easy to see from this formula that F cannot be analytically continued past 1. ∎

As a consequence of our last proposition,

$$S_n^{p,q}(r) = r^{p+q} \frac{F(p, q, p + q + n; r^2)}{F(p, q, p + q + n; 1)}$$

is well defined and C^{∞} when $0 \leq r < 1$.

THEOREM 6.8.9
Let $f \in \mathcal{H}^{p,q}$. Then the solution of the Dirichlet problem

$$\begin{cases} \Delta_B u = 0 & \text{on } B \\ u = f & \text{on } \partial B \equiv \Sigma \end{cases}$$

is given by

$$u(r\zeta) = f(\zeta) S_n^{p,q}(r)$$

for $\zeta \in \Sigma$ and $0 \leq r \leq 1$.

PROOF To simplify the calculations, we shall prove the theorem only in dimension $n = 2$.

Let $F_0(z) = z_1^p \bar{z}_2^q$ and $f_0 = F_0|_{\partial B}$. Then the ordinary Laplacian

$$\Delta \equiv 4 \left(\frac{\partial^2}{\partial z_1 \partial \bar{z}_1} + \frac{\partial^2}{\partial z_2 \partial \bar{z}_2} \right)$$

annihilates F_0. Recall that $\mathcal{H}^{p,q}$ is irreducible for $U(2)$. This means that $\{f \circ \sigma\}_{\sigma \in U(2)}$ spans all of $\mathcal{H}^{p,q}$ (for if it did not it would generate a nontrivial invariant subspace, and these do not exist by definition of irreducibility). Furthermore, Δ_B commutes with $U(2)$ so if we prove the assertion for f_0, F_0 then the full result follows.

For $z \in B$ we set $r = |z|$. Then $r^2 = z_1 \bar{z}_1 + z_2 \bar{z}_2$. We will seek a solution of our Dirichlet problem of the form

$$u(z) = g(r^2) z_1^p \bar{z}_2^q.$$

Recall that

$$\Delta_B = \frac{4}{n+1} (1 - |z|^2) \sum_{i,j=1}^n (\delta_{ij} - z_i \bar{z}_j) \frac{\partial^2}{\partial z_i \partial \bar{z}_j}.$$

We calculate $\Delta_B u$.

Now

$$\frac{\partial}{\partial \bar{z}_j} u = z_j g'(r^2) \left[z_1^p \bar{z}_2^q \right] + g(r^2) z_1^p \left(q \bar{z}_2^{q-1} \right) \delta_{2j}$$

and

$$\frac{\partial^2}{\partial z_i \bar{z}_j} u = g''(r^2) \bar{z}_i z_j \left[z_1^p \bar{z}_2^q \right] + \delta_{ij} g'(r^2) \left[z_1^p \bar{z}_2^q \right]$$

$$+ z_j g'(r^2) \left[p z_1^{p-1} \bar{z}_2^q \delta_{i1} \right]$$

$$+ \bar{z}_i g'(r^2) \left[z_1^p q \bar{z}_2^{q-1} \delta_{2j} \right]$$

$$+ g(r^2) \left[(p z_1^{p-1} \delta_{1i})(q \bar{z}_2^{q-1} \delta_{2j}) \right].$$

Therefore

$$\sum_{i=1}^2 \frac{\partial^2 u}{\partial z_i \partial \bar{z}_i} = \left\{ \sum_{i=1}^2 [g''(r^2)|z_i|^2 + g'(r^2)] z_1^p \bar{z}_2^q \right\} + g'(r^2) p z_1^p \bar{z}_2^q + g'(r^2) q z_1^p \bar{z}_2^q$$

$$= z_1^p \bar{z}_2^q [g''(r^2)r^2 + (2 + p + q)g'(r^2)].$$

By a similar calculation we find that

$$\sum_{i,j=1}^2 z_i \bar{z}_j \frac{\partial^2 u}{\partial z_i \bar{z}_j} = z_1^p \bar{z}_2^q [r^4 g''(r^2) + (p + q + 1)r^2 g'(r^2) + pq\, g(r^2)].$$

Substituting these two calculations into the equation $\triangle_B u = 0$ (and remembering that $n = 2$), we find that

$$0 = \triangle_B u = \frac{4}{2+1}(1-r^2)z_1^p\bar{z}_2^q\left[g''(r^2)r^2 + (2+p+q)g'(r^2)\right]$$

$$-\frac{4}{2+1}(1-r^2)z_1^p\bar{z}_2^q\left[g''(r^2)r^4 + (p+q+1)r^2g'(r^2) + pq\,g(r^2)\right]$$

$$=\frac{4}{2+1}(1-r^2)z_1^p\bar{z}_2^q\Big\{r^2(1-r^2)g''(r^2)$$

$$+\left[(p+q+2) - (p+q+1)r^2\right]g'(r^2) - pq\,g(r^2)\Big\}.$$

Therefore, if a solution of our Dirichlet problem of the form of $u(z) = g(r^2)z_1^p\bar{z}_2^q$ exists, then g must satisfy the following ordinary differential equation:

$$r^2(1-r^2)g''(r^2) + \left[(p+q+2) - (p+q+1)r^2\right]g'(r^2) - pq\,g(r^2) = 0.$$

We may bring the essential nature of this equation to the surface with the changes of variables $t = r^2$, $a = p$, $b = q$, $c = p+q+2$. Then the equation becomes

$$t(1-t)g'' + \left[c - (a+b+1)t\right]g' - ab\,g = 0.$$

This, of course, is a hypergeometric equation. Since u is the solution of an elliptic problem, it must be C^∞ on the interior. Thus g must be C^∞ on $[0,1)$. Given the solutions that we have found of the hypergeometric equation, we conclude that

$$g(t) = F(p,q,p+q+n;t).$$

Consequently,

$$u(z) = \frac{F(p,q,p+q+n;r^2)}{F(p,q,p+q+n;1)}z_1^p\bar{z}_2^q$$

$$= S_n^{p,q}(r)r^{p+q}f(\zeta).\qquad\blacksquare$$

THEOREM 6.8.10
Let $0 \le r < 1$ and $\eta,\zeta \in \partial B$. Then the Poisson–Szegő kernel for the ball $B \subseteq \mathbb{C}^n$ is given by the formula

$$\mathcal{P}(r\eta,\zeta) = \sum_{p,q=0}^{\infty} S_n^{p,q}(r)H_n^{p,q}(\eta,\zeta).$$

PROOF Recall that if $g \in C(\partial B)$, then

$$G(z) = \begin{cases} \int_{\partial B} \mathcal{P}(z,\zeta)g(\zeta)\,d\sigma(\zeta) & \text{on } B \\ g(z) & \text{on } \partial B \end{cases}$$

solves the Dirichlet problem for $\triangle_B$ with data g. Recall also that $H_n^{p,q}(\eta, \zeta)$ is the zonal harmonic for $\mathcal{H}^{p,q}$.

Let us first prove that the series in the statement of the theorem converges. An argument similar to the one we gave for real spherical harmonics shows that

$$|H_n^{p,q}(\eta, \zeta)| \leq C \cdot D(p, q; n).$$

Here $D(p, q; n)$ is the dimension of $\mathcal{H}^{p,q}$. Clearly,

$$D(p, q; n) \leq \dim \mathcal{H}_{2n}^{p+q} = \binom{2n + (p+q) - 1}{p+q} - \binom{2n + (p+q) - 3}{p+q-2}$$

$$\leq C \cdot (p + q + 1)^{2n}.$$

Recall that

$$S_n^{p,q}(r) = r^{p+q} \frac{F(p, q, p + q + n; r^2)}{F(p, q, p + q + n; 1)}$$

and observe that $F(p, q, p + q + n; r^2)$ is an increasing function of r. Thus

$$S_n^{p,q}(r) \leq r^{p+q} \cdot 1.$$

Putting together all of our estimates, we find that

$$S_n^{p,q}(r) \cdot H_n^{p,q}(\eta, \zeta) \leq C \cdot r^{p+q}(p + q + 1)^{2n}.$$

Summing on p and q for $0 \leq r < 1$ we see that our series converges absolutely.

It remains to show that the sum of the series is actually the Poisson–Szegö kernel. What we will in fact show is that for $\eta \in \partial B$ and $0 < r < 1$ we have

$$\int_{\partial B} \mathcal{P}(r\eta, \zeta) f(\zeta) \, d\sigma(\zeta) = \int_{\partial B} \sum_{p,q} S_n^{p,q}(r) H_n^{p,q}(\eta, \zeta) f(\zeta) \, d\sigma(\zeta)$$

for every $f \in C(\partial B)$. But we already know that this identity holds for $f \in \mathcal{H}^{p,q}$. Finite linear combinations of $\cup_{p,q} \mathcal{H}^{p,q}$ are dense in $C(\partial B)$. Hence the result follows. ∎

Now we return to the question that has motivated all of our work. Namely, we want to understand the lack of boundary regularity for the Dirichlet problem for the Laplace–Beltrami operator on the ball. As a preliminary, we must introduce a new piece of terminology.

DEFINITION 6.8.11 *Let $U \subseteq \mathbb{C}^n$ be an open set and suppose that f is a continuous function defined on U. We say that f is **pluriharmonic** on U if for* *every $a \in U$ and every $b \in \mathbb{C}^n$, it holds that the function*

$$\zeta \mapsto f(a + \zeta b)$$

is harmonic on the open set (in $\mathbb{C}$) of those ζ such that $a + \zeta b \in U$.

A function is pluriharmonic if and only if it is harmonic in the classical sense on every complex line $\zeta \mapsto a + \zeta b$. Pluriharmonic functions arise naturally because they are (locally) the real parts of holomorphic functions of several complex variables (see [KR1, Ch. 2] for a detailed treatment of these matters).

Remark that a C^2 function v is pluriharmonic if and only if we have $(\partial^2 / \partial z_j \partial \bar{z}_k) v \equiv 0$ for all j, k. In the notation of differential forms, this condition is conveniently written as $\partial \bar{\partial} v \equiv 0$.

Now we have

THEOREM 6.8.12
Let $f \in C^\infty(\partial B)$. Consider the Dirichlet problem

$$\begin{cases} \Delta_B u = 0 & \text{on } B \\ u\big|_{\partial B} = f & \text{on } \partial B. \end{cases}$$

Suppose that the solution u of this problem (given in Proposition 6.5.2) lies in $C^\infty(\bar{B})$. Then u must be of the form

$$\sum_\alpha c_\alpha z^\alpha + \sum_\beta d_\beta \bar{z}^\beta.$$

That is, u must be pluriharmonic. The converse statement holds as well: if f is the boundary function of a pluriharmonic function u that is continuous on $\bar{B}$ and if f is C^∞ on the boundary, then $U \in C^\infty(\bar{B})$.

PROOF Now let $v \in C(\bar{B})$ and suppose that v is pluriharmonic on B. Let $v\big|_{\partial B} \equiv f$. Then the solution to the Dirichlet problem for Δ_B with data f is in fact the function v (exercise). But then v is also the ordinary Poisson integral of f. Thus if $f \in C^\infty(\partial B)$ then $v \in C^\infty(\bar{B})$. This proves the converse (the least interesting) direction of the theorem.

For the forward direction, let $f \in C^\infty(\partial B)$ and suppose that the solution u of the Dirichlet problem for Δ_B with data f is C^∞ on $\bar{B}$. We write

$$f = \sum_{p,q} Y_{p,q},$$

where each $Y_{p,q} \in \mathcal{H}^{p,q}$. We proved above that

$$\mathcal{P}(r\eta, \zeta) = \sum_{p,q} S_n^{p,q}(r) H_n^{p,q}(\eta, \zeta)$$

and also that the solution to the Dirichlet problem for $\triangle_B$ is given by

$$
\begin{aligned}
u(r\eta) \quad &= \quad \int \mathcal{P}(r\eta,\zeta)f(\zeta)\,d\sigma(\zeta) \\[2mm]
&= \quad \sum_{p',q'}\sum_{p,q}\int S_n^{p',q'}(r)H_n^{p',q'}(\eta,\zeta)Y_{p,q}(\zeta)\,d\sigma(\zeta) \\[2mm]
&\overset{\text{(orthogonality)}}{=\!=\!=} \sum_{p,q} S_n^{p,q}(r)\int H_n^{p,q}(\eta,\zeta)Y_{p,q}(\zeta)\,d\sigma(\zeta) \\[2mm]
&= \quad \sum_{p,q} S_n^{p,q}(r)Y_{p,q}(\eta).
\end{aligned}
$$

Therefore if $\mathcal{P}f = u$ is smooth on $\bar{B}$ then we may define for each p,q the function

$$
\begin{aligned}
\mathcal{Q}_{p,q}(r) &= \int_{\partial B}(\mathcal{P}f)(r\zeta)\overline{Y_{p,q}(\zeta)}\,d\sigma(\zeta) \\
&= S_n^{p,q}(r)\|Y_{p,q}\|^2.
\end{aligned}
$$

Thus if $\mathcal{P}f$ is C^∞ up to the boundary then, by differentiation under the integral sign, $\mathcal{Q}_{p,q}(r)$ is C^∞ up to $r = 1$. But recall that

$$
S_n^{p,q}(r) = r^{p+q}\frac{F(p,q,p+q+n;r^2)}{F(p,q,p+q+n;1)}
$$

is smooth at $r = 1$ if and only if either $p = 0$ or $q = 0$. So the only nonvanishing terms in the expansion of f are elements of $\mathcal{H}^{p,0}$ or $\mathcal{H}^{0,q}$. That is what we wanted to prove. ∎

We leave it to the reader to prove the refined statement that if a solution u to the Dirichlet problem is C^n up to the closure, then u must be pluriharmonic.

The analysis of the Poisson–Szegö kernel using bigraded spherical harmonics is due to G. B. Folland [FOL]. We thank Folland for useful conversations and correspondence regarding this material.

An analysis of boundary regularity for the Dirichlet problem of the Laplace–Beltrami operator on strongly pseudoconvex domains was begun in [GRL]. Interestingly, these authors uncovered a difference between the case of dimension 2 and the case of dimensions 3 and higher.

7

The $\bar{\partial}$-Neumann Problem

Introductory Remarks

The $\bar{\partial}$-Neumann problem is a generalization to several complex variables of the Cauchy–Riemann equations of one complex variable. While the groundwork for this problem was laid by D. C. Spencer, Charles Morrey, P. Conner, and others, it was J. J. Kohn [KOH1] who tamed the problem. It is the key to many of the important techniques of the function theory of several complex variables.

The $\bar{\partial}$-Neumann problem was also one of the first non-elliptic problems for which a regularity theory was established. Whereas in an elliptic problem of order m we have learned that the solution is m degrees smoother than the data, the $\bar{\partial}$-Neumann problem does not exhibit the maximal degree of smoothing. Indeed the $\bar{\partial}$-Neumann problem is *subelliptic* rather than elliptic; roughly speaking, this means that the solution gains a predictable number of derivatives, but that number is less than the degree of the partial differential operator in question.

Subelliptic regularity was quite unexpected in the early 1960s when it was first discovered (see [KOH1]). It cannot be treated by a naive application of the theory of pseudodifferential operators or by other classical techniques. The monograph [GRS] describes a special calculus of pseudodifferential operators designed for the study of the $\bar{\partial}$-Neumann problem on an important special class of domains; indeed, it is the same special class that we shall study here.

The method that we present here is not Kohn's original, which is rather complicated, but is a simpler method, called *elliptic regularization*, that he developed later in collaboration with Nirenberg (see [KO2], [FOK] and references therein). The monograph [FOK] is the canonical reference for matters related to the $\bar{\partial}$-Neumann problem. However, the presentation there is perhaps too complicated for our purposes, since it is in the setting of an arbitarary metric structure on an almost complex manifold. In an effort to maintain simplicity, we shall formulate and solve the $\bar{\partial}$-Neumann problem only on a smoothly bounded strongly pseudoconvex domain in $\mathbb{C}^n$. We do follow the basic steps in [FOK], but by specializing we can provide considerably more detail and context.

Those interested in the most general setting for the $\bar{\partial}$-Neumann problem will, after reading the material here, be well prepared to consult [FOK] and other more recent sources (see, for instance, [RAN] and [CAT1]–[CAT3]) on the $\bar{\partial}$-Neumann problem.

7.1   Introduction to Hermitian Analysis

As previously mentioned, we shall be working strictly on domains in $\mathbb{C}^n$. We may thus bypass a certain amount of formalism by relying on the standard coordinates in space.

If $P \in \mathbb{C}^n \cong \mathbb{R}^{2n}$ then the tangent space to $\mathbb{R}^{2n}$ at P is spanned by

$$\frac{\partial}{\partial x_1}, \frac{\partial}{\partial y_1}, \ldots, \frac{\partial}{\partial x_n}, \frac{\partial}{\partial y_n}.$$

It clearly has (real) dimension $2n$. We set

$$\frac{\partial}{\partial z_j} = \frac{1}{2}\left[\frac{\partial}{\partial x_j} - i\frac{\partial}{\partial y_j}\right]$$

$$\frac{\partial}{\partial \bar{z}_j} = \frac{1}{2}\left[\frac{\partial}{\partial x_j} + i\frac{\partial}{\partial y_j}\right].$$

Then we let $T^{1,0}$ denote the complex linear space spanned by

$$\frac{\partial}{\partial z_1}, \ldots, \frac{\partial}{\partial z_n}$$

and $T^{0,1} \equiv \overline{T^{1,0}}$ denote the complex linear space spanned by

$$\frac{\partial}{\partial \bar{z}_1}, \ldots, \frac{\partial}{\partial \bar{z}_n}.$$

The *complexified tangent space* is then

$$CT_P(\mathbb{C}^n) \equiv T^{1,0} \oplus_{\mathbb{C}} T^{0,1}.$$

It obviously has complex dimension $2n$. The complexified tangent space is, in a natural sense, the tensor product of the real tangent space with $\mathbb{C}$. However, we shall have no use for this formalism (see [WEL] for more on this point of view).

There is associated to the complexified tangent space a complexified cotangent space. We set

$$dz_j = dx_j + idy_j$$

$$d\bar{z}_j = dx_j - idy_j.$$

A trivial calculation shows that

$$\left\langle dz_j, \frac{\partial}{\partial z_j} \right\rangle = 1, \qquad \left\langle dz_j, \frac{\partial}{\partial \bar{z}_j} \right\rangle = 0,$$

$$\left\langle d\bar{z}_j, \frac{\partial}{\partial z_j} \right\rangle = 0, \qquad \left\langle d\bar{z}_j, \frac{\partial}{\partial \bar{z}_j} \right\rangle = 1.$$

If $\alpha = (a_1, \ldots, a_s)$ is a multiindex, then we set

$$dz^\alpha \equiv dz_{a_1} \wedge dz_{a_2} \wedge \ldots \wedge dz_{a_s},$$

and

$$d\bar{z}^\alpha \equiv d\bar{z}_{a_1} \wedge d\bar{z}_{a_2} \wedge \ldots \wedge d\bar{z}_{a_s}.$$

In this context the *magnitude* of the multiindex is $|\alpha| \equiv s$. (The use of multiindices here is rather at odds with some earlier uses in this book, but should lead to no confusion.) Then $\bigwedge^{p,q}$ denotes the space of forms of the type

$$\omega = \sum_{|\alpha|=p, |\beta|=q} a_{\alpha\beta} dz^\alpha \wedge d\bar{z}^\beta.$$

It is an exercise in linear algebra to see that if $\bigwedge^r$ denotes the space of all classical (real variable) r-forms on $\mathbf{C}^n \cong \mathbf{R}^{2n}$, then

$$\bigwedge\nolimits^r = \bigoplus_{p+q=r} \bigwedge\nolimits^{p,q}.$$

Next we turn to the exterior derivative. Let

$$\omega = \sum_{\alpha,\beta} a_{\alpha,\beta} dz^\alpha \wedge d\bar{z}^\beta$$

be a differential form. The exterior derivative d may be written as $d = \partial + \bar{\partial}$, where

$$\partial \omega = \sum_{\alpha,\beta} \sum_{j=1}^n \frac{\partial a_{\alpha,\beta}}{\partial z_j} dz_j \wedge dz^\alpha \wedge d\bar{z}^\beta$$

and

$$\bar{\partial} \omega = \sum_{\alpha,\beta} \sum_{j=1}^n \frac{\partial a_{\alpha,\beta}}{\partial \bar{z}_j} d\bar{z}_j \wedge dz^\alpha \wedge d\bar{z}^\beta.$$

Clearly,

$$\partial : \bigwedge\nolimits^{p,q} \to \bigwedge\nolimits^{p+1,q} \qquad \text{and} \qquad \bar{\partial} : \bigwedge\nolimits^{p,q} \to \bigwedge\nolimits^{p,q+1}.$$

Since

$$0 = d^2 = (\partial + \bar{\partial})(\partial + \bar{\partial}),$$

we see by counting degrees that

$$\partial^2 = 0, \qquad \bar{\partial}^2 = 0, \qquad \partial\bar{\partial} = -\bar{\partial}\partial.$$

We define a Hermitian inner product on $\mathbb{C}T_P(\mathbb{C}^n)$ as follows:

$$\left\langle \frac{\partial}{\partial z_j}, \frac{\partial}{\partial z_k} \right\rangle_P = \left\langle \frac{\partial}{\partial \bar{z}_j}, \frac{\partial}{\partial \bar{z}_k} \right\rangle_P = \frac{1}{2}\delta_{jk}$$

and

$$\left\langle \frac{\partial}{\partial z_j}, \frac{\partial}{\partial \bar{z}_k} \right\rangle_P = 0 \qquad \text{for all } j, k.$$

In particular, we see that

$$\mathbb{C}T_P(\mathbb{C}^n) = T_P^{1,0} \oplus T_P^{0,1}$$

is an orthogonal decomposition.

There is a corresponding inner product on covectors. We have

$$\langle dz_j, dz_k \rangle_P = \langle d\bar{z}_j, d\bar{z}_k \rangle_P = 2\delta_{jk}$$

and

$$\langle dz_j, d\bar{z}_k \rangle_P = 0 \qquad \text{for all } j, k.$$

By functoriality, if ϕ, ψ are both (p, q)-forms then

$$\phi = \sum_{\substack{|I|=p \\ |J|=q}} \phi_{IJ} dz^I \wedge d\bar{z}^J \qquad \text{and} \qquad \psi = \sum_{\substack{|I|=p \\ |J|=q}} \psi_{IJ} dz^I \wedge d\bar{z}^J,$$

then

$$\langle \phi, \psi \rangle_P = 2^{p+q} \sum_{I,J} \phi_{IJ}(P) \cdot \bar{\psi}_{IJ}(P). \tag{7.1.1}$$

It follows in particular that if ϕ is a form of type (p, q) and ψ is a form of type (p', q') with $(p, q) \neq (p', q')$, then ϕ is orthogonal to ψ.

When $\mathbb{C}^n$ is identified with $\mathbb{R}^{2n}$ then the volume form, in real coordinates, is

$$dV = dx_1 \wedge dy_1 \wedge \cdots \wedge dx_n \wedge dy_n.$$

It is more convenient in complex analysis to use complex coordinates: since $dx_j \wedge dy_j = (i/2)(dz_j \wedge d\bar{z}_j)$, we find that

$$dV = \left(\frac{i}{2}\right)^n dz_1 \wedge d\bar{z}_1 \wedge \cdots \wedge dz_n \wedge d\bar{z}_n.$$

We now use the volume form to produce an inner product on *forms* that is consistent with the inner product that we have defined on covectors. Let $\Omega \subseteq \mathbf{C}^n$ be a domain. Then, with ϕ, ψ as above,

$$\langle \phi, \psi \rangle_\Omega = \int_\Omega \langle \phi, \psi \rangle_P \, dV.$$

More explicitly,

$$\langle \phi, \psi \rangle_\Omega = 2^{p+q} \int_\Omega \sum_{I,J} \phi_{IJ} \bar{\psi}_{IJ} \, dV.$$

We next define certain linear spaces of forms with coefficients that satisfy regularity properties. Set

$$\bigwedge^{p,q} = \bigwedge^{p,q}(\Omega) \equiv \left\{ \phi = \sum_{\substack{|I|=p \\ |J|=q}} \phi_{IJ} \, dz^I \wedge d\bar{z}^J : \phi_{IJ} \in C^\infty(\bar{\Omega}) \text{ for all } I, J \right\};$$

$$\bigwedge_c^{p,q} \equiv \left\{ \phi = \sum_{\substack{|I|=p \\ |J|=q}} \phi_{IJ} \, dz^I \wedge d\bar{z}^J : \phi_{IJ} \in C_c^\infty(\Omega) \text{ for all } I, J \right\};$$

$$H_s^{p,q} \equiv \left\{ \phi = \sum_{\substack{|I|=p \\ |J|=q}} \phi_{IJ} \, dz^I d\bar{z}^J : \phi_{IJ} \in H_s(\Omega) \text{ for all } I, J \right\}.$$

Here $H_s = H^s$ is the standard Sobolev space of functions. In what follows, when we write $\bigwedge_c^{p,q}(\bar{\Omega})$ or $\bigwedge_c^{p,q}(U \cap \bar{\Omega})$ then we interpret the closure bar on the domain to mean that the support of the form is allowed to intersect the boundary. Thus the support is compact *in the closure*.

The final piece of elementary mathematics that we need is the Hodge star operator. First consider the context of real analysis on $\mathbf{R}^N$. Let

$$\gamma = dx_1 \wedge \cdots \wedge dx_N$$

be the volume form, $\bigwedge^k$ the space of k-alternating forms, and $\bigwedge_P^k$ the k-covectors at $P \in \mathbf{R}^N$. Define

$$* : \bigwedge_P^k \rightarrow \bigwedge_P^{N-k}$$

by the equality

$$\langle \psi, \phi \rangle_P \gamma = \psi_P \wedge (*\phi_P) \tag{7.1.2}$$

for all $\psi \in \Lambda_P^k$. Now that we have defined the $*$ operation (pointwise) on covectors, we extend it to forms by setting

$$\langle \psi, \phi \rangle \, \gamma = \psi \wedge (*\phi)$$

for all $\psi \in \Lambda^k$.

Example: Here is an example of the utility of the $*$ operator in a classical setting. Let

$$\Omega = \{x \in \mathbb{R}^N : r(x) < 0\}$$

be a smooth domain. We assume that $dr \neq 0$ on $\partial\Omega$. We may replace r by $r/|dr|$ so that $|dr| = 1$ on $\partial\Omega$. Then the area form on $\partial\Omega$ is $*dr$. (Exercise: use Green's theorem.)   ▯

Now let us turn our attention to $\mathbb{C}^n$. Define

$$* : \Lambda_P^{p,q} \to \Lambda_P^{n-q,n-p}$$

by the identity

$$\langle \psi, \phi \rangle_P \, \gamma = \psi_P \wedge \overline{(*\phi_P)}$$

for $\psi \in \Lambda_P^{p,q}$. Likewise

$$* : \Lambda^{p,q} \to \Lambda^{n-p,n-q}$$

is defined by

$$\langle \psi, \phi \rangle \, \gamma = \psi \wedge \overline{(*\phi)},$$

where $\langle \psi, \phi \rangle$ is the function of z given by

$$\langle \phi, \psi \rangle \, (z) = 2^{p+q} \sum_{|I|=p,|J|=q} \phi_{IJ}(z) \overline{\psi_{IJ}}(z).$$

7.2   The Formalism of the $\bar{\partial}$ Problem

We want to study existence and regularity for the partial differential equation

$$\bar{\partial} v = f. \tag{7.2.1}$$

We shall restrict attention to the case that v is a function and f a $(0,1)$-form. If v exists then we may apply $\bar{\partial}$ to both sides of the equation to obtain

$$0 = \bar{\partial}^2 v = \bar{\partial} f.$$

Thus *a necessary condition for the equation* (7.2.1) *to have a solution is that*
$\bar{\partial} f = 0$. This point bears a moment's discussion.

In $\mathbb{C}^1$, the Cauchy–Riemann equations for C^1 functions $v = \xi + i\eta$ and
$f = u + iv$ can be written either as

$$\frac{1}{2}\left(\frac{\partial \xi}{\partial x} - \frac{\partial \eta}{\partial y}\right) = u \quad \text{and} \quad \frac{1}{2}\left(\frac{\partial \eta}{\partial x} + \frac{\partial \xi}{\partial y}\right) = v$$

or as

$$\frac{\partial v}{\partial \bar{z}} = f.$$

In either notation, the number of unknown functions is in balance with the
number of equations. However, in dimensions two and greater the situation is
different. Write $f = f_1 dz_1 + \cdots + f_n dz_n$. Then our system (7.2.1) is

$$\frac{\partial v}{\partial \bar{z}_j} = f_j, \quad j = 1, \ldots, n,$$

and the number of equations (n) exceeds the number of unknown functions
(1). Passing to real notation results in no improvement. We call such a system
overdetermined. Algebraic considerations (or dimension) suggest that some
compatibility conditions will be needed on the data in order for the system to
be solvable. And indeed that is what we have discovered. (A formal theory of
overdetermined systems of partial differential equations has been developed by
D. C. Spencer; see for instance [SPE].)

There is also a problem with uniqueness for solutions of (7.2.1). For if u is
one solution to this equation and h is any holomorphic function, then $u + h$ is
also a solution. In order to obtain a workable theory for this partial differential
equation, we shall need a canonical method for choosing a "good" solution. For
this we shall exploit the metric structure introduced in the last section. First let
us see that the $\bar{\partial}$ operator is elliptic in a natural sense. When we are dealing
with an operator on *forms*, a certain amount of formalism is ultimately necessary.
However, the $\bar{\partial}$ operator acting on functions may be thought of as an n-tuple
comprised of the operators $\partial/\partial \bar{z}_1, \ldots, \partial/\partial \bar{z}_n$. Each of these is plainly elliptic
on a suitable copy of $\mathbb{C}$, and they span all directions. Hence so is the operator
$\bar{\partial}$ itself elliptic.

Now we turn to the question of ellipticity on forms. Let $\Omega \subseteq \mathbb{C}^n$ be a domain.
Think of $E = E^{p,q}$ as the vector bundle of (p,q) covectors over Ω. Then a
differential form ω of type (p,q) is a section of this vector bundle: we write
$\omega \in \Gamma(E)$. Letting $F = E^{p,q+1}$, we then have

$$\bar{\partial} : \Gamma(E) \to \Gamma(F).$$

Fix $z \in \Omega$. The *symbol* of $\bar{\partial}$ assigns to each covector η at z a linear mapping

$$\sigma(\bar{\partial}, \eta) : E_z \to F_z.$$

This is done as follows: It is elementary to construct a scalar-valued function ρ such that $\rho(z) = 0$ and $d\rho(z) = \eta$ (just use Tayor series—or even the fundamental theorem of calculus). If $\mu \in E_z$, we let $\tilde{\mu}$ be a local section of E such that $\tilde{\mu}(z) = \mu$ (just work in a neighborhood of z over which E is trivial). We define

$$\sigma(\bar{\partial}, \eta)\mu = \bar{\partial}(\rho \cdot \tilde{\mu})\big|_z.$$

The reader should check that for a classical first-order partial differential operator acting on functions, this definition is consistent with the one discussed in Chapters 3 and 4. Of course, in the case of functions, the vector bundles E and F are just $\Omega \times \mathbb{C}$.

With this definition of symbol, we now explicitly calculate the symbol of $\bar{\partial}$. With η, ρ, μ as in the definition we have

$$\sigma(\bar{\partial}, \eta)\mu_z = \bar{\partial}(\rho\tilde{\mu})_z = \left[\bar{\partial}\rho \wedge \tilde{\mu} + \rho\bar{\partial}\tilde{\mu}\right]_z = \{\Pi_{0,1}\eta\} \wedge \mu_z.$$

Here $\Pi_{0,1}$ is the projection of a 1-form into $\bigwedge^{0,1}$. We see that

$$\sigma(\bar{\partial}, \eta)(\,\cdot\,) = \Pi_{0,1}\eta \wedge (\,\cdot\,). \tag{7.2.2}$$

Now, in the present general setting, an operator is said to be elliptic if the complex

$$0 \to E_z \xrightarrow{\sigma(\bar{\partial}, \eta)} F_z$$

is exact, that is, if $\sigma(\bar{\partial}, \eta)$ is injective. In the case that $p = 0, q = 1$, that is, when $\bar{\partial}$ is acting on functions, it is then clear that $\bar{\partial}$ is elliptic. When $\bar{\partial}$ is acting on forms of higher order, we must speak of *ellipticity of a complex*. Namely, a sequence

$$\Gamma(E) \xrightarrow{D_1} \Gamma(F) \xrightarrow{D_2} \Gamma(G)$$

is called elliptic if the induced sequence

$$E_z \xrightarrow{\sigma(D_1, \eta)} F_z \xrightarrow{\sigma(D_2, \eta)} G_z$$

is exact at F_z—that is, if the kernel of the second mapping equals the image of the first. We invite the reader to check that the $\bar{\partial}$ complex

$$\bigwedge^{p,q} \xrightarrow{\bar{\partial}} \bigwedge^{p,q+1} \xrightarrow{\bar{\partial}} \bigwedge^{p,q+2}$$

is elliptic (or see [FOK, pp. 10–11] for further details).

Let us now discuss the formal adjoint of the $\bar{\partial}$ operator. We work on a domain $\Omega \subseteq \mathbb{C}^n$. If we think of $\bar{\partial} : \bigwedge^{p,q} \to \bigwedge^{p,q+1}$ then the formal adjoint $\vartheta : \bigwedge_c^{p,q+1} \to \bigwedge_c^{p,q}$ is defined by the relation

$$\langle \vartheta\phi, \psi \rangle = \langle \phi, \bar{\partial}\psi \rangle$$

for all $\psi \in \bigwedge_c^{p,q}$. Here the inner product is the Hermitian one for forms that we introduced in the last section. It is easy to see that the operator ϑ is well defined.

Let us calculate ϑ when $p = q = 0$. Thus we are looking at $\bar{\partial}$ acting on functions and sending them to $(0,1)$-forms. If ψ is a function then of course

$$\bar{\partial}\psi = \sum_{j=1}^{n} \frac{\partial\psi}{\partial\bar{z}_j} \, d\bar{z}_j.$$

Then, for ϕ a $(0,1)$-form, we have

$$\langle \vartheta\phi, \psi \rangle = \langle \phi, \bar{\partial}\psi \rangle$$

$$= \left\langle \sum_j \phi_j d\bar{z}_j, \sum_j \frac{\partial}{\partial\bar{z}_j}\psi \, d\bar{z}_j \right\rangle$$

$$= \sum_j \left\langle \phi_j d\bar{z}_j, \frac{\partial}{\partial\bar{z}_j}\psi d\bar{z}_j \right\rangle$$

$$= \sum_j 2 \int_\Omega \phi_j \overline{\frac{\partial}{\partial\bar{z}_j}\psi} \, dV.$$

Now we can perform integration by parts in each of these integrals; *because the function ψ is compactly supported, the boundary terms in the integration by parts vanish.* The last line therefore equals

$$-\sum_j 2 \int \frac{\partial\phi_j}{\partial z_j} \, \bar{\psi} \, dV = \int \left(-2 \sum_j \frac{\partial\phi_j}{\partial z_j} \right) \bar{\psi} \, dV.$$

Comparing the far left and far right sides of our calculations, and using the fact that C_c^∞ functions are dense in $L^2(\Omega)$, we find that

$$\vartheta\phi = -2 \sum_j \frac{\partial\phi_j}{\partial z_j} \, .$$

Exercise: Check that for $\bar{\partial}$ operating on general (p,q)-forms, the adjoint ϑ is

$$\vartheta\phi = 2(-1)^{p+1} \sum_{I,H,J,k} \epsilon_{kH}^J \frac{\partial\phi_{IJ}}{\partial z_k} \, dz^I \wedge d\bar{z}^H,$$

where ϵ_{kH}^J is the sign of the permutation changing $(k, h_1, \ldots, h_q)$ into $J = (j_1, \ldots, j_{q+1})$.

One checks that $\bar{\partial}\vartheta + \vartheta\bar{\partial}$ is elliptic as an operator just because the $\bar{\partial}$ complex is elliptic.

In setting up the $\bar{\partial}$-Neumann problem, a principal task for us is to compare the formal adjoint ϑ of $\bar{\partial}$ with the the Hilbert space adjoint $\bar{\partial}^*$. Recall that if $\mathcal{H}^1, \mathcal{H}^2$ are Hilbert spaces and $L : \mathcal{H}^1 \to \mathcal{H}^2$ is a (not necessarily bounded) linear operator defined on a dense subset of $\mathcal{H}^1$, then the Hilbert space adjoint L^* is defined on the set

$$\text{dom } L^* \equiv \left\{ \phi \in \mathcal{H}^2 : \exists c > 0 \text{ with } |\langle \phi, L\psi \rangle_{\mathcal{H}^2}| \leq c\|\psi\|_{\mathcal{H}^1} \ \forall \ \psi \in \text{dom } L \right\}.$$

Then the map

$$\mathcal{H}^1 \ni \psi \mapsto \langle \phi, L\psi \rangle_{\mathcal{H}^2}$$

is linear and satisfies $|\langle \phi, L\psi \rangle_{\mathcal{H}^2}| \leq c\|\psi\|_{\mathcal{H}^1}$. By the Hahn–Banach theorem, the functional extends to a bounded linear functional on all of $\mathcal{H}^1$ with the same bound. Therefore the Riesz representation theorem guarantees that there is an element $\alpha_\phi \in \mathcal{H}^1$ such that

$$\langle \phi, L\psi \rangle_{\mathcal{H}^2} = \langle \alpha_\phi, \psi \rangle_{\mathcal{H}^1}$$

for all $\psi \in \text{dom } L$. We set $L^*\phi \equiv \alpha_\phi$.

If our Hilbert spaces are L^2 spaces, then the operator $\bar{\partial}$ is densely defined on $\bigwedge_c^{p,q}$. We wish to determine the domain of $\bar{\partial}^*$, and to relate $\bar{\partial}^*$ to ϑ. As a first step we prove the following lemma:

LEMMA 7.2.3

Let $\phi, \psi \in C_c^\infty(\overline{\mathbf{R}_+^{N+1}})$ (that is, the supports of these functions may intersect the boundary). Let the partial differential operator L be given by

$$L = \sum_{j=1}^N a_j \frac{\partial}{\partial t_j} + b\frac{\partial}{\partial r},$$

where r is the downward (negative) normal coordinate to $\mathbf{R}^N = \partial\mathbf{R}_+^{N+1}$ and a_j, b are functions. Then

$$\langle L\phi, \psi \rangle = \langle \phi, L'\psi \rangle + \int_{\mathbf{R}^N} \left[\phi b\overline{\psi} \right]_{r=0} dt,$$

where L' is the formal adjoint for L that is determined, as usual, by inner product with functions that are compactly supported in $\mathbf{R}_+^{N+1}$.

PROOF Now

$$\langle L\phi, \psi \rangle = \int_{\mathbf{R}_+^{N+1}} \left(\sum_{j=1}^{N} a_j \frac{\partial \phi}{\partial t_j} + b \frac{\partial \phi}{\partial r} \right) \overline{\psi} \, dt \, dr$$

$$= \sum_{j=1}^{N} \int_{\mathbf{R}_+^{N+1}} a_j \frac{\partial \phi}{\partial t_j} \overline{\psi} \, dt \, dr + \int_{\mathbf{R}_+^{N+1}} b \frac{\partial \phi}{\partial r} \overline{\psi} \, dt \, dr$$

$$= -\sum_{j=1}^{N} \int_{\mathbf{R}_+^{N+1}} \phi \frac{\partial}{\partial t_j} (a_j \overline{\psi}) \, dt \, dr + \int_{\mathbf{R}^N} \phi b \overline{\psi} \Big|_{r=-\infty}^{0} \, dt$$

$$- \int_{\mathbf{R}_+^{N+1}} \phi \frac{\partial}{\partial r} (b\overline{\psi}) \, dt \, dr.$$

Notice that in the first group of integrals we have used the fact that the t_j directions are *tangential*, together with the compact support of ψ, to see that no boundary terms result from the integrations by parts. Now the last line is

$$\int_{\mathbf{R}_+^{N+1}} \phi \left[-\sum_{j=1}^{N} \frac{\partial}{\partial t_j} (\overline{a_j}\psi) - \frac{\partial}{\partial r} (\overline{b}\psi) \right] \, dt \, dr + \int_{\mathbf{R}^N} [\phi b \overline{\psi}]_{r=0} \, dt$$

$$\equiv \langle \phi, L'\psi \rangle + \int_{\mathbf{R}^N} [\phi b \overline{\psi}]_{r=0} \, dt.$$

That completes the proof. ∎

A simple computation shows that

$$\sigma(L, \sum_{j=1}^{N} \eta_j dt_j + \eta_r dr) = \sum_{j=1}^{N} \eta_j a_j(t, r) + \eta_r b(t, r).$$

In particular, $\sigma(L, dr) = b$ and the integral from our lemma satisfies

$$\int_{\mathbf{R}^N} [\phi b \overline{\psi}]_{r=0} \, dt = \int_{\partial \mathbf{R}_+^{N+1}} \langle \sigma(L, dr)\phi, \psi \rangle \, dt.$$

The arguments that we have presented are easily adapted from the half-space to a smoothly bounded domain (just use local boundary coordinate patches). The result is that

$$\langle Lf, g \rangle = \langle f, L'g \rangle + \int_{\partial \Omega} \langle \sigma(L, dr)f, g \rangle.$$

The arguments also are easily applied to partial differential operators on a vector bundle (we leave details to the interested reader). Thus, in particular,

$$\langle \bar{\partial}\phi, \psi \rangle = \langle \phi, \vartheta\psi \rangle + \int_{\partial\Omega} \langle \sigma(\bar{\partial}, dr)\phi, \psi \rangle \qquad (7.2.3.1)$$

and

$$\langle \vartheta\phi, \psi \rangle = \langle \phi, \bar{\partial}\psi \rangle + \int_{\partial\Omega} \langle \sigma(\vartheta, dr)\phi, \psi \rangle . \qquad (7.2.3.2)$$

Set

$$\mathcal{D}^{p,q} \equiv \bigwedge\nolimits^{p,q}(\bar{\Omega}) \cap \operatorname{dom} \bar{\partial}^* .$$

Then we have

PROPOSITION 7.2.4
The linear space $\mathcal{D}^{p,q}$ is equal to the space of those $\phi \in \bigwedge^{p,q}(\bar{\Omega})$ such that $\sigma(\vartheta, dr)\phi = 0$ on $\partial\Omega$. Moreover, $\bar{\partial}^* = \vartheta$ on $\mathcal{D}^{p,q}$.

PROOF Let $\phi \in \mathcal{D}^{p,q}$ and $\psi \in \bigwedge_c^{p,q-1}(\Omega)$. According to formula (7.2.3.2) above we have

$$\begin{aligned}
\langle \bar{\partial}^*\phi, \psi \rangle &= \langle \phi, \bar{\partial}\psi \rangle \\
&= \langle \vartheta\phi, \psi \rangle - \int_{\partial\Omega} \langle \sigma(\vartheta, dr)\phi, \psi \rangle \\
&= \langle \vartheta\phi, \psi \rangle .
\end{aligned} \qquad (7.2.4.1)$$

In the last equality we have used the fact that ψ vanishes on $\partial\Omega$ to make the integral vanish.

But $\bigwedge_c^{p,q-1}$ is dense in $H_0^{p,q-1}$. Therefore

$$\langle \bar{\partial}^*\phi, \psi \rangle = \langle \vartheta\phi, \psi \rangle$$

for all $\psi \in H_0^{p,q-1}$; that is,

$$\bar{\partial}^*\phi = \vartheta\phi$$

for all $\phi \in \mathcal{D}^{p,q}$. But then (7.2.3.1) implies that

$$\int_{\partial\Omega} \langle \sigma(\vartheta, dr)\phi, \psi \rangle = 0$$

for $\phi \in \mathcal{D}^{p,q}, \psi \in \bigwedge^{p,q-1}$. Therefore $\sigma(\vartheta, dr)\phi = 0$ on $\partial\Omega$. We have proved that

$$\mathcal{D}^{p,q} \subseteq \left\{ \phi \in \bigwedge\nolimits^{p,q}(\bar{\Omega}) : \sigma(\vartheta, dr)\phi = 0 \text{ on } \partial\Omega \right\} .$$

Conversely, let $\phi \in \bigwedge^{p,q}(\bar{\Omega})$ satisfy $\sigma(\vartheta, dr)\phi = 0$ on $\partial\Omega$. From the equation

$$\langle \phi, \bar{\partial}\psi \rangle = \langle \vartheta\phi, \psi \rangle - \int_{\partial\Omega} \langle \sigma(\vartheta, dr)\phi, \psi \rangle$$

for $\psi \in \bigwedge_c^{p,q-1}(\bar{\Omega})$, we learn that

$$\langle \phi, \bar{\partial}\psi \rangle = \langle \vartheta\phi, \psi \rangle$$

and therefore

$$|\langle \phi, \bar{\partial}\psi \rangle| \leq \|\vartheta\phi\|_{L^2}\|\psi\|_{L^2}.$$

Therefore $\phi \in \text{dom}\,\bar{\partial}^*$ and we see that $\bar{\partial}^*\phi = \vartheta\phi$. That proves the proposition. ∎

7.3   Formulation of the $\bar{\partial}$-Neumann Problem

Let L be a uniformly elliptic partial differential operator of order m such as we studied in Chapters 4 and 5. Classically, the heart of the study of an elliptic boundary value problem for L is a *coercive estimate* of the form

$$\sum_{|\alpha| \leq m} \|D^\alpha u\|_0 \leq C \left(\|Lu\|_0 + \|u\|_0 \right).$$

However, the condition that we discovered in the last proposition of Section 7.2, namely that $\sigma(\vartheta, dr)\phi = 0$ on $\partial\Omega$, is not a coercive boundary condition—it does not satisfy the Lopatinski criterion. We shall need to develop a substitute for the classical approach, and that will require building up some machinery. We begin with some elementary functional analysis:

LEMMA 7.3.1   FRIEDRICHS
Let H be a Hilbert space equipped with the inner product $\langle \cdot, \cdot \rangle$ and corresponding norm $\| \ \|$. Let $Q(\cdot, \cdot)$ be a densely defined hermitian form on H. Let $\mathcal{D}$ be the domain of Q and assume that

$$Q(\phi, \phi) \geq c \cdot \|\phi\|^2$$

for all $\phi \in \mathcal{D}$. Assume also that $\mathcal{D}$ itself is a Hilbert space when it is equipped with the inner product Q.

Then there exists a canonical self-adjoint map F on H associated with Q such that

1. *dom $F \subseteq \mathcal{D}$;*
2. *$Q(\phi, \psi) = (F\phi, \psi)$ for all $\phi \in \text{dom}\,F, \psi \in \mathcal{D}$.*

PROOF   After rescaling, we may assume that $c = 1$. Let $\alpha \in H$ and consider the linear functional on $\mathcal{D}$ given by

$$\mu_\alpha : \mathcal{D} \ni \psi \mapsto \langle \psi, \alpha \rangle.$$

We have

$$|\mu_\alpha(\psi)| \leq \|\alpha\| \cdot \|\psi\|$$
$$\leq \|\alpha\| \cdot Q(\psi,\psi)^{1/2}.$$

Therefore the Riesz representation theorem implies the existence of an element $\phi_\alpha \in \mathcal{D}$ such that

$$\langle \psi, \alpha \rangle = \mu_\alpha(\psi)$$
$$= Q(\psi, \phi_\alpha),$$

that is,

$$\langle \alpha, \psi \rangle = Q(\phi_\alpha, \psi).$$

Define

$$T : H \ni \alpha \mapsto \phi_\alpha \in \mathcal{D}.$$

Then

$$\|T\alpha\|^2 \leq Q(T\alpha, T\alpha)$$
$$= \langle \alpha, T\alpha \rangle$$
$$\leq \|\alpha\| \cdot \|T\alpha\|.$$

It follows that $\|T\alpha\| \leq \|\alpha\|$ and T is bounded as an operator from H to H.

The operator T is injective since $T\alpha = 0$ implies that $\langle \alpha, \psi \rangle = Q(T\alpha, \psi) = 0$ for every $\psi \in \mathcal{D}$ and $\mathcal{D}$ is dense in H. Furthermore, T is self-adjoint because

$$\langle T\alpha, \beta \rangle = \overline{\langle \beta, T\alpha \rangle}$$
$$= \overline{Q(T\beta, T\alpha)}$$
$$= Q(T\alpha, T\beta)$$
$$= \langle \alpha, T\beta \rangle.$$

Next, set $U = \text{range } T \subseteq \mathcal{D}$ and define

$$F = T^{-1} : U \to H.$$

By the equality $Q(T\alpha, \psi) = \langle \alpha, \psi \rangle$ we obtain

$$Q(\beta, \psi) = \langle F\beta, \psi \rangle$$

for $\psi \in \mathcal{D}, \beta \in U$. That completes the proof. ∎

Exercise: Show that U is dense both in $\mathcal{D}$ and in H. (Hint: Use the fact that T is self-adjoint.)

We intend to apply the Friedrichs lemma to the $\bar{\partial}$-Neumann problem. To this end, we introduce the following notation:

1. Let $H = H_0^{p,q} = \{(p,q)\text{-forms with } L^2 \text{ coefficients on } \Omega\}$.
2. Let $Q(\phi, \psi) = (\bar{\partial}\phi, \bar{\partial}\psi) + (\vartheta\phi, \vartheta\psi) + (\phi, \psi)$.
3. Let $\mathcal{D} = \tilde{\mathcal{D}}^{p,q} \equiv$ the closure of $\mathcal{D}^{p,q}$ in the Q-topology.

Now we have to do some formal checking: There is a natural continuous inclusion of $\mathcal{D}^{p,q}$ in $H_0^{p,q}$. We will see that this induces an inclusion of $\tilde{\mathcal{D}}^{p,q}$ in $H_0^{p,q}$.

Let $\{\phi_n\}$ be a Q-Cauchy sequence in $\mathcal{D}^{p,q}$. By definition of Q, we see that $\{\phi_n\}, \{\vartheta\phi_n\}$, and $\{\bar{\partial}\phi_n\}$ are Cauchy sequences in $H_0^{p,q}$ (i.e., in the L^2 topology). Let ϕ be the L^2 limit of $\{\phi_n\}$. When interpreted in the weak or distribution sense, $\bar{\partial}$ and ϑ are closed operators. Hence we have

$$\bar{\partial}\phi_n \to \bar{\partial}\phi$$

and

$$\vartheta\phi_n \to \vartheta\phi.$$

If $\phi = 0$ in L^2 then

$$Q(\phi, \phi) = \lim_n Q(\phi_n, \phi_n) = \lim_{n\to\infty} \left[\langle \phi_n, \phi_n \rangle + \langle \bar{\partial}\phi_n, \bar{\partial}\phi_n \rangle + \langle \vartheta\phi_n, \vartheta\phi_n \rangle \right]$$
$$= \langle \phi, \phi \rangle + \langle \bar{\partial}\phi, \bar{\partial}\phi \rangle + \langle \vartheta\phi, \vartheta\phi \rangle = 0.$$

Therefore $\phi_n \to 0$ in the Q-topology. It follows that the inclusion $\mathcal{D}^{p,q} \hookrightarrow H_0^{p,q}$ extends to the inclusion $\tilde{\mathcal{D}}^{p,q} \hookrightarrow H_0^{p,q}$.

Now we may apply the Friedrichs theorem to obtain a (canonical) self-adjoint map

$$F : \operatorname{dom} F \left(\underset{\text{dense}}{\subseteq} \tilde{\mathcal{D}}^{p,q} \right) \to H_0^{p,q}$$

such that

$$Q(\phi, \psi) = \langle F\phi, \psi \rangle$$

for all $\phi \in \operatorname{dom} F, \psi \in \mathcal{D}$. Now we need to identify F and determine its domain.

Set

$$\mathcal{D}_c^{p,q} \equiv \bigwedge_c^{p,q} \cap \operatorname{dom} \bar{\partial}^* = \{\rho \in \mathcal{D}^{p,q} : \operatorname{supp} \rho \text{ is compact}\}.$$

If $\phi, \psi \in \mathcal{D}^{p,q}_c$ then

$$
\begin{aligned}
Q(\phi, \psi) &= \langle \bar{\partial}\phi, \bar{\partial}\psi \rangle + \langle \vartheta\phi, \vartheta\psi \rangle + \langle \phi, \psi \rangle \\
&= \langle \vartheta\bar{\partial}\phi, \psi \rangle + \langle \bar{\partial}\vartheta\phi, \psi \rangle + \langle \phi, \psi \rangle \\
&= \langle ((\vartheta\bar{\partial} + \bar{\partial}\vartheta) + I)\,\phi, \psi \rangle \\
&\equiv \langle (\Box + I)\phi, \psi \rangle.
\end{aligned}
$$

Notice that passing ϑ and $\bar{\partial}$ from side to side in the inner products is justified here because ϕ, ψ are compactly supported in Ω. The reader may also check that, because we are working in $\mathbb{C}^n$ with the standard Hermitian inner product, it turns out that $\Box = -\Delta$. We use the notation $\Box$ by convention.

PROPOSITION 7.3.2
Let $\phi \in \mathcal{D}^{p,q}$. Then $\phi \in \operatorname{dom} F$ if and only if $\bar{\partial}\phi \in \mathcal{D}^{p,q+1}$. In this case it holds that

$$
F\phi = (\Box + I)\phi.
$$

PROOF Let $\phi \in \mathcal{D}^{p,q}$ lie in $\operatorname{dom} F$. We know that, for $\psi \in \bigwedge^{p,q}_c$,

$$
\langle F\phi, \psi \rangle = \langle (\Box + I)\phi, \psi \rangle. \tag{7.3.2.1}
$$

Since $\bigwedge^{p,q}_c$ is dense in $H^{p,q}_0$, it follows that (7.3.2.1) holds for $\psi \in \mathcal{D}^{p,q}$. By (7.2.3.1), (7.2.3.2), and 7.2.4 we have that

$$
\begin{aligned}
\langle \vartheta\phi, \vartheta\psi \rangle &= \langle \bar{\partial}\vartheta\phi, \psi \rangle - \int_{\partial\Omega} \langle \sigma(\bar{\partial}, dr)\vartheta\phi, \psi \rangle \\
&= \langle \bar{\partial}\vartheta\phi, \psi \rangle - \int_{\partial\Omega} \langle \vartheta\phi, \sigma(\vartheta, dr)\psi \rangle \\
&= \langle \bar{\partial}\vartheta\phi, \psi \rangle \tag{7.3.2.2}
\end{aligned}
$$

since $\psi \in \mathcal{D}^{p,q}$ (hence $\sigma(\vartheta, dr)\psi = 0$ on $\partial\Omega$). Also,

$$
\langle \bar{\partial}\phi, \bar{\partial}\psi \rangle = \langle \vartheta\bar{\partial}\phi, \psi \rangle - \int_{\partial\Omega} \langle \sigma(\vartheta, dr)\bar{\partial}\phi, \psi \rangle.
$$

We cannot argue as before since we have no control over the support of ϕ. Instead we proceed as follows:

$$
\begin{aligned}
Q(\phi, \psi) &= \langle \bar{\partial}\phi, \bar{\partial}\psi \rangle + \langle \vartheta\phi, \vartheta\psi \rangle + \langle \phi, \psi \rangle \\
&= \langle \vartheta\bar{\partial}\phi, \psi \rangle - \int_{\partial\Omega} \langle \sigma(\vartheta, dr)\bar{\partial}\phi, \psi \rangle + \langle \bar{\partial}\vartheta\phi, \psi \rangle + \langle \phi, \psi \rangle \\
&= \langle (\Box + I)\phi, \psi \rangle - \int_{\partial\Omega} \langle \sigma(\vartheta, dr)\bar{\partial}\phi, \psi \rangle.
\end{aligned}
$$

Here we have used (7.3.2.2). Since $\phi \in \text{dom } F$, we know that

$$Q(\phi, \psi) = \langle F\phi, \psi \rangle = \langle (\square + I)\phi, \psi \rangle.$$

Therefore

$$\int_{\partial \Omega} \langle \sigma(\vartheta, dr)\bar{\partial}\phi, \psi \rangle = 0 \qquad (7.3.2.3)$$

for all $\psi \in \Lambda_c^{p,q}$; by density, (7.3.2.3) persists for $\psi \in \tilde{\mathcal{D}}^{p,q}$. We wish to conclude that $\sigma(\vartheta, dr)\bar{\partial}\phi = 0$ on $\partial \Omega$. If we choose ψ to be $\sigma(\vartheta, dr)\bar{\partial}\phi$ then

$$\sigma(\vartheta, dr)\psi = \sigma(\vartheta, dr)\left(\sigma(\vartheta, dr)\bar{\partial}\phi\right)$$
$$= \sigma(\vartheta^2, dr)\bar{\partial}\phi$$
$$= 0.$$

Hence $\psi = \sigma(\vartheta, dr)\bar{\partial}\phi \in \mathcal{D}^{p,q}$ and putting this choice of ψ into (7.3.2.3) yields

$$\int_{\partial \Omega} \langle \sigma(\vartheta, dr)\bar{\partial}\phi, \sigma(\vartheta, dr)\bar{\partial}\phi \rangle = 0.$$

Therefore

$$\sigma(\vartheta, dr)\bar{\partial}\phi = 0 \quad \text{on} \quad \partial \Omega.$$

We conclude that $\bar{\partial}\phi \in \tilde{\mathcal{D}}^{p,q+1}$. That $F\phi = (\square + I)\phi$ is then automatic from what has gone before.

Conversely, if $\phi \in \mathcal{D}^{p,q}$ and $\bar{\partial}\phi \in \mathcal{D}^{p,q+1}$ we let $\psi \in \mathcal{D}^{p,q}$. Then

$$Q(\phi, \psi) = \langle (\square + I)\phi, \psi \rangle$$

(just calculate), hence $\phi \in \text{dom } F$ and $F\phi = (\square + I)\phi$. ∎

We have made the following important discovery:

The $\bar{\partial}$-Neumann Boundary Conditions

Let $\phi \in \Lambda^{p,q}(\bar{\Omega})$. Then $\phi \in \text{dom } F$ if and only if

1. $\phi \in \mathcal{D}^{p,q}$ (that is, $\sigma(\vartheta, dr)\phi = 0$ on $\partial \Omega$);
2. $\bar{\partial}\phi \in \mathcal{D}^{p,q+1}$ (that is, $\sigma(\vartheta, dr)\bar{\partial}\phi = 0$ on $\partial \Omega$).

In studying the $\bar{\partial}$-Neumann problem, we will follow the paradigm already laid out in the study of elliptic boundary value problems in Chapter 5. Namely, we will prove an *a priori* estimate that will in turn lead to a sharp existence and regularity result.

We remark in passing that the $\bar{\partial}$-Neumann problem as formulated and solved here is in the Hilbert space $L^2 = H_0$. It would be of some interest to solve the problem in other Hilbert spaces, such as the Sobolev space H^s. The groundwork for studying those Hilbert spaces has been laid in [BO1]. However, as of this writing, it is not clear what the Neumann boundary conditions should be in that context.

7.4 The Main Estimate

If $\alpha \in H_0^{p,q}$, then the construction of the operator F guarantees the existence of a unique $\phi \in \tilde{\mathcal{D}}^{p,q}$ such that $F\phi = \alpha$. We are interested in the regularity up to the boundary of this solution ϕ. If we had a classical Gårding type, or coercive, inequality

$$Q(\phi, \phi) \geq c\|\phi\|_1^2,$$

then it would be straightforward to prove the desired estimates. We do not have such an inequality, but will find a substitute that will do the job.

LEMMA 7.4.1
Let $\Omega \subseteq \mathbf{C}^n$ be a smoothly bounded domain with defining function r: $\Omega = \{z \in \mathbf{C}^n : r(z) < 0\}$ and $|\nabla r| = 1$ on $\partial\Omega$. Then

1. *If $\phi \in \bigwedge^{0,1}$, then $\phi \in \mathcal{D}^{0,1}$ if and only if $\sum_j (\partial r/\partial z_j)\phi_j = 0$ on $\partial\Omega$.*
2.

$$\frac{\partial r}{\partial z_j}(*dr) = \left(-\frac{1}{2i}\right)^n dz_1 \wedge d\bar{z}_1 \wedge \cdots \wedge \widehat{dz_j} \wedge \cdots \wedge dz_n \wedge d\bar{z}_n$$

   on $\partial\Omega$.
3. *If $f, g \in C^\infty(\bar{\Omega})$ then we have*

$$d\left[f\bar{g}\left(-\frac{1}{2i}\right)^n dz_1 \wedge d\bar{z}_1 \wedge \cdots \wedge \widehat{dz_j} \wedge \cdots \wedge dz_n \wedge \bar{z}_n\right] = \left(\frac{\partial f}{\partial z_j}\bar{g} + f\frac{\partial \bar{g}}{\partial z_j}\right)\gamma,$$

   where γ is the volume element.
4.

$$\left(-\frac{1}{2i}\right)^n \sum_j \int_{\partial\Omega} f\bar{g}\frac{\partial r}{\partial z_j}\, d\sigma = \sum_j \int_\Omega \left(\frac{\partial f}{\partial z_j}\bar{g} + f\frac{\partial \bar{g}}{\partial z_j}\right)\gamma.$$

5. *The analog of (4) with $\partial/\partial z_j$ replaced by $\partial/\partial \bar{z}_j$.*

PROOF   (1) This is just definition checking: Let $p \in \partial\Omega$ and let r be a defining function with $|dr| = 1$ on $\partial\Omega$. Of course $r(p) = 0$. Let $\phi = \sum_{i=1}^{n} \phi_i d\bar{z}_i$. Then $\phi \in \mathcal{D}^{0,1}$ if and only if

$$
\begin{aligned}
0 &= \sigma(\vartheta, dr)\phi\big|_p \\
&= \vartheta(r\phi)\big|_p \\
&= (\vartheta r)_p(\phi_p) \\
&= -\sum_{j=1}^{n} \frac{\partial r}{\partial z_j}(p)\phi_j(p).
\end{aligned}
$$

If we apply $\wedge dr$ to both sides of the equation in the statement of (2), then the result is $\partial r / \partial z_j \cdot \gamma = \partial r / \partial z_j \cdot \gamma$. Result (2) follows.

To prove (3), we calculate that

$$
d\left[f\bar{g}\left(-\frac{1}{2i}\right)^n dz_1 \wedge d\bar{z}_1 \wedge \cdots \widehat{dz_j} \wedge \cdots \wedge dz_n \wedge d\bar{z}_n \right]
$$

$$
= \left[\frac{\partial}{\partial z_j}(f\bar{g})\left(-\frac{1}{2i}\right)^n dz_j \wedge dz_1 \wedge d\bar{z}_1 \wedge \cdots \wedge \widehat{dz_j} \wedge \cdots \wedge dz_n \wedge d\bar{z}_n \right]
$$

$$
= \left(-\frac{1}{2i}\right)^n \left(\frac{\partial f}{\partial z_j}\bar{g} + f\frac{\partial \bar{g}}{\partial z_j} \right)\gamma.
$$

To prove (4), we apply Stokes's theorem to (3). Thus

$$
\sum_{j=1}^{n} \int_{\partial\Omega} f\bar{g}\left(-\frac{1}{2i}\right)^n dz_1 \wedge d\bar{z}_1 \wedge \ldots \wedge \widehat{dz_j} \wedge \cdots \wedge dz_n \wedge d\bar{z}_n
$$

$$
= \sum_{j=1}^{n} \int_{\Omega} d\left[f\bar{g}\left(-\frac{1}{2i}\right)^n dz_1 \wedge d\bar{z}_1 \wedge \cdots \wedge \widehat{dz_j} \wedge \cdots \wedge dz_n \wedge d\bar{z}_n \right]
$$

$$
= \sum_{j=1}^{n} \int_{\Omega} \left(\frac{\partial f}{\partial z_j}\bar{g} + f\frac{\partial \bar{g}}{\partial z_j} \right)\gamma.
$$

Now we use (2) and the fact that $*dr = d\sigma$ to see that the left side is equal to

$$
\left(-\frac{1}{2i}\right)^n \sum_{j=1}^{n} \int_{\partial\Omega} f\bar{g}\frac{\partial r}{\partial z_j}\, d\sigma. \qquad\qquad ∎
$$

PROPOSITION 7.4.2
Let $\Omega \subseteq \mathbb{C}^n$ be a smoothly bounded domain in $\mathbb{C}^n$ with defining function r. Let $\phi \in \mathcal{D}^{0,1}$. Then

$$Q(\phi, \phi) = \sum_{j,k} \left\| \frac{\partial \phi_j}{\partial \bar{z}_k} \right\|^2 + \sum_{j,k} \int_{\partial \Omega} \frac{\partial^2 r}{\partial z_j \partial \bar{z}_k} \phi_j \bar{\phi}_k + \|\phi\|^2.$$

(Here $\| \cdot \|$ denotes the L^2-norm.)

PROOF   If $\phi \in \bigwedge^{0,1}(\bar{\Omega})$ then

$$\bar{\partial}\phi = \bar{\partial} \left(\sum_{j=1}^{n} \phi_j \, d\bar{z}_j \right)$$

$$= \sum_j \sum_k \frac{\partial \phi_j}{\partial \bar{z}_k} \, d\bar{z}_k \wedge d\bar{z}_j$$

$$= \sum_{j>k} \left(\frac{\partial \phi_j}{\partial \bar{z}_k} - \frac{\partial \phi_k}{\partial \bar{z}_j} \right) d\bar{z}_k \wedge d\bar{z}_j.$$

Now

$$\left\| \frac{\partial \phi_j}{\partial \bar{z}_k} - \frac{\partial \phi_k}{\partial \bar{z}_j} \right\|^2 = \left\langle \frac{\partial \phi_j}{\partial \bar{z}_k} - \frac{\partial \phi_k}{\partial \bar{z}_j}, \frac{\partial \phi_j}{\partial \bar{z}_k} - \frac{\partial \phi_k}{\partial \bar{z}_j} \right\rangle$$

$$= \left\| \frac{\partial \phi_k}{\partial \bar{z}_j} \right\|^2 + \left\| \frac{\partial \phi_j}{\partial \bar{z}_k} \right\|^2 - \left\langle \frac{\partial \phi_k}{\partial \bar{z}_j}, \frac{\partial \phi_j}{\partial \bar{z}_k} \right\rangle - \left\langle \frac{\partial \phi_j}{\partial \bar{z}_k}, \frac{\partial \phi_k}{\partial \bar{z}_j} \right\rangle.$$

Therefore

$$\|\bar{\partial}\phi\|^2 = \sum_{j>k} \left\| \frac{\partial \phi_j}{\partial \bar{z}_k} - \frac{\partial \phi_k}{\partial \bar{z}_j} \right\|^2$$

$$= \frac{1}{2} \sum_{j,k=1}^{n} \left\| \frac{\partial \phi_j}{\partial \bar{z}_k} - \frac{\partial \phi_k}{\partial \bar{z}_j} \right\|^2$$

$$= \sum_{j,k=1}^{n} \left\| \frac{\partial \phi_j}{\partial \bar{z}_k} \right\|_0^2 - \sum_{j,k=1}^{n} \left\langle \frac{\partial \phi_j}{\partial \bar{z}_k}, \frac{\partial \phi_k}{\partial \bar{z}_j} \right\rangle. \qquad (7.4.2.1)$$

Now

$$\sum_{j,k=1}^{n} \left\langle \frac{\partial \phi_j}{\partial \bar{z}_k}, \frac{\partial \phi_k}{\partial \bar{z}_j} \right\rangle = \sum_{j,k=1}^{n} \int_{\Omega} \frac{\partial \phi_j}{\partial \bar{z}_k} \frac{\overline{\partial \phi_k}}{\partial z_j} \gamma$$

$$= -\sum_{j,k=1}^{n} \int_{\Omega} \frac{\partial^2 \phi_j}{\partial \bar{z}_k \partial z_j} \overline{\phi_k} \, \gamma + \sum_{j,k=1}^{n} \int_{\partial \Omega} \frac{\partial \phi_j}{\partial \bar{z}_k} \overline{\phi_k} \frac{\partial r}{\partial z_j} \, d\sigma$$

$$= \sum_{j,k=1}^{n} \int_{\Omega} \frac{\partial \phi_j}{\partial z_j} \frac{\overline{\partial \phi_k}}{\partial \bar{z}_k} \gamma - \sum_{j,k=1}^{n} \int_{\partial \Omega} \frac{\partial \phi_j}{\partial z_j} \overline{\phi_k} \frac{\partial r}{\partial \bar{z}_k} \, d\sigma$$

$$+ \sum_{j,k=1}^{n} \int_{\partial \Omega} \frac{\partial \phi_j}{\partial \bar{z}_k} \overline{\phi_k} \frac{\partial r}{\partial z_j} \, d\sigma, \tag{7.4.2.2}$$

where we have used part (4) of the last lemma twice. By part (1) of that lemma,

$$\sum_{k} \frac{\partial r}{\partial \bar{z}_k} \bar{\phi}_k = \overline{\sum_{k} \left(\frac{\partial r}{\partial z_k} \phi_k \right)} = 0$$

on $\partial \Omega$. Hence the second term on the right-hand side of (7.4.2.2) vanishes. Notice now that the condition

$$\sum_{j} \frac{\partial r}{\partial z_j} \phi_j = 0$$

on $\partial \Omega$ (since $\phi \in \mathcal{D}^{0,1}$) implies that any tangential derivative of this expression vanishes on $\partial \Omega$. Further observe that

$$\sum_{k} \bar{\phi}_k \frac{\partial}{\partial z_k}$$

is a tangential derivative. Therefore, on $\partial \Omega$,

$$\sum_{k} \bar{\phi}_k \frac{\partial}{\partial \bar{z}_k} \left(\sum_{j} \phi_j \frac{\partial r}{\partial z_j} \right) = 0.$$

This means that on $\partial \Omega$ we have

$$\sum_{j,k} \frac{\partial^2 r}{\partial z_j \partial \bar{z}_k} \phi_j \overline{\phi_k} = -\sum_{j,k} \frac{\partial r}{\partial z_j} \frac{\partial \phi_j}{\partial \bar{z}_k} \overline{\phi_k}. \tag{7.4.2.3}$$

Now equations (7.4.2.2) and (7.4.2.3) yield that

$$\sum_{j,k} \left\langle \frac{\partial \phi_j}{\partial \bar{z}_k}, \frac{\partial \phi_k}{\partial \bar{z}_j} \right\rangle = \sum_{.j,k} \left\langle \frac{\partial \phi_j}{\partial z_j}, \frac{\partial \phi_k}{\partial z_k} \right\rangle - \sum_{j,k} \int_{\partial \Omega} \frac{\partial^2 r}{\partial z_j \partial \bar{z}_k} \phi_j \overline{\phi_k} \, d\sigma. \tag{7.4.2.4}$$

Substituting (7.4.2.4) into (7.4.2.1) yields that

$$\|\bar{\partial}\phi\|^2 = \sum_{j,k} \left\| \frac{\partial \phi_j}{\partial \bar{z}_k} \right\|^2 - \sum_{j,k} \left\langle \frac{\partial \phi_j}{\partial z_j}, \frac{\partial \phi_k}{\partial z_k} \right\rangle + \int_{\partial\Omega} \sum_{j,k} \frac{\partial^2 r}{\partial z_j \partial \bar{z}_k} \phi_j \overline{\phi_k} \, d\sigma$$

$$= \sum_{j,k} \left\| \frac{\partial \phi_j}{\partial \bar{z}_k} \right\|^2 - \|\vartheta\phi\|^2 + \int_{\partial\Omega} \sum_{j,k} \frac{\partial^2 r}{\partial z_j \partial \bar{z}_k} \phi_j \overline{\phi_k} \, d\sigma.$$

As a result,

$$Q(\phi, \phi) = \|\bar{\partial}\phi\|^2 + \|\vartheta\phi\|^2 + \|\phi\|^2$$

$$= \sum_{j,k} \left\| \frac{\partial \phi_j}{\partial \bar{z}_k} \right\|^2 + \int_{\partial\Omega} \sum_{j,k} \frac{\partial^2 r}{\partial z_j \partial \bar{z}_k} \phi_j \overline{\phi_k} \, d\sigma + \|\phi\|^2.$$

That completes the proof. ∎

DEFINITION 7.4.3 Let $\Omega \equiv \{z \in \mathbb{C}^n : r(z) < 0\}$ be a smoothly bounded domain. If $p \in \partial\Omega$ then let a be a complex tangent vector to $\partial\Omega$ at p, i.e.,

$$\sum_{j=1}^{n} \frac{\partial r}{\partial z_j}(p) a_j = 0.$$

The Levi form at p is defined to be the quadratic form

$$L_p(a) = \sum_{j,k=1}^{n} \frac{\partial^2 r}{\partial z_j \partial \bar{z}_k}(p) a_j \bar{a}_k.$$

We say that Ω is **pseudoconvex** at p if L_p is positive semidefinite on the space of complex tangent vectors. We say that Ω is **strongly pseudoconvex** at p if L_p is positive definite on the space of complex tangent vectors. We say that Ω is pseudoconvex (resp. strongly pseudoconvex) if every boundary point is pseudoconvex (resp. strongly pseudoconvex).

For emphasis, a pseudoconvex domain is sometimes called *weakly pseudoconvex*.

The geometric meaning of strong pseudoconvexity does not lie near the surface. It is a biholomorphically invariant version of strong convexity (which concept is not biholomorphically invariant). A detailed discussion of pseudoconvexity and strong pseudoconvexity appears in [KR1]. In particular, it is proved in that reference that if p is a point of strong pseudoconvexity, then there is a biholomorphic change of coordinates in a neighborhood of p so that p becomes strongly convex.

DEFINITION 7.4.4 *We define a norm $E(\,\cdot\,)$ on $\bigwedge^{0,1}(\bar{\Omega})$ by setting*

$$E(\phi)^2 \equiv \sum_{j,k=1}^{n} \left\| \frac{\partial \phi_j}{\partial \bar{z}_k} \right\|^2 + \int_{\partial\Omega} |\phi|^2 \, d\sigma + \|\phi\|^2.$$

PROPOSITION 7.4.5
Let Ω be a smoothly bounded domain. Then there exists a constant $c > 0$ such that for all $\phi \in \mathcal{D}^{0,1}$,

$$Q(\phi, \phi) \leq cE(\phi)^2. \tag{7.4.5.1}$$

Moreover, if Ω is strongly pseudoconvex then there is a constant $c' > 0$ such that for $\phi \in \mathcal{D}^{0,1}$ we have

$$Q(\phi, \phi) \geq c' E(\phi)^2. \tag{7.4.5.2}$$

PROOF   Given our identity

$$Q(\phi, \phi) = \sum_{j,k} \left\| \frac{\partial \phi_j}{\partial \bar{z}_k} \right\|^2 + \int_{\partial\Omega} \sum_{j,k} \frac{\partial^2 r}{\partial z_j \partial \bar{z}_k} \phi_j \overline{\phi_k} \, d\sigma + \|\phi\|^2,$$

part (1) is obvious. Part (2) follows also from this identity and the definitions of $E(\phi)$, the Levi form, and strong pseudoconvexity.   ∎

REMARK   Inequality (7.4.5.2) is our substitute for the coercive estimate (compare them at this time). Notice that it was necessary for us to give up some regularity. Indeed, $E(\phi)$ contains no information about derivatives of the type $\partial \phi_j / \partial z_k$.

Notice, moreover, that (7.4.5.2) has the *form* of the hypothesis of the Friedrichs lemma. However, it is not the same since $E(\phi)$ is not comparable with $\|\phi\|$.   ∎

DEFINITION 7.4.6   THE BASIC ESTIMATE   *Let Ω be a smoothly bounded domain in $\mathbb{C}^n$ and let $E(\phi)$ be defined as above. We say that the **basic estimate** holds for elements of $\mathcal{D}^{p,q}(\Omega)$ provided that there is a constant $c > 0$ such that*

$$Q(\phi, \phi) \geq cE(\phi)^2$$

for all $\phi \in \mathcal{D}^{p,q}(\Omega)$.

Putting our definitions together, we see that the basic estimate holds for elements of $\mathcal{D}^{p,q}(\Omega)$ when Ω is strongly pseudoconvex.

Exercise: Show that for $\psi \in \Lambda_c^{p,q}(U)$, $U \subset\subset \Omega$, we have $Q(\psi, \psi) \geq c\|\psi\|_1^2$ (hint: integrate by parts). Then on $\Lambda_c^{p,q}(U)$ we have a classical coercive estimate. The lack of full regularity in some directions for the $\bar{\partial}$-Neumann problem is due to the complex geometry of the boundary.

Exercise: Show that on any smoothly bounded domain Ω the expression $E(\,\cdot\,)$ satisfies

$$E(\phi) \leq C \cdot \|\phi\|_1^2$$

for all $\phi \in \Lambda^{p,q}(\bar{\Omega})$, but that in general there is no constant $C' > 0$ such that

$$E(\phi) \geq C'\|\phi\|_1^2$$

for all $\phi \in \Lambda^{p,q}(\bar{\Omega})$.

Now we are ready to formulate the main theorem of this chapter. Recall that, by construction, the equation $F\phi = \alpha$ admits a unique solution $\phi \in \operatorname{dom}(F)$ for every $\alpha \in H_0^{p,q}(\Omega)$.

THEOREM 7.4.7   THE MAIN ESTIMATE
Let $\Omega \subseteq \mathbb{C}^n$ be a smoothly bounded domain. Assume that the basic estimate holds for elements of $\mathcal{D}^{p,q}(\Omega)$. For $\alpha \in H_0^{p,q}$, we let ϕ denote the unique solution to the equation $F\phi = \alpha$. Then we have:

1. *If W is a relatively open subset of $\bar{\Omega}$ and if $\alpha|_W \in \Lambda^{p,q}(W)$, then $\phi|_W \in \Lambda^{p,q}(W)$.*
2. *Let ρ, ρ_1 be smooth functions with $\operatorname{supp} \rho \subseteq \operatorname{supp} \rho_1 \subseteq W$ and $\rho_1 \equiv 1$ on $\operatorname{supp} \rho$. Then:*

   (a)   If $W \cap \partial\Omega = \emptyset$ then $\forall s \geq 0$ there is a constant $c_s > 0$ (depending on ρ, ρ_1 but independent of α) such that

   $$\|\rho\phi\|_{s+2}^2 \leq c_s \left(\|\rho_1\alpha\|_s^2 + \|\alpha\|_0^2\right).$$

   (b)   If $W \cap \partial\Omega \neq \emptyset$ then $\forall s \geq 0$ there exists a constant $c_s \geq 0$ such that

   $$\|\rho\phi\|_{s+1}^2 \leq c_s \left(\|\rho_1\alpha\|_s^2 + \|\alpha\|_0^2\right).$$

REMARK   Observe that (1) states that F is hypoelliptic. Statement (2a) asserts that, in the interior of Ω, the operator F enjoys the regularity of a strongly elliptic operator. That is, F is of order 2 and the solution of $F\phi = \alpha$ exhibits a gain of two derivatives.

On the other hand, (2b) states that at the boundary F enjoys only subelliptic regularity—the solution enjoys a gain of only one derivative. Examples ([FOK], [GRE], [KR4]) show that this estimate is sharp.   ∎

The proof of the Main Estimate is quite elaborate and will take up most of the remainder of the chapter. We will begin by building up some technical machinery. Then we show how to derive (2a) from the results of Chapter 4. Next, and what is of most interest, we study the boundary estimate. Of course (1) follows from (2a), (2b). All the hard work goes into proving statement 2(b). The tradeoff between existence and regularity is rather delicate in this context. To address this issue we shall use the technique, developed by Kohn and Nirenberg, of elliptic regularization (see [KON2]).

Before we end the section, we wish to stress that part (2a) is the least interesting of all the parts of the Main Estimate. For notice that if $\phi, \psi \in \Lambda_0^{p,q}$ then

$$\langle F\phi, \psi \rangle = ((\Box + I)\phi, \psi).$$

Now $\Box + I$ is elliptic, so the regularity statement follows from Theorem 4.2.4.

7.5   Special Boundary Charts, Finite Differences, and Other Technical Matters

The proof of the Main Estimate has at its heart a number of sophisticated applications of the method of integration by parts. As a preliminary exercise we record here some elementary but useful facts that will be used along the way.

LEMMA 7.5.1
For every $\epsilon > 0$ there is a $K > 0$ such that for any $a, b \in \mathbb{R}$ we have

$$ab \le \epsilon a^2 + K b^2.$$

PROOF   Recall that $2\alpha\beta \le \alpha^2 + \beta^2$ for all $\alpha, \beta \in \mathbb{R}$. Hence

$$2ab = 2\left(\sqrt{2\epsilon}a\right)\left(\frac{1}{\sqrt{2\epsilon}}b\right)$$

$$\le 2\epsilon a^2 + \frac{1}{2\epsilon}b^2.$$

Thus $K = 1/4\epsilon$ does the job.   ∎

LEMMA 7.5.2
If D_1, D_2 are partial differential operators of degrees k_1, k_2, respectively, then

$$[D_1, D_2] \equiv D_1 D_2 - D_2 D_1$$

has degree not exceeding $k_1 + k_2 - 1$.

PROOF Exercise: write it out. ∎

DEFINITION 7.5.3 If A and B are numerical quantities, then we shall use Landau's notation

$$A = \mathcal{O}(B)$$

to indicate that

$$A \leq C \cdot B$$

for some constant C. We will sometimes write $A \lesssim B$, which has the same meaning.

We write $A \approx B$ to mean both $A \lesssim B$ and $B \lesssim A$.

Special Boundary Charts

For the moment let us identify $\mathbb{C}^n$ with $\mathbb{R}^{2n}$. We consider a domain $\Omega = \{x \in \mathbb{R}^{2n} : r(x) < 0\}$ with $|\nabla r| = 1$ on $\partial\Omega$. Let $U \subseteq \bar{\Omega}$ be a relatively open set that has nontrivial intersection with $\partial\Omega$. Coordinates $(t_1, \ldots, t_{2n-1}, r)$ constitute a special boundary chart if r is the defining function for Ω and $(t_1, \ldots, t_{2n-1})$ form coordinates for $\partial\Omega \cap U$. (This construct is quite standard in differential geometry and is called "giving U a product structure.") We associate to a special boundary chart an orthonormal basis $\omega_1, \ldots, \omega_n$ for $\bigwedge^{1,0}$ such that $\omega_n = \sqrt{2}\partial r$. It is frequently convenient to take the function r to be (signed) Euclidean distance to the boundary. Obviously $\partial r / \partial z_n \neq 0$ on U.

If U is a special boundary chart then we set

$$L_j = \frac{\partial r}{\partial z_j}\frac{\partial}{\partial z_n} - \frac{\partial r}{\partial z_n}\frac{\partial}{\partial z_j}, \qquad j = 1, \ldots, n-1,$$

$$L_n = \frac{\partial}{\partial z_n}.$$

Then $L_1, \ldots, L_n$ generate $T_P^{1,0}(\Omega)$ for $P \in \Omega$ and $L_1, \ldots, L_{n-1}$ are tangential; this last statement means that $L_j r = 0$ on U, $j = 1, \ldots, n-1$.

Exercise: Assume that $|\nabla r| \equiv 1$. Set

$$\nu = \sum_{j=1}^{n} \frac{\partial r}{\partial \bar{z}_j}\frac{\partial}{\partial z_j}.$$

Then ν is normal to $U \cap \partial\Omega$. Complete ν to an orthonormal basis $\tilde{L}_1, \ldots, \tilde{L}_{n-1}, \nu$ for $T_P^{1,0}(\Omega), P \in U$. The canonical dual basis $\tilde{\omega}_1, \ldots, \tilde{\omega}_n \in \bigwedge^{1,0}(\Omega)$ satisfies $\omega_n = \sqrt{2}\partial r$.

Check that $\tilde{L}_1, \ldots \tilde{L}_{n-1}, \bar{\tilde{L}}_1, \ldots, \bar{\tilde{L}}_{n-1}, \operatorname{Im} \nu$ form a basis for $T_q(\partial\Omega), q \in \partial\Omega$.

On a special boundary chart the $\bar{\partial}$-Neumann boundary conditions can be easily expressed in terms of coordinates:

LEMMA 7.5.4
If U is a special boundary chart and

$$\phi = \sum_{I,J} \phi_{IJ}\omega^I \wedge \bar{\omega}^J$$

on U then $\phi \in \mathcal{D}^{p,q}$ if and only if $\phi \in \bigwedge^{p,q}(\bar{\Omega})$ and $\phi_{IJ} = 0$ on $\partial\Omega$ whenever $n \in J$.

PROOF   Exercise. This is just definition chasing (or see [FOK, p. 33]).   ∎

Finite Differences

In learning about the calculus of finite differences we shall use $\mathbb{R}^N$ as our setting. If u is a function on $\mathbb{R}^N, j \in \{1, \ldots, N\}$, and $h \in \mathbb{R}$, then we define

$$\Delta_h^j u(x_1, \ldots, x_N) = \frac{1}{2ih}\Big[u(x_1, \ldots, x_{j-1}, x_j + h, x_{j+1}, \ldots, x_N) \\ - u(x_1, \ldots, x_{j-1}, x_j - h, x_{j+1}, \ldots, x_N) \Big].$$

Also if $\beta = (\beta_1, \ldots, \beta_N)$ is a multiindex and

$$H = \left(h_{jk} \right)_{\substack{j=1,\ldots,N \\ k=1,\ldots,\beta_j}}$$

is an array of real numbers then we define

$$\Delta_H^\beta = \prod_{j=1}^N \prod_{k=1}^{\beta_j} \Delta_{h_{jk}}^j.$$

Here the symbol $\prod$ is to be interpreted as a *composition* of the Δ operators.

Example: Let $N = 1$, $\beta = (2)$, and $H = (h, h)$ with $h > 0$. If u is a function then

$$\Delta_H^\beta u(x) = \Delta_h^1 \left(\Delta_h^1 u(x) \right)$$

$$= \Delta_h^1 \left(u(x + h) - u(x - h) \right)$$

$$= u(x + 2h) + u(x - 2h) - 2u(x).$$

This is a standard second difference operator such as one encounters in [ZYG] (see also Chapters 4 and 5).   ⬚

A comprehensive consideration of finite difference operators may be found in [KR2]. We shall limit ourselves here to a few special facts.

LEMMA 7.5.5
If $u \in L^2(\mathbb{R}^N)$ then

$$\left(\Delta_h^j u \right)^\wedge (\xi) = -\frac{\sin(h\xi_j)}{h} \widehat{u}(\xi).$$

Here $^\wedge$ denotes the standard Fourier transform.

PROOF   We calculate that

$$\left(\Delta_h^j u \right)^\wedge (\xi) = \frac{1}{2ih} \int e^{ix\xi} [u(x_1, \ldots, x_j + h, \ldots, x_N)$$

$$- u(x_1, \ldots, x_j - h, \ldots, x_N)] \, dx$$

$$= \frac{1}{2ih} \left[e^{-ih\xi_j} - e^{ih\xi_j} \right] \widehat{u}(\xi)$$

$$= -\frac{\sin(h\xi_j)}{h} \widehat{u}(\xi). \qquad\qquad \blacksquare$$

Notice that

$$\left| \frac{\sin(h\xi_j)}{h} \right| \approx |\xi_j|$$

as $h \to 0$.

LEMMA 7.5.6
Let $u \in H^{s'}(\mathbb{R}^N)$, β a multiindex, $|\beta| = s$. If $r \leq s' - s$ then for any H we have

$$\| \Delta_H^\beta u \|_r \leq \| D^\beta u \|_r.$$

(Here the norms are Sobolev space norms.)

PROOF   We calculate that

$$\| \Delta_H^\beta u \|_r^2 = \int (1 + |\xi|^2)^r \left| (\Delta_H^\beta u)\widehat{\ }\ (\xi) \right|^2 d\xi$$

$$= \int (1 + |\xi|^2)^r \prod_{j=1}^N \prod_{k=1}^{\beta_j} \left| \frac{\sin h_{j_k} \xi_j}{h_{j_k}} \right|^2 |\widehat{u}(\xi)|^2 d\xi$$

$$\lesssim \int (1 + |\xi|^2)^r \prod_{j=1}^N |\xi_j|^{2\beta_j} |\widehat{u}(\xi)|^2 d\xi$$

$$= \int (1 + |\xi|^2)^r \left| (D^\beta u)\widehat{\ }\ (\xi) \right|^2 d\xi$$

$$= \| D^\beta u \|_r^2. \qquad\blacksquare$$

LEMMA 7.5.7   SCHUR
Let $(X, \mu), (Y, \nu)$ be measure spaces. Let $K : X \times Y \to \mathbb{C}$ be a jointly measurable function such that

$$\int |K(x, y)| \, d\mu(x) \le C,$$

uniformly in $y \in Y$, and

$$\int |K(x, y)| \, d\nu(y) \le C,$$

uniformly in $x \in X$. Then the operator

$$f \mapsto \int K(x, y) f(y) \, d\nu(y)$$

maps $L^p(Y, \nu)$ to $L^p(X, \mu), 1 \le p \le \infty$.

PROOF   For $p = \infty$ the assertion is obvious. For $1 \le p < \infty$, use Jensen's inequality from measure theory (or the generalized Marcinkiewicz inequality, for which see [STSI]). $\blacksquare$

We now introduce an important analytic tool that is useful in studying smoothness of functions. We saw a version of it earlier in Chapter 5.

DEFINITION 7.5.8   *The **Bessel potential** of order r is defined by the Fourier analytic expression*

$$\Lambda^r : \phi \mapsto \left((1 + |\xi|^2)^{r/2} \widehat{\phi} \right)^{\vee},$$

where $\phi \in C_c^\infty$. This operation extends to H^r in a natural way.

Observe that $\phi \in H^r$ if and only if $\Lambda^r \phi \in L^2 (\equiv H^0)$. Notice also that $\Lambda^r : H^t \to H^{t-r}$ for any $t, r \in \mathbb{R}$.

LEMMA 7.5.9
If $a \in C_c^\infty(\mathbb{R}^N)$ and $r \le s' - s$, then for all $u \in H^{s'-1}$ and multiindices β with $|\beta| = s$, we have

$$\left\| [a, \Delta_H^\beta] u \right\|_r \lesssim \|u\|_{r+s-1},$$

that is, $[a, \Delta_H^\beta]$ behaves like an operator of order $s - 1$.

PROOF   First we consider the case $|\beta| = 1$. We want to show that

$$\| [a, \Delta_h^j] u \|_r \lesssim \|u\|_r$$

for any r and any $u \in H^{s'}$, $s' \ge r$. Using Bessel potentials, this inequality is seen to be equivalent to

$$\| \Lambda^r [a, \Delta_h^j] \Lambda^{-r} v \| \lesssim \|v\|_0.$$

Now, if $\mathcal{F}$ denotes the Fourier transform then

$$\mathcal{F}\left(\Lambda^r [a, \Delta_h^j] \Lambda^{-r} v \right)(\eta) = \int K(\xi, \eta) \hat{v}(\xi)\, d\xi, \tag{7.5.9.1}$$

where

$$K(\xi, \eta) = \left(\frac{1 + |\eta|^2}{1 + |\xi|^2} \right)^{r/2} \cdot \left[\frac{\sin(h\eta_j)}{h} - \frac{\sin(h\xi_j)}{h} \right] \hat{a}(\eta - \xi).$$

In fact, let us do the calculation that justifies this assertion: According to the definition of the Bessel potential we have

$$\mathcal{F}\left(\Lambda^r [a, \Delta_h^j] \Lambda^{-r} v \right)(\eta) = (1 + |\eta|^2)^{r/2} \mathcal{F}\left([a, \Delta_h^j] \Lambda^{-r} v \right)(\eta). \tag{7.5.9.2}$$

Next,

$$\mathcal{F}([a, \Delta_h^j] \Lambda^{-r} v)(\eta) = \mathcal{F}\left(a\, \Delta_h^j\, \Lambda^{-r}(v) - \Delta_h^j a \Lambda^{-r}(v) \right)(\eta)$$

$$= \hat{a} * (\Delta_h^j \Lambda^{-r} v)\hat{\ }\,(\eta) + \frac{\sin(h\eta_j)}{h} \left(\hat{a} * (\Lambda^{-r} v)\hat{\ } \right)(\eta)$$

$$= -\left(\hat{a} * \left[\frac{\sin(h \cdot_j)}{h} (1 + |\cdot|^2)^{-r/2} \hat{v}(\cdot) \right] \right)(\eta)$$

$$+ \frac{\sin(h\eta_j)}{h} \left(\hat{a} * (1 + |\cdot|^2)^{-r/2} \hat{v}(\cdot) \right)(\eta)$$

$$= \int \hat{a}(\eta - \xi) \left[\frac{\sin(h\eta_j)}{h} - \frac{\sin(h\xi_j)}{h} \right] (1 + |\xi|^2)^{-r/2} \hat{v}(\xi)\, d\xi.$$

The last line, combined with (7.5.9.2), gives (7.5.9.1).

Now we will verify that the kernel $K(\xi, \eta)$ satisfies the hypotheses of Schur's lemma. Note that

$$\left(\frac{1 + |\eta|^2}{1 + |\xi|^2}\right)^{r/2} \leq C \cdot \left(\frac{1 + |\eta|}{1 + |\xi|}\right)^r$$

$$\leq C \cdot (1 + |\eta - \xi|)^{|r|}$$

$$\leq C(1 + |\eta - \xi|^2)^{|r|/2}.$$

Here we are using Proposition 3.2.7. Then

$$|K(\xi, \eta)| \leq C \left(1 + |\eta - \xi|^2\right)^{|r|/2} |\xi - \eta| \cdot |\hat{a}(\eta - \xi)|$$

so that $K(\xi, \eta)$ is uniformly integrable in ξ and η. By Schur's lemma we have

$$\|\Lambda^r[a, \Delta_h^j]\Lambda^{-r}v\|_0 = \left\|\mathcal{F}\left(\Lambda^r[a, \Delta_h^j]\Lambda^{-r}v\right)\right\|_0$$

$$= \left\|\int K(\xi, \eta)\hat{v}(\xi)\,d\xi\right\|_0$$

$$\leq C\|v\|_0.$$

This proves the desired inequality, and we have handled the case $|\beta| = 1$.

If now $|\beta| = s > 1$, we claim that $[a, \Delta_H^\beta]$ is a sum of terms of the form

$$\Delta_{H'}^{\beta'}[a, \Delta_h^j]\Delta_{H''}^{\beta''}$$

with $|\beta'| + |\beta''| = s - 1$. To see this assertion in case $s = 2$, we calculate that

$$[a, \Delta_H^\beta] = [a, \Delta_h^j \Delta_k^\ell]$$

$$= a\,\Delta_h^j\,\Delta_k^\ell - \Delta_h^j\,\Delta_k^\ell\,a$$

$$= \Delta_h^j a\,\Delta_k^\ell + [a, \Delta_h^j]\,\Delta_k^\ell - \Delta_h^j\,\Delta_k^\ell a$$

$$= \Delta_h^j[a, \Delta_k^\ell] + [a, \Delta_h^j]\,\Delta_k^\ell.$$

The claim now follows easily by induction.

Finally, using induction, (7.5.9.2), and the claim we have

$$
\begin{aligned}
\left\| [a, \triangle_H^\beta] u \right\|_r &\leq \sum \left\| \left(\triangle_{H'}^{\beta'} [a, \triangle_h^j] \triangle_{H'}^{\beta''} \right) u \right\|_r \\
&\leq \sum \left\| [a, \triangle_h^j] \triangle_{H'}^{\beta''} u \right\|_{r+|\beta'|} \\
&\leq \sum \left\| \triangle_{H''}^{\beta''} u \right\|_{r+|\beta'|} \\
&\leq C \cdot \|u\|_{r+|\beta''|+|\beta'|} \\
&= C \cdot \|u\|_{r+s-1}.
\end{aligned}
$$

LEMMA 7.5.10
Let $u, v \in H^s, 0 \leq r \leq s$. Then

$$
\langle \triangle_h^j u, v \rangle = \langle u, \triangle_h^j v \rangle.
$$

PROOF   This is just an elementary change of variable.   ∎

Next we have

LEMMA 7.5.11   THE GENERALIZED SCHWARZ INEQUALITY
Let f, g be L^2 functions and $s \in \mathbb{R}$. Then

$$
\langle f, g \rangle \leq \|f\|_s \|g\|_{-s}.
$$

PROOF   Look at the Fourier transform side and use the standard Schwarz inequality.   ∎

PROPOSITION 7.5.12
Let $K \subseteq \mathbb{R}^N$ be compact and let $u, v \in L^2(K)$. Let D be a first-order differential operator and let β be a multiindex with $|\beta| = s$. Then

1.   If $u \in H^s$ then $\| \triangle_H^\beta u \|_0 \leq \|u\|_s$ uniformly in H as $|H| \to 0$.
2.   If $u \in H^s$ then $\|[D, \triangle_H^\beta] u\|_0 \lesssim \|u\|_s$ uniformly in H as $|H| \to 0$.
3.   We have

$$
\langle \triangle_H^\beta u, v \rangle = \langle u, \triangle_H^\beta v \rangle.
$$

4.   If $u \in H^{s-1}, v \in H^1$ then

$$
\left| \langle u, [\triangle_H^\beta, D] v \rangle \right| \leq \|u\|_{s-1} \|v\|_1,
$$

uniformly as $|H| \to 0$.
5.   If $u \in H^s$ and $\| \triangle_H^\beta u \|_s$ is bounded as $|H| \to 0$, then $D^\beta u \in H^s$.

PROOF (1) By Lemma 7.5.6 we have

$$\| \Delta_H^\beta u \|_0 \le \| D^\beta u \|_0$$

$$\le \| u \|_s$$

uniformly in H as $|H| \to 0$.

(2) Write $D = \sum_{j=1}^N a_j(x) D^j$. Assume that $\operatorname{supp} a_j \subseteq K$. Notice that the finite difference operators commute with D^j. Then $[D, \Delta_H^\beta] = \sum_{j=1}^N [a_j, \Delta_H^\beta] D^j$. Thus

$$\left\| [D, \Delta_H^\beta] u \right\|_0 \le \sum_{j=1}^N \left\| [a_j, \Delta_H^\beta] D^j u \right\|_0$$

$$\le \sum_{=1}^N \| D^j u \|_{s-1}$$

$$\le C \cdot \| u \|_s.$$

Note that we have used 7.5.9 in the penultimate inequality.

(3) This is an **elementary change of variable**.

(4) Write $D = \sum_{j=1}^N a_j(x) D^j$. Then $[D, \Delta_H^\beta] = \sum_{j=1}^B [a_j, \Delta_H^\beta] D^j$. Therefore

$$\left| \langle u, [\Delta_H^\beta, D] v \rangle \right| \le \sum_{j=1}^N \left| \langle u, [a_j, \Delta_H^\beta] D^j v \rangle \right|$$

$$\lesssim \sum_{j=1}^N \| u \|_{s-1} \left\| [a_j, \Delta_H^\beta] D^j v \right\|_{1-s}$$

$$\lesssim \sum_{j=1}^N \| u \|_{s-1} \| D^j v \|_0$$

$$\lesssim \| u \|_{s-1} \| v \|_1.$$

Here we have used the **generalized Schwarz inequality**.

(5) If $u \in H^s$ and $\| \Delta_H^\beta u \|_s$ is bounded as $|H| \to 0$, then we want to show that $D^\beta u \in H^s$. It suffices to prove the result for $|\beta| = 1$, i.e., $\Delta_H^\beta = \Delta_h^j$. Saying that $\| \Delta_h^j u \|_s$ is bounded as $h \to 0$ means that

$$\int \left| \frac{\sin(\xi_j h)}{h} (1 + |\xi|^2)^{s/2} \hat{u}(\xi) \right|^2 \, d\xi$$

is bounded as $h \to 0$. Thus the dominated convergence theorem gives the result. ∎

Tangential Sobolev Spaces

One of the important features of the analysis of the $\bar{\partial}$-Neumann problem is that it is *nonisotropic*. This means that in different directions the analysis is different. In particular, a normal derivative behaves like two tangential derivatives. It turns out that the reason for this is that, when the boundary is strongly pseudoconvex, the (complex) normal derivative is a commutator of tangential derivatives (exercise: calculate $[L_j, \bar{L}_j]$ in the boundary of the ball to see this). This assertion will be brought to the surface in the course of our calculations.

Our nonisotropic analysis will be facilitated by the introduction of some specialized function spaces known as the tangential Sobolev spaces. Consider $\mathbb{R}_+^{N+1}$ with coordinates $(t_1, \ldots, t_N, r), r < 0$. Define the *partial Fourier transform* (PFT) by

$$\tilde{u}(\tau, r) = \int_{\mathbb{R}^N} e^{-it\tau} u(t, r) \, dt.$$

[For convenience here we define the Fourier transform with a minus sign.] Then the *tangential Bessel potential* is defined to be

$$\left(\Lambda_t^s u \right)^{\tilde{}} (\tau, r) = (1 + |\tau|^2)^{s/2} \tilde{u}(\tau, r), \qquad (\tau, r) \in \mathbb{R}_+^{N+1}.$$

We then define *tangential Sobolev norms* by

$$\| u \|_s^2 \equiv \| \Lambda_t^s u \|_0^2 = \int_{\mathbb{R}^N} \int_{-\infty}^0 (1 + |\tau|^2)^s |\tilde{u}(\tau, r)|^2 \, dr \, d\tau.$$

For the remainder of the section, we will write $\tilde{\Delta}_H^\beta$ to denote a finite difference operator acting on the first N variables only. Let $K \subseteq \mathbb{R}_+^{N+1}$. We have

(1) If $u \in H^{s'}, |\beta| = s$, and $r \leq s' - s$ then

$$\| \tilde{\Delta}_H^\beta u \|_r \lesssim \| D^\beta u \|_r.$$

This is proved by imitating the proof of the isotropic result 7.5.6, using the tangential Fourier transform and integrating out in the r variable.

(2) If $\rho \in C_c^\infty, u \in H^{s'-1}, |\beta| = s$, and $r \leq s' - s$ then

$$\left\| [\rho, \tilde{\Delta}_H^\beta] u \right\|_r \lesssim \| u \|_{r+s-1}.$$

We prove this by first reducing to the case of a single difference (as we have done before). Then we express the left-hand side as an integral with kernel (as in the isotropic case) and use the Schwarz inequality.

(3) We have

$$\left\langle \tilde{\Delta}_h^j u, v \right\rangle = \left\langle u, \tilde{\Delta}_h^j v \right\rangle.$$

From (1), (2), and (3) we can obtain that, for $u \in H^s$ and $|\beta| = s$,

$$\|\tilde{\Delta}_H^\beta u\|_0 \lesssim |u|_s.$$

Also, for D a first-order partial differential operator, $|\beta| = s$,

$$\|[D, \tilde{\Delta}_H^\beta]u\|_0 \lesssim \sum_{j=1}^N |D^j u|_{s-1}$$

and

$$\left\langle \tilde{\Delta}_H^\beta u, v \right\rangle = \left\langle u, \tilde{\Delta}_H^\beta v \right\rangle.$$

Finally,

$$|\langle u, v \rangle| \leq |u|_{s-1} |v|_{1-s}.$$

We invite the reader to fill in the details in the proofs of these assertions about the tangential Sobolev norms. See also [FOK].

7.6   First Steps in the Proof of the Main Estimate

The method of proof presented here is not the one in Kohn's original work [KOH1]. In fact, the method presented here was developed in the later work [KON2]. In the latter paper, a method of elliptic regularizaion is developed that allows us to exploit the elliptic regularity theory developed in Chapters 4 and 5 to avoid the nasty question of existence for the $\bar{\partial}$-Neumann problem. The idea is to add to the (degenerate) quadratic form Q an expression consisting of δ times the quadratic form for the classical Laplacian. Certain estimates are proved, uniformly in δ, and then δ is allowed to tend to 0. We begin with a definition:

DEFINITION 7.6.1   *Let $\Omega \subseteq \mathbb{C}^n$ be a smoothly bounded domain. (We do not assume that the basic estimate holds in $\mathcal{D}^{p,q}$.) For $\delta > 0$ and $D^j = i\partial/\partial x_j$, we define*

$$Q^\delta(\phi, \psi) = Q(\phi, \psi) + \delta \sum_{j=1}^{2n} \langle D^j \phi, D^j \psi \rangle.$$

Observe that

$$Q^\delta(\phi, \phi) \geq \delta \|\phi\|_1^2. \tag{7.6.2}$$

The point of creating Q^δ is that it has better properties at the boundary than does Q. Indeed, in the interior we have $Q(\ \cdot\ ,\ \cdot\ ) = \langle(\square + I)\ \cdot\ ,\ \cdot\ \rangle$ and this is perfectly suited to our purposes. But Q does not satisfy a coercive estimate at the boundary. Since Q^δ *does* satisfy such an estimate, it is much more useful.

Recall that we defined $\tilde{\mathcal{D}}^{p,q}$ to be the closure of $\mathcal{D}^{p,q}$ (which we know equals dom $\bar{\partial}^* \cap \bigwedge^{p,q}$) in the Q-topology. We let $\tilde{\mathcal{D}}_\delta^{p,q}$ be the closure of $\mathcal{D}^{p,q}$ in the Q^δ-topology. This setup would be intractable if $\tilde{\mathcal{D}}_\delta^{p,q}$ varied with different values of δ. Fortunately, that is not the case:

LEMMA 7.6.3
For all $\delta, \delta' > 0$ it holds that $\tilde{\mathcal{D}}_\delta^{p,q} = \tilde{\mathcal{D}}_{\delta'}^{p,q}$. All of the spaces $\tilde{\mathcal{D}}_\delta^{p,q}$ are contained in $\tilde{\mathcal{D}}^{p,q} \cap H_1^{p,q}$.

PROOF   A sequence $\{\phi_k\}$ is Cauchy in the Q^δ-topology if and only if $\{\phi_k\}$ and $\{D^j \phi_k\}$ are Cauchy in $L^2, j = 1, \ldots, 2n$. And this statement does not depend on δ. Thus the notion of closure is independent of δ.

The second statement is now obvious.   ∎

Notice that we are *not* saying that $\tilde{\mathcal{D}}^{p,q} = \tilde{\mathcal{D}}_\delta^{p,q}$. In fact, this equality is not true. To see this, suppose that the two spaces *were* equal. Then the open mapping principal implies that the Q-norm and the Q^δ-norm are equivalent. This would imply that Q contains information (as does Q^δ) about the L^2-norm of $\partial\phi_j/\partial z_k$ when $\phi \in \bigwedge^{0,1}$. But this is clearly not the case. That gives a contradiction.

Next we wish to apply the Friedrichs theory to the Q^δ's in $\tilde{\mathcal{D}}^{p,q}$. Note that $Q^\delta \geq Q \geq \|\cdot\|_0^2$. Also, by construction, $\tilde{\mathcal{D}}_\delta^{p,q}$ is complete in the Q^δ-topology. Thus, by the Friedrichs theory, there is a self-adjoint F^δ with dom $F^\delta \subseteq H_0^{p,q}$ and

$$F^\delta : \text{dom } F^\delta \to H_0^{p,q}$$

univalently and surjectively. Evidently F^δ will correspond to $\square + I + \delta \sum_j D^{j'} D^j$.

We see that, given a form $\alpha \in H_0^{p,q}$, there exists a unique $\phi^\delta \in \tilde{\mathcal{D}}_\delta^{p,q}$ such that

$$F^\delta \phi^\delta = \alpha.$$

Moreover, ϕ^δ satisfies interior estimates uniformly in δ since

$$\sigma_P(F^\delta) = \begin{pmatrix} (1+\delta)|\xi|^2 & & 0 \\ & \ddots & \\ 0 & & (1+\delta)|\xi|^2 \end{pmatrix}$$

and we see that the eigenvalues of the principal symbol are bounded from below, uniformly in δ.

THEOREM 7.6.4

Let U be a special boundary chart and $V \subseteq \bar{V} \subseteq U$. Let ρ_1 be a smooth, real-valued function with support in U and $\rho_1 \equiv 1$ on $\bar{V}$. If $\phi \in \operatorname{dom} F^\delta$ and $\rho_1 F^\delta \phi \in \bigwedge^{p,q}(\bar{\Omega})$, then for every smooth ρ that is supported in $\bar{V}$ we have $\rho\phi \subseteq \bigwedge^{p,q}(\bar{\Omega})$.

REMARK Recall that $F^\delta = I + \Box + \delta \sum_j D^{j'} D^j$. The theorem says that F^δ is hypoelliptic. From the proof (below) we will see that

$$\|\rho\phi\|_{s+2}^2 \lesssim \|\rho_1 F^\delta \phi\|_s + \|\phi\|_0.$$

However, the constant in the inequality will depend heavily on δ. It will, in fact, be of size δ^{-s-2}, so it blows up as $\delta \to 0$. This means in particular that the uniformity that we need will not come cheaply. ∎

PROOF OF THE THEOREM We know that $\phi \in \operatorname{dom} F^\delta$ implies that $\phi \in H_1^{p,q}$. We will prove that if $\rho\phi \in H_s^{p,q}$ for all cutoff functions ρ supported in V then $\rho\phi \in H_{s+1}^{p,q}$. The theorem will then follow from the Sobolev lemma.

We begin by proving the following claim:

Claim 1: If $\rho\phi \in H_s^{p,q}$ for all ρ, then $D_t^\beta \rho\phi \in H_1^{p,q}$ whenever $|\beta| = s$.

Notice that this claim is not the full statement that we are proving: it gives us information about the derivatives of order $s+1$ only when all but one of the derivatives is in the tangential directions. We deal with tangential derivatives first since tangential derivatives and tangential differences preserve $\mathcal{D}^{p,q}$ and tangential differences preserve $\tilde{\mathcal{D}}^{p,q}$.

For the claim, it suffices to show that

$$\|\tilde{\triangle}_H^\beta \rho\phi\|_1^2$$

is bounded as $|H| \to 0$ when $|\beta| = s$.

Now $\|\psi\|_1^2 \lesssim Q^\delta(\psi, \psi)$ (where the constant depends on δ); therefore it suffices to estimate $Q^\delta(\triangle_H^\beta \rho\phi, \triangle_H^\beta \rho\phi)$. In order to perform this estimate, we need to analyze:

1. $\langle \bar{\partial} \tilde{\triangle}_H^\beta \rho\phi, \bar{\partial} \tilde{\triangle}_H^\beta \rho\phi \rangle$

2. $\langle \vartheta \tilde{\triangle}_H^\beta \rho\phi, \vartheta \tilde{\triangle}_H^\beta \rho\phi \rangle$

3. $\langle \tilde{\triangle}_H^\beta \rho\phi, \tilde{\triangle}_H^\beta \rho\phi \rangle$

4. $\langle D^j \tilde{\triangle}_H^\beta \rho\phi, D^j \tilde{\triangle}_H^\beta \rho\phi \rangle$.

Analysis of (1): Now

$$\langle \bar{\partial} \tilde{\triangle}_H^\beta \rho\phi, \bar{\partial} \tilde{\triangle}_H^\beta \rho\phi \rangle = \langle \tilde{\triangle}_H^\beta \bar{\partial} \rho\phi, \bar{\partial} \tilde{\triangle}_H^\beta \rho\phi \rangle + \langle [\bar{\partial}, \tilde{\triangle}_H^\beta] \rho\phi, \bar{\partial} \tilde{\triangle}_H^\beta \rho\phi \rangle.$$

Recall that $[\bar{\partial}, \tilde{\Delta}_H^\beta]$ acts like an operator of order s because of part (2) of 7.5.12. Therefore

$$\left| \langle [\bar{\partial}, \tilde{\Delta}_H^\beta] \rho\phi, \bar{\partial}\tilde{\Delta}_H^\beta \rho\phi \rangle \right| \lesssim \left\| [\bar{\partial}, \tilde{\Delta}_H^\beta] \rho\phi \right\|_0 \cdot \left\| \bar{\partial}\tilde{\Delta}_H^\beta \rho\phi \right\|_0$$

$$\lesssim \|\rho\phi\|_s \left\| \tilde{\Delta}_H^\beta \rho\phi \right\|_1 .$$

Next,

$$\langle \tilde{\Delta}_H^\beta \bar{\partial}\rho\phi, \bar{\partial}\tilde{\Delta}_H^\beta \rho\phi \rangle = \langle \tilde{\Delta}_H^\beta \rho\bar{\partial}\phi, \bar{\partial}\tilde{\Delta}_H^\beta \rho\phi \rangle + \langle \tilde{\Delta}_H^\beta [\bar{\partial}, \rho]\phi, \bar{\partial}\tilde{\Delta}_H^\beta \rho\phi \rangle.$$

Obviously $[\bar{\partial}, \rho]$ is an operator of order 0, i.e. it consists of multiplication by a smooth function with support in V: call it ρ'. In fact, the support of ρ' lies in the support of ρ. Then

$$\left| \langle \tilde{\Delta}_H^\beta [\bar{\partial}, \rho]\phi, \bar{\partial}\tilde{\Delta}_H^\beta \rho\phi \rangle \right| \lesssim \left\| \tilde{\Delta}_H^\beta \rho'\phi \right\|_0 \left\| \bar{\partial}\tilde{\Delta}_H^\beta \rho\phi \right\|_0$$

$$\lesssim \|\rho'\phi\|_s \| \tilde{\Delta}_H^\beta \rho\phi \|_1.$$

It follows that

$$\langle \bar{\partial}\tilde{\Delta}_H^\beta \rho\phi, \bar{\partial}\tilde{\Delta}_H^\beta \rho\phi \rangle = \langle \tilde{\Delta}_H^\beta \rho\bar{\partial}\phi, \bar{\partial}\tilde{\Delta}_H^\beta \rho\phi \rangle + \mathcal{O}\left(\|\rho'\phi\|_s \cdot \| \tilde{\Delta}_H^\beta \rho\phi \|_1 \right).$$

Here and in what follows, we use ρ' to denote some smooth function with support a subset of the support of ρ—in particular, the support of ρ' lies in V.

Now we use part (3) of 7.5.12 and the generalized Schwarz inequality to see that

$$\langle \bar{\partial}\tilde{\Delta}_H^\beta \rho\phi, \bar{\partial}\tilde{\Delta}_H^\beta \rho\phi \rangle = \langle \rho\bar{\partial}\phi, \tilde{\Delta}_H^\beta \bar{\partial}\tilde{\Delta}_H^\beta \rho\phi \rangle + \mathcal{O}(\|\rho'\phi\|_s \cdot \| \tilde{\Delta}_H^\beta \rho\phi \|_1)$$

$$= \langle \rho\bar{\partial}\phi, \bar{\partial}\tilde{\Delta}_H^\beta \tilde{\Delta}_H^\beta \rho\phi \rangle + \langle \rho\bar{\partial}\phi, [\tilde{\Delta}_H^\beta, \bar{\partial}]\tilde{\Delta}_H^\beta \rho\phi \rangle$$

$$\quad + \mathcal{O}(\|\rho'\phi\|_s \cdot \| \tilde{\Delta}_H^\beta \rho\phi \|_1)$$

$$= \langle \rho\bar{\partial}\phi, \bar{\partial}\tilde{\Delta}_H^\beta \tilde{\Delta}_H^\beta \rho\phi \rangle$$

$$\quad + \mathcal{O}(\|\rho\bar{\partial}\phi\|_{s-1} \| [\tilde{\Delta}_H^\beta, \bar{\partial}]\tilde{\Delta}_H^\beta \rho\phi \|_{1-s})$$

$$\quad + \mathcal{O}(\|\rho'\phi\|_s \| \tilde{\Delta}_H^\beta \rho\phi \|_1)$$

$$= \langle \rho\bar{\partial}\phi, \bar{\partial}\tilde{\Delta}_H^\beta \tilde{\Delta}_H^\beta \rho\phi \rangle + \mathcal{O}(\|\rho'\phi\|_s \| \tilde{\Delta}_H^\beta \rho\phi \|_1).$$

Write $[\bar{\partial}, \rho] = \rho'$ and $\tilde{\Delta}_H^{\beta} = \tilde{\Delta}_{H'}^{\beta'}\tilde{\Delta}_h^{j}$. Then the last line equals

$$
\begin{aligned}
&= \langle \bar{\partial}\phi, \bar{\partial}\rho\tilde{\Delta}_H^{\beta}\tilde{\Delta}_H^{\beta}\rho\phi\rangle + \langle \bar{\partial}\phi, \rho'\tilde{\Delta}_H^{\beta}\tilde{\Delta}_H^{\beta}\rho\phi\rangle \\
&\quad + \mathcal{O}(\|\rho'\phi\|_{\bullet} \cdot \|\tilde{\Delta}_H^{\beta}\rho\phi\|_1) \\
&= \langle \bar{\partial}\phi, \bar{\partial}\rho\tilde{\Delta}_H^{\beta}\tilde{\Delta}_H^{\beta}\rho\phi\rangle + \langle \rho'\bar{\partial}\phi, \tilde{\Delta}_{H'}^{\beta'}\cdot\tilde{\Delta}_h^{j}\tilde{\Delta}_H^{\beta}\rho\phi\rangle \\
&\quad + \mathcal{O}(\|\rho'\phi\|_{\bullet}\|\tilde{\Delta}_H^{\beta}\rho\phi\|_1) \\
&= \langle \bar{\partial}\phi, \bar{\partial}\rho\tilde{\Delta}_H^{\beta}\tilde{\Delta}_H^{\beta}\rho\phi\rangle + \langle \tilde{\Delta}_{H'}^{\beta'}\rho'\bar{\partial}\phi, \tilde{\Delta}_h^{j}\tilde{\Delta}_H^{\beta}\rho\phi\rangle \\
&\quad + \mathcal{O}(\|\rho'\phi\|_{\bullet}\|\tilde{\Delta}_H^{\beta}\rho\phi\|_1) \\
&= \langle \bar{\partial}\phi, \bar{\partial}\rho\tilde{\Delta}_H^{\beta}\tilde{\Delta}_H^{\beta}\rho\phi\rangle + \mathcal{O}(\|\tilde{\Delta}_{H'}^{\beta'}\rho'\bar{\partial}\phi\|_0\|\tilde{\Delta}_h^{j}\tilde{\Delta}_H^{\beta}\rho\phi\|_0) \\
&\quad + \mathcal{O}(\|\rho'\phi\|_{\bullet}\|\tilde{\Delta}_H^{\beta}\rho\phi\|_1) \\
&= \langle \bar{\partial}\phi, \bar{\partial}\rho\tilde{\Delta}_H^{\beta}\tilde{\Delta}_H^{\beta}\rho\phi\rangle + \mathcal{O}(\|\rho'\phi\|_{\bullet}\|\tilde{\Delta}_H^{\beta}\rho\phi\|_1).
\end{aligned}
$$

We have thus proved that

$$
\langle \bar{\partial}\tilde{\Delta}_H^{\beta}\rho\phi, \bar{\partial}\tilde{\Delta}_H^{\beta}\rho\phi\rangle = \langle \bar{\partial}\phi, \bar{\partial}\rho\tilde{\Delta}_H^{\beta}\tilde{\Delta}_H^{\beta}\rho\phi\rangle + \mathcal{O}\left(\|\rho'\phi\|_{\bullet}\|\tilde{\Delta}_H^{\beta}\rho\phi\|_1\right).
$$

The very same argument may be used to prove that

$$
\langle \vartheta\tilde{\Delta}_H^{\beta}\rho\phi, \vartheta\tilde{\Delta}_H^{\beta}\rho\phi\rangle = \langle \vartheta\phi, \vartheta\rho\tilde{\Delta}_H^{\beta}\tilde{\Delta}_H^{\beta}\rho\phi\rangle + \mathcal{O}\left(\|\rho'\phi\|_{\bullet}\|\tilde{\Delta}_H^{\beta}\rho\phi\|_1\right).
$$

As an exercise, the reader may apply similar arguments to

$$
\langle \tilde{\Delta}_H^{\beta}\rho\phi, \tilde{\Delta}_H^{\beta}\rho\phi\rangle \qquad \text{and} \qquad \delta\sum_j \langle D^j\tilde{\Delta}_H^{\beta}\rho\phi, D^j\tilde{\Delta}_H^{\beta}\rho\phi\rangle
$$

to obtain

$$
\begin{aligned}
Q^{\delta}\left(\tilde{\Delta}_H^{\beta}\rho\phi, \tilde{\Delta}_H^{\beta}\rho\phi\right) &= Q^{\delta}\left(\phi, \rho\tilde{\Delta}_H^{\beta}\tilde{\Delta}_H^{\beta}\rho\phi\right) \\
&\quad + \mathcal{O}\left(\|\rho'\phi\|_{\bullet}\|\tilde{\Delta}_H^{\beta}\rho\phi\|_1\right).
\end{aligned} \tag{7.6.4.1}
$$

Let $\rho_1 \in C_c^{\infty}$ with $\rho_1 \equiv 1$ on $\bar{V}$. Since $\phi \in \mathrm{dom}\, F^{\delta}$, we may rewrite the right-hand side of (7.6.4.1) as

$$
\begin{aligned}
&\langle F^{\delta}\phi, \rho\tilde{\Delta}_H^{\beta}\tilde{\Delta}_H^{\beta}\rho\phi\rangle + \mathcal{O}\left(\|\rho'\phi\|_{\bullet}\|\tilde{\Delta}_H^{\beta}\rho\phi\|_1\right) \\
&= \langle \rho\rho_1 F^{\delta}\phi, \tilde{\Delta}_H^{\beta}\tilde{\Delta}_H^{\beta}\rho\phi\rangle + \mathcal{O}\left(\|\rho'\phi\|_{\bullet}\|\tilde{\Delta}_H^{\beta}\rho\phi\|_1\right) \\
&= \langle \tilde{\Delta}_{H'}^{\beta'}\rho\rho_1 F^{\delta}\phi, \tilde{\Delta}_h^{j}\tilde{\Delta}_H^{\beta}\rho\phi\rangle + \mathcal{O}\left(\|\rho'\phi\|_{\bullet}\|\tilde{\Delta}_H^{\beta}\rho\phi\|_1\right).
\end{aligned}
$$

Now we apply the Schwarz inequality to see that the right-hand side in modulus does not exceed

$$\lesssim \|\tilde{\Delta}_{H'}^{\beta'} \rho \rho_1 F^{\delta} \phi\|_0 \|\tilde{\Delta}_h^j \tilde{\Delta}_H^{\beta} \rho\phi\|_0 + \mathcal{O}\left(\|\rho'\phi\|_s \|\tilde{\Delta}_H^{\beta}\rho\phi\|_1\right)$$

$$\lesssim \|\rho_1 F^{\delta}\phi\|_{s-1} \|\tilde{\Delta}_H^{\beta}\rho\phi\|_1 + \mathcal{O}\left(\|\rho'\phi\|_s \|\tilde{\Delta}_H^{\beta}\rho\phi\|_1\right)$$

$$\lesssim \frac{2}{\epsilon}\|\rho_1 F^{\delta}\phi\|_{s-1}^2 + \frac{\epsilon}{2}\|\tilde{\Delta}_H^{\beta}\rho\phi\|_1^2 + \frac{2}{\epsilon}\|\rho'\phi\|_s^2 + \frac{\epsilon}{2}\|\tilde{\Delta}_H^{\beta}\rho\phi\|_1^2.$$

Here $\epsilon > 0$ is to be specified. Thus, using (7.6.2) and (7.6.4.1), we have

$$\|\tilde{\Delta}_H^{\beta}\rho\phi\|_1 \lesssim Q^{\delta}\left(\tilde{\Delta}_H^{\beta}\rho\phi, \tilde{\Delta}_H^{\beta}\rho\phi\right)$$

$$\lesssim \frac{2}{\epsilon}\|\rho_1 F^{\delta}\phi\|_{s-1}^2 + \frac{2}{\epsilon}\|\rho'\phi\|_s^2 + \epsilon\|\tilde{\Delta}_H^{\beta}\rho\phi\|_1^2.$$

If we select ϵ to be positive and smaller than $1/4$, then we find that

$$\|\tilde{\Delta}_H^{\beta}\rho\phi\|_1^2 \lesssim \|\rho_1 F^{\delta}\phi\|_{s-1}^2 + \|\rho'\phi\|_s.$$

Applying part (5) of 7.5.12 now yields that

$$\|D_t^{\beta}\rho\phi\|_1 < \infty.$$

We have proved the claim.
Now we will prove:

Claim 2: If $m \geq 2$ and $|\beta| + m = s + 1$ then

$$D_t^{\beta} D_r^m(\rho\phi) \in H_0^{p,q}.$$

(Equivalently, $\rho D_t^{\beta} D_r^m \phi \in H_0^{p,q}$.)

The case $m = 1$ of the claim has already been covered in Claim 1. Now let $m = 2$. In local coordinates the operator F^{δ} looks like

$$F^{\delta} = A_{2n}D_r^2 + \sum_{j=1}^{2n-1} A_j D_t^j D_r + \sum_{j,k=1}^{2n-1} A_{j,k}D_t^j D_t^k,$$

where the $A_1, \ldots, A_{2n}, A_{j,k}, B_j, C$ are matrices of smooth functions. The ellipticity of F^{δ} is expressed by the invertibility of the matrix of its symbols $A_1, \ldots, A_{2n}$. In particular, A_{2n} is invertible. Therefore, recalling that $\rho_1 \equiv 1$ on supp ρ, we see that

$$\rho_1 D_r^2 \phi = \rho_1 A_{2n}^{-1}\left[\rho_1 F^{\delta}\phi - \sum_{j=1}^{2n-1} A_j D_t^j D_r\phi - \sum_{j,k=1}^{2n-1} A_{j,k}D_t^j D_t^k\phi\right].$$

Applying ρD_t^β with $|\beta| = s - 1$ to both sides yields that

$$\rho D_t^\beta D_r^2 \phi = \rho D_t^\beta \rho_1 A_{2n}^{-1} \left[\rho_1 F^\delta \phi - \sum_{j=1}^{2n-1} A_j D_t^j D_r \phi - \sum_{j,k=1}^{2n-1} A_{j,k} D_t^j D_t^k \phi \right].$$

(Here we use explicitly the fact that $\rho_1 = 1$ on the support of ρ.) Thus we see that we can express $\rho D_t^\beta D_r^2 \phi$, $|\beta| = s - 1$, in terms of two types of expressions:

1. $(s - 1)$ tangential derivatives of the expression $\rho_1 F^\delta \phi$;
2. $(s + 1)$ derivatives, at most one of which is in the normal direction, of ϕ.

These two types of expressions are both elements of $H_0^{p,q}$. Hence $\rho D_t^\beta D_r^2 \phi \in H_0^{p,q}$. This proves the case $m = 2$.

Proceeding by induction on m we find that $\rho D_t^\beta D_r^2 \phi$ is expressed in terms of (1) $(s - 1)$ tangential derivatives of $\rho_1 F^\delta \phi$ and (2) $(s + 1)$ derivatives of $\rho \phi$ of which at most $(m - 1)$ are in the normal direction.

This concludes the proof. ∎

7.7   Estimates in the Sobolev $-1/2$ Norm

We begin this section by doing some calculations in the tangential Sobolev norms. Recall that

$$|\phi|_s \equiv \|\Lambda_t^s \phi\|_0.$$

If D is any first-order linear differential operator then

$$|D\phi|_s^2 = \left| \sum_{j=1}^{2n} a_j D^j \phi + a_0 \phi \right|_s^2$$

$$\approx \sum_{j=1}^{2n} |a_j D^j \phi|_s^2 + \|\phi\|_0^2$$

$$\approx |\phi|_{s+1}^2 + |D_r \phi|_s^2.$$

Let $A\phi \equiv A_k \phi \equiv \rho_1 \Lambda_t^k(\rho\phi)$ (which in turn equals $\Lambda_t^k(\rho_1 \rho\phi) + [\rho_1, \Lambda_t^k](\rho\phi)$). Let A' denote the formal adjoint of A; that is, if $\phi, \psi \in \Lambda_c^{p,q}(U \cap \bar{\Omega})$ then $\langle A'\phi, \psi \rangle \equiv \langle \phi, A\psi \rangle$. Fix a special boundary chart U.

LEMMA 7.7.1
For all real s and $\phi \in \bigwedge_c^{p,q}(U \cap \bar{\Omega})$, we have

1. $\|A\phi\|_s \lesssim \|\phi\|_{s+k}.$

1'. $\|A'\phi\|_s \lesssim \|\phi\|_{s+k}.$

2. $\|(A - A')\phi\|_s \lesssim \|\phi\|_{s+k-1}.$

3. *If D is any first-order linear differential operator, then*

    (a) $\|[A, D]\phi\|_s \lesssim \|D\phi\|_{s+k-1}.$

    (b) $\|[A - A', D]\phi\|_s \lesssim \|D\phi\|_{s+k-2}.$

    (c) $\|[A, [A, D]]\phi\|_s \lesssim \|D\phi\|_{s+2k-2}.$

PROOF The proof is straightforward using techniques that we have already presented. We leave the details to the reader. ∎

Observe that A preserves $\mathcal{D}^{p,q}$.

LEMMA 7.7.2
It holds that

$$Q\langle A\phi, A\phi \rangle - Re\, Q\langle \phi, A'A\phi \rangle = \mathcal{O}\left(\|\nabla\phi\|_{k-1}^2\right),$$

uniformly in $\phi \in \mathcal{D}^{p,q} \cap \bigwedge_c^{p,q}(U \cap \bar{\Omega})$; here ∇ denotes the gradient of ϕ.

PROOF Recall that $Q(\phi, \psi) = \langle \bar{\partial}\phi, \bar{\partial}\psi \rangle + \langle \vartheta\phi, \vartheta\psi \rangle + \langle \phi, \psi \rangle$. Consider the expression

$$\langle \bar{\partial}A\phi, \bar{\partial}A\phi \rangle - Re\,\langle \bar{\partial}\phi, \bar{\partial}A'A\phi \rangle = \frac{1}{2}\left\{ 2\langle \bar{\partial}A\phi, \bar{\partial}A\phi \rangle - \langle \bar{\partial}\phi, \bar{\partial}A'A\phi \rangle \right.$$
$$\left. - \langle \bar{\partial}A'A\phi, \bar{\partial}\phi \rangle \right\}.$$

We write

$$\langle \bar{\partial}\phi, \bar{\partial}A'A\phi \rangle = \langle \bar{\partial}\phi, A'\bar{\partial}A\phi \rangle + \langle \bar{\partial}\phi, [\bar{\partial}, A']A\phi \rangle$$
$$= \langle A\bar{\partial}\phi, \bar{\partial}A\phi \rangle + \langle \bar{\partial}\phi, [\bar{\partial}, A']A\phi \rangle$$
$$= \langle \bar{\partial}A\phi, \bar{\partial}A\phi \rangle + \langle [A, \bar{\partial}]\phi, \bar{\partial}A\phi \rangle + \langle \bar{\partial}\phi, [\bar{\partial}, A']A\phi \rangle.$$

Also

$$\langle \bar{\partial}A'A\phi, \bar{\partial}\phi \rangle = \langle \bar{\partial}A\phi, \bar{\partial}A\phi \rangle + \langle \bar{\partial}A\phi, [A, \bar{\partial}]\phi \rangle + \langle [\bar{\partial}, A']A\phi, \bar{\partial}\phi \rangle.$$

As a result,

$$
\begin{aligned}
\langle \bar{\partial} A\phi, \bar{\partial} A\phi \rangle - \mathrm{Re}\, \langle \bar{\partial}\phi, \bar{\partial} A'A\phi \rangle &= -\frac{1}{2}\{\langle \bar{\partial}\phi, [\bar{\partial}, A']A\phi \rangle + \langle [A, \bar{\partial}]\phi, \bar{\partial} A\phi \rangle \\
&\quad + \langle [\bar{\partial}, A']A\phi, \bar{\partial}\phi \rangle + \langle \bar{\partial} A\phi, [A, \bar{\partial}]\phi \rangle \} \\
&= -\frac{1}{2}\{I + II + III + IV\}.
\end{aligned}
$$

We will estimate $II + III$; the corresponding estimate for $I + IV$ will then follow easily.

Notice that

$$
\begin{aligned}
&\langle [\bar{\partial}, A' - A]A\phi, \bar{\partial}\phi \rangle + \langle [[\bar{\partial}, A], A]\phi, \bar{\partial}\phi \rangle + \langle [\bar{\partial}, A]\phi, (A' - A)\bar{\partial}\phi \rangle \\
&\quad + \langle [\bar{\partial}, A]\phi, [A, \bar{\partial}]\phi \rangle \\
&= \langle [\bar{\partial}, A']A\phi, \bar{\partial}\phi \rangle - \langle [\bar{\partial}, A]A\phi, \bar{\partial}\phi \rangle + \langle [\bar{\partial}, A]A\phi, \bar{\partial}\phi \rangle - \langle A[\bar{\partial}, A]\phi, \bar{\partial}\phi \rangle \\
&\quad + \langle [\bar{\partial}, A]\phi, A'\bar{\partial}\phi \rangle - \langle [\bar{\partial}, A]\phi, A\bar{\partial}\phi \rangle + \langle [\bar{\partial}, A]\phi, A\bar{\partial}\phi \rangle - \langle [\bar{\partial}, A]\phi, \bar{\partial} A\phi \rangle \\
&= \langle [\bar{\partial}, A']A\phi, \bar{\partial}\phi \rangle - \langle [\bar{\partial}, A]\phi, \bar{\partial} A\phi \rangle \\
&= \langle [\bar{\partial}, A']A\phi, \bar{\partial}\phi \rangle + \langle [A, \bar{\partial}]\phi, \bar{\partial} A\phi \rangle. \qquad\qquad (7.7.2.1)
\end{aligned}
$$

Therefore

$$
\begin{aligned}
|II + III| &\equiv |\langle [A, \bar{\partial}]\phi, \bar{\partial} A\phi \rangle + \langle [\bar{\partial}, A']A\phi, \bar{\partial}\phi \rangle| \\
&= |\langle [\bar{\partial}, A' - A]A\phi, \bar{\partial}\phi \rangle + \langle [[\bar{\partial}, A], A]\phi, \bar{\partial}\phi \rangle + \langle [\bar{\partial}, A]\phi, (A' - A)\bar{\partial}\phi \rangle \\
&\quad + \langle [\bar{\partial}, A]\phi, [A, \bar{\partial}]\phi \rangle|. \qquad\qquad (7.7.2.2)
\end{aligned}
$$

Now, using (7.7.2.2) and 7.7.1, we have the estimates

$$
\begin{aligned}
|II + III| &\lesssim \|[\bar{\partial}, A' - A]A\phi\|_{1-k} \cdot \|\bar{\partial}\phi\|_{k-1} + \|[[\bar{\partial}, A], A]\phi\|_{1-k}\|\bar{\partial}\phi\|_{k-1} \\
&\quad + \|[\bar{\partial}, A]\phi\|_0 \cdot \|(A' - A)\bar{\partial}\phi\|_0 + \|[\bar{\partial}, A]\phi\|_0^2 \\
&\lesssim \|\nabla A\phi\|_{-1}\|\nabla\phi\|_{k-1} + \|\nabla\phi\|_{k-1}^2.
\end{aligned}
$$

But, by 7.7.1 again,

$$
\begin{aligned}
\|\nabla A\phi\|_{-1}^2 &\lesssim \sum_j \|AD^j\phi\|_{-1}^2 + \sum_j \|[D^j, A]\phi\|_{-1}^2 + \|A\phi\|_{-1}^2 \\
&\lesssim \sum_j \|D^j\phi\|_{k-1}^2 + \sum_j \|D^j\phi\|_{k-2}^2 + \|\phi\|_{k-1}^2 \\
&\lesssim \|\nabla\phi\|_{k-1}^2.
\end{aligned}
$$

Therefore

$$|II + III| \lesssim \|\nabla\phi\|_{k-1}^2.$$

Since $I + IV = \overline{II + III}$, we may also conclude that

$$|I + IV| \lesssim \|\nabla\phi\|_{k-1}^2.$$

Together these yield

$$\left| \langle \bar{\partial} A\phi, \bar{\partial} A\phi \rangle - \text{Re}\, \langle \bar{\partial}\phi, \bar{\partial} A' A\phi \rangle \right| \lesssim \|\nabla\phi\|_{k-1}^2.$$

Similarly, it may be shown that

$$\langle \vartheta A\phi, \vartheta A\phi \rangle - \text{Re}\, \langle \vartheta\phi, \vartheta A' A\phi \rangle \lesssim \|\nabla\phi\|_{k-1}^2$$

and

$$\langle A\phi, A\phi \rangle - \text{Re}\, \langle \phi, A' A\phi \rangle = 0.$$

This concludes the proof of the lemma. ∎

It is convenient to think of the preceding lemma as a sophisticated exercise in integration by parts. In the case $k = 0$ we shall now derive a slightly strengthened result:

LEMMA 7.7.3
We have the estimate

$$Q\langle \rho\phi, \rho\phi \rangle - \text{Re}\, Q(\phi, \rho^2\phi) = \mathcal{O}(\|\phi\|_0^2).$$

PROOF We take A in the preceding lemma to be the operator corresponding to multiplication by ρ (this is just the case $k = 0$). Assuming as we may that ρ is real-valued, we know that $A = A'$. By (7.7.2.1) in the proof of 7.7.2 we have

$$\langle [\bar{\partial}, A']A\phi, \bar{\partial}\phi \rangle + \langle [A, \bar{\partial}]\phi, \bar{\partial} A\phi \rangle$$

$$= \langle [\bar{\partial}, A - A']A\phi, \bar{\partial}\phi \rangle + \langle [[\bar{\partial}, A], A]\phi, \bar{\partial}\phi \rangle$$

$$+ \langle [\bar{\partial}, A]\phi, (A - A')\bar{\partial}\phi \rangle + \langle [\bar{\partial}, A]\phi, [A, \bar{\partial}]\phi \rangle.$$

Since $A = \rho$, the operator $[\bar{\partial}, A]$ is simply multiplication by a matrix of functions. Thus $[\bar{\partial}, A]$ and A commute and $[[\bar{\partial}, A], A] = 0$ and of course $A - A' = 0$. The result follows. ∎

Recall the quadratic form

$$E(\phi)^2 \equiv \sum_{j,k} \left\| \frac{\partial \phi_j}{\partial \bar{z}_k} \right\|_0^2 + \int_{\partial\Omega} |\phi|^2 + \|\phi\|_0^2.$$

The basic estimate is

$$E(\phi)^2 \lesssim Q(\phi, \phi),$$

and it is a standing hypothesis that this estimate holds on the domain under study. *We know that the basic estimate holds, for instance, on any strongly pseudoconvex domain.* The next result contains the key estimate in our proof of regularity for the $\bar{\partial}$-Neumann problem up to the boundary.

THEOREM 7.7.4
For each $P \in \partial\Omega$ there exists a special boundary chart V about P such that

$$\|\nabla\phi\|_{-1/2}^2 \lesssim E(\phi)^2$$

for all $\phi \in \bigwedge_c^{p,q}(V \cap \bar{\Omega})$.

The result follows quickly from the following lemma:

LEMMA 7.7.5
Let U be a special boundary chart for Ω and let $M_1, \ldots, M_N$ be first-order, homogeneous differential operators on U. Write

$$M_k = \sum_{j=1}^{2n} a_{jk} D^j.$$

Assume that there exists no $0 \neq \eta \in T^*(U)$ such that $\sigma(M_k, \eta) = 0$ for all $k \in \{1, \ldots, N\}$. Then for all $P \in \partial\Omega \cap U$ there exists a neighborhood $V \subseteq U$ of P such that

$$\sum_{j=1}^{2n} \|D^j \phi\|_{-1/2}^2 \lesssim \sum_{k=1}^{N} \|M_k \phi\|_{-1/2}^2 + \int_{\partial\Omega} |\phi|^2 \qquad (7.7.5.1)$$

for all $\phi \in \bigwedge_c^{p,q}(V \cap \bar{\Omega})$.

Assuming the lemma for the moment, let us prove the theorem.

PROOF OF THEOREM 7.7.4   The vector fields

$$\frac{\partial}{\partial \bar{z}_1}, \ldots, \frac{\partial}{\partial \bar{z}_n}$$

satisfy the conditions of the lemma so that

$$\sum_{j=1}^{2n} |D^j \phi|^2_{-1/2} = |\nabla \phi|^2_{-1/2}$$

$$\lesssim \sum_{j,k=1}^{n} \left\| \frac{\partial \phi_k}{\partial \bar{z}_j} \right\|^2_{-1/2} + \int_{\partial \Omega} |\phi|^2$$

$$\leq \sum_{j,k=1}^{n} \left\| \frac{\partial \phi_k}{\partial \bar{z}_j} \right\|^2_{0} + \int_{\partial \Omega} |\phi|^2 + \|\phi\|^2_0$$

$$= E(\phi)^2. \qquad \blacksquare$$

Note that the M_k's in the lemma cannot be too few, for they have to span all possible directions. Elementary considerations of dimensionality show that $N \geq n$ (the base field is the complex numbers). The $\{D^j\}$ are the simplest example of a collection of vector fields that satisfies the symbol condition in the lemma.

Before proving Lemma 7.7.5, we need a preliminary result:

LEMMA 7.7.6
Let $s > t$. Then for any $\epsilon > 0$ there is a neighborhood V of $P \in \Omega$ such that

$$\|u\|_t \leq \epsilon \|u\|_s$$

whenever u is supported in $V \cap \bar{\Omega}$.

REMARK Results of this sort are commonly used in elliptic theory. They are a form of the Rellich lemma. We shall have to do a little extra work when V intersects the boundary. $\blacksquare$

PROOF OF THE LEMMA We first prove the statement in the case when there is no intervention of the boundary, that is, when $u \in C_c^\infty(V)$ and $V \subset\subset \Omega$. Also assume at first that $s > t \geq 0$. If the assertion were false then there would exist an $\epsilon > 0$ and a sequence $\{u_k\}$ of functions with $\operatorname{supp} u_k \searrow \{P\}$, $\|u_k\|_t = 1$, and $\|u_k\|_s < 1/\epsilon$. By Rellich's lemma there is a subsequence converging in H_t to a function u (it is a function since we have assumed that $t \geq 0$). But $\operatorname{supp} u = \{P\}$, which is impossible since $\|u\|_t = 1$.

If $s \leq 0$ but P is still in the interior of Ω, we let V' be a neighborhood of P for which $\|u\|_{-s} \leq \epsilon \|u\|_{-t}$ for all $u \in C_c^\infty(V')$. Let $V \subseteq \bar{V} \subseteq V'$. Then we may exploit the duality between H_t and H_{-t} to see that if $u \in C_c^\infty(V)$ then

$$\|u\|_t = \sup_{\substack{v \in C_c^\infty(V) \\ v \neq 0}} \frac{|\langle u, v \rangle|}{\|v\|_{-t}}$$

$$\leq \epsilon \sup_{\substack{v \in C_c^\infty(V) \\ v \neq 0}} \frac{|\langle u, v \rangle|}{\|v\|_{-s}}$$

$$= \epsilon \|u\|_s.$$

Notice that the case $t < 0 \leq s$ follows from these first two cases.

Now let us consider the case that $P \in \partial\Omega$. It is enough to consider the problem on the half-space in $\mathbf{R}^{2n}$ with boundary $\mathbf{R}^{2n-1}$. Let $V' \subseteq \mathbf{R}^{2n-1}$ be a relative neighborhood of $P \in \mathbf{R}^{2n-1}$ such that

$$\|u\|_t \leq \epsilon \|u\|_s$$

for all u supported in V' (note here that P is in the *interior* of $\mathbf{R}^{2n-1}$ so our preceding result applies). Let $V \equiv V' \times I$, where I is any interval in $(0, \infty)$. Then

$$|u|_t^2 = \int_{-\infty}^0 \|u(\cdot, r)\|_t^2 \, dr$$

$$\leq \epsilon^2 \int_{-\infty}^0 \|u(\cdot, r)\|_s^2 \, dr$$

$$= \epsilon^2 |u|_s^2. \qquad\qquad \blacksquare$$

PROOF OF LEMMA 7.7.5   The first step is to reduce to the case in which the M_k's are operators with constant coefficients. Let $V \subset\subset U$ be a neighborhood of P, $\rho \in C_c^\infty$, $0 \leq \rho \leq 1$. Suppose also that $\rho \equiv 1$ on V and $W \equiv \operatorname{supp}\rho \subseteq U$. It suffices to prove our result for functions since the Sobolev norm on forms is defined componentwise.

Let $u \in \Lambda_c^{0,0}(V \cap \bar\Omega)$. Now we freeze the coefficients of the M_k, that is, we set

$$N_k = \sum_{j=1}^{2n} a_{j,k}(P) D^j,$$

$k = 1, \ldots, N$. Let $b_{j,k}(x) \equiv a_{j,k}(x) - a_{j,k}(P)$. Then

$$|(M_k - N_k)u|_{-1/2} = \left| \sum_j \left(a_{j,k}(x) - a_{j,k}(P) \right) D^j u \right|_{-1/2}$$

$$\lesssim \sum_j |b_{j,k} D^j u|_{-1/2}$$

$$= \sum_j \left\| \rho b_{j,k} D^j u \right\|_{-1/2}$$

$$= \sum_j \left\| \Lambda_t^{-1/2} \left(\rho b_{j,k} D^j u \right) \right\|_0$$

$$= \sum_j \left\| \rho b_{j,k} \Lambda_t^{-1/2} D^j u + [\Lambda_t^{-1/2}, \rho b_{j,k}] D^j u \right\|_0$$

$$\lesssim \sum_j \sup_W |\rho b_{j,k}| \cdot \left\| \Lambda_t^{-1/2} D^j u \right\|_0$$

$$+ \sum_j \left\| [\Lambda_t^{-1/2}, \rho b_{j,k}] D^j u \right\|_0.$$

Note that if W is small enough then $\sup_W |\rho b_{j,k}| < \epsilon$ since $|b_{j,k}(P)| = 0$. Now the commutator in the second sum is of tangential order $-3/2$. Thus the last line is

$$\lesssim \epsilon \sum_j \left\| D^j u \right\|_{-1/2} + \sum_j \left\| D^j u \right\|_{-3/2}$$

$$\lesssim \epsilon \left\| D^j u \right\|_{-1/2},$$

where (shrinking V if necessary) we have applied Lemma 7.7.6. As a result,

$$\left\| N_k u \right\|_{-1/2} \leq \left\| M_k u \right\|_{-1/2} + \left\| (M_k - N_k) u \right\|_{-1/2}$$

$$\lesssim \left\| M_k u \right\|_{-1/2} + \epsilon \sum_j \left\| D^j u \right\|_{-1/2}. \tag{7.7.5.2}$$

If we can prove the constant coefficient case of our inequality then we would have

$$\sum_j \left\| D^j u \right\|_{-1/2}^2 \lesssim \sum_{k=1}^N \left\| N_k u \right\|_{-1/2}^2 + \int_{\partial \Omega} |u|^2.$$

Coupling this with (7.7.5.2) would yield

$$\sum_{j=1}^{2n} \left\| D^j u \right\|_{-1/2}^2 \lesssim \sum_{k=1}^N \left(\left\| M_k u \right\|_{-1/2} + \epsilon \sum_j \left\| D^j u \right\|_{-1/2} \right)^2$$

$$+ \int_{\partial \Omega} |u|^2$$

$$\lesssim \sum_{k=1}^N \left\| M_k u \right\|_{-1/2}^2 + \epsilon \sum_j \left\| D^j u \right\|_{-1/2}^2 + \int_{\partial \Omega} |u|^2.$$

The full result then follows.

So we have reduced matters to the case of the M_k's having constant coefficients. Assume for the moment that $u\big|_{\partial\Omega} \equiv 0$. Then we can extend u to be zero outside Ω. The extended function will be continuous on all of space. We may suppose that V is a special boundary chart. Notice that, on V, the $D_t^j u$ are continuous and $D_r u$ has only a jump discontinuity. Therefore all of these first derivative functions are square integrable.

Write $\xi = (\tau, \zeta)$ with $\tau \in \mathbb{R}^{2n-1}, \zeta \in \mathbb{R}$. We use the symbol condition on the M_k's to see that

$$
\sum_{k=1}^{N} |M_k u|^2_{-1/2} \quad \equiv \quad \sum_{k=1}^{N} \left\| (1 + |\tau|^2)^{-1/4} \widetilde{M_k u}(\tau, r) \right\|_0^2
$$

$$
\overset{\left(\substack{\text{Plancherel in} \\ \text{last var.}}\right)}{=} \quad \sum_{k=1}^{N} \left\| (1 + |\tau|^2)^{-1/4} \left[\sum_{j=1}^{2n-1} a_{j,k}(P)\tau_j \right. \right.
$$

$$
\left. \left. + a_{2n,k}(P)\zeta \right] \hat{u}(\xi) \right\|_0^2
$$

$$
= \quad \sum_{k=1}^{N} \int_{\mathbb{R}^{2n}} (1 + |\tau|^2)^{-1/2} |\sigma(N_k, \xi)|^2 |\hat{u}(\xi)|^2 \, d\xi
$$

$$
\gtrsim \quad \int_{\mathbb{R}^{2n}} (1 + |\tau|^2)^{-1/2} |\xi|^2 |\hat{u}(\xi)|^2 \, d\xi
$$

$$
\gtrsim \quad \int_{\mathbb{R}^{2n}} (1 + |\tau|^2)^{-1/2} \left(\widehat{\sum_j |D^j u|^2} \right)(\xi) \, d\xi
$$

$$
\overset{\left(\substack{\text{Plancherel in} \\ \text{last var.}}\right)}{=} \quad \int_{\mathbb{R}^{2n-1}} (1 + |\tau|^2)^{-1/2} \int_{\mathbb{R}} \sum_j |\widetilde{D^j u}(\tau, r)|^2 \, dr \, d\tau
$$

$$
\equiv \quad \sum_j |D^j u|^2_{-1/2} .
$$

Therefore

$$
\sum_{k=1}^{N} |M_k u|^2_{-1/2} \gtrsim \sum_{j=1}^{2n} |D^j u|^2_{-1/2}
$$

in the constant coefficient case provided that u vanishes on $\partial\Omega$.

Next suppose that u may or may not vanish at the boundary. Let

$$
\tilde{w}(\tau, r) \equiv \exp\left[(1 + |\tau|^2)^{1/2} r \right] \tilde{u}(\tau, 0).
$$

This is a regular extension of $\tilde{u}(\tau, 0)$ to $(\tau, r), r \leq 0$. By the Fourier inversion theorem we have

$$
w(t, 0) = u(t, 0).
$$

Set $v = u - w$. Then v vanishes on the boundary so that the previous result applies to v. We then have

$$\sum_{j=1}^{2n} |D^j v|^2_{-1/2} \lesssim \sum_{k=1}^{N} |M_k v|^2_{-1/2}.$$

Therefore

$$\sum_{j=1}^{2n} |D^j u|^2_{-1/2} \lesssim \sum_{j=1}^{2n} |D^j v|^2_{-1/2} + |D^j w|^2_{-1/2}$$

$$\lesssim \sum_{k=1}^{N} |M_k v|^2_{-1/2} + \sum_{j=1}^{2n} |D^j w|^2_{-1/2}$$

$$\lesssim \sum_{k=1}^{N} \left[|M_k u|^2_{-1/2} + |M_k w|^2_{-1/2} \right] + \sum_{j=1}^{2n} |D^j w|^2_{-1/2}$$

$$\lesssim \sum_{k=1}^{N} |M_k u|^2_{-1/2} + \sum_{j=1}^{2n} |D^j w|^2_{-1/2},$$

since the M_k's are linear combinations of the D^j's. Observe that we are finished if we can show that

$$|D^j w|^2_{-1/2} \lesssim \int_{\partial \Omega} |u|^2 \, d\sigma,$$

$j = 1, \ldots, 2n$.

First suppose that $j \in \{1, \ldots, 2n - 1\}$. Then

$$|D^j w|^2_{-1/2} = \int_{\mathbb{R}^{2n-1}} \int_{-\infty}^{0} (1 + |\tau|^2)^{-1/2} |\tau_j|^2 \exp \left[2(1 + |\tau|^2)^{1/2} r \right]$$

$$\times |\tilde{u}(\tau, 0)|^2 \, dr \, d\tau$$

$$\lesssim \int_{\mathbb{R}^{2n-1}} \int_{-\infty}^{0} (1 + |\tau|^2)^{-1/2} (1 + |\tau|^2) \exp \left[2(1 + |\tau|^2)^{1/2} r \right]$$

$$\times |\tilde{u}(\tau, 0)|^2 \, dr \, d\tau$$

$$= \int_{\mathbb{R}^{2n-1}} |\tilde{u}(\tau, 0)|^2 \left(\frac{1}{2} \exp \left[2(1 + |\tau|^2)^{1/2} r \right] \Big|_{-\infty}^{0} \right) d\tau$$

$$= \frac{1}{2} \int_{\mathbb{R}^{2n-1}} |\tilde{u}(\tau, 0)|^2 \, d\tau$$

$$= \frac{1}{2} \int_{\partial \Omega} |u|^2 \, d\sigma.$$

If instead $j = 2n$, so that D^j is the usual normal derivative, then

$$\|D^{2n}w\|^2_{-1/2} = \int_{\mathbb{R}^{2n}_+} (1 + |\tau|^2)^{-1/2} |\widetilde{D_r w}(\tau, r)|^2 \, d\tau \, dr.$$

Since the derivative does not affect the variables in which we take the Fourier transform, the two operations commute. Hence

$$\widetilde{D_r w}(\tau, r) = (1 + |\tau|^2)^{1/2} \exp[(1 + |\tau|^2)^{1/2} r] \tilde{u}(\tau, 0).$$

Therefore

$$
\begin{aligned}
\|D_r w\|^2_{-1/2} &= \int_{\mathbb{R}^{2n-1}} \int_{-\infty}^0 (1 + |\tau|^2)^{-1/2}(1 + |\tau|^2) \exp\left[2(1 + |\tau|^2)^{1/2} r\right] \\
&\qquad \times |\tilde{u}(\tau, 0)|^2 \, dr \, d\tau \\
&= \int_{\mathbb{R}^{2n-1}} \int_{-\infty}^0 (1 + |\tau|^2)^{1/2} \exp[2(1 + |\tau|^2)^{1/2} r] |\tilde{u}(\tau, 0)|^2 \, dr \, d\tau \\
&= \frac{1}{2} \int_{\mathbb{R}^{2n-1}} |\tilde{u}(\tau, 0)|^2 \, d\tau \\
&= \frac{1}{2} \int_{\partial\Omega} |u|^2 \, d\sigma.
\end{aligned}
$$

This concludes the second case and the proof. ∎

7.8 Conclusion of the Proof of the Main Estimate

Now we pass from the Sobolev $-1/2$ norm to higher order norms.

LEMMA 7.8.1
Suppose that the basic estimate $E(\phi)^2 \lesssim Q(\phi, \phi), \phi \in D^{p,q}$, holds on a special boundary chart V. Let $\{\rho_k\}$ be a sequence of cutoff functions in $\Lambda^{0,0}_c(V \cap \bar\Omega)$ such that

$$\rho_k \equiv 1 \qquad on \quad supp \, \rho_{k+1}.$$

Then for each $k > 0$ we have an a priori estimate

$$\|\nabla \rho_k \phi\|^2_{(k-2)/2} \lesssim \|\rho_1 F \phi\|^2_{(k-2)/2} + \|F\phi\|^2_0.$$

PROOF We proceed by induction on k. For $k = 1$ we have

$$\|\nabla \rho_1 \phi\|^2_{-1/2} \overset{(7.7.4)}{\lesssim} E(\rho_1 \phi)^2$$

$$\overset{\text{(basic est.)}}{\lesssim} Q(\rho_1 \phi, \rho_1 \phi)$$

$$\overset{(7.7.3)}{=} \operatorname{Re} Q(\phi, \rho_1^2 \phi) + \mathcal{O}(\|\phi\|_0^2)$$

$$\overset{\text{(Friedrichs)}}{=} \operatorname{Re} \langle F\phi, \rho_1^2 \phi \rangle + \mathcal{O}(\|\phi\|_0^2)$$

$$\lesssim \|F\phi\|_0 \|\rho_1^2 \phi\|_0 + \mathcal{O}(\|\phi\|_0^2)$$

$$\lesssim \|F\phi\|_0 \|\phi\|_0 + \mathcal{O}(\|\phi\|_0^2)$$

$$\lesssim \|F\phi\|_0^2 + \mathcal{O}(\|\phi\|_0^2)$$

$$\lesssim \|F\phi\|_0^2$$

$$\lesssim \|\rho_1 F\phi\|^2_{-1/2} + \|F\phi\|_0^2.$$

In the last two lines but one we have used the fact that $T = F^{-1}$ is bounded on L^2. This is the first step of the induction.

Now take $k > 1$ and assume that we have proved the result for $k - 1$. Set $\Lambda = \Lambda_t^{(k-1)/2}$. Note that Λ commutes with $D^j, j = 1, \ldots, 2n - 1$, because Λ is a convolution operator in the tangential variables. If $j = 2n$ then it is even easier to see that D^j commutes with Λ since Λ does not act in the r variable. Thus

$$\|\nabla \rho_k \phi\|^2_{(k-2)/2} = \|\Lambda(\nabla \rho_k \phi)\|^2_{-1/2}$$

$$= \|\nabla \Lambda(\rho_k \phi)\|^2_{-1/2}$$

$$= \|\nabla \Lambda(\rho_1 \rho_k \phi)\|^2_{-1/2}. \tag{7.8.1.1}$$

Observe that

$$[\Lambda, [D^j, \rho_1]] \rho_k \rho_{k-1} \phi + [\Lambda, \rho_1][D^j, \rho_k] \rho_{k-1} \phi + [\Lambda, \rho_1] \rho_k D^j \rho_{k-1} \phi$$

$$= [\Lambda, [D^j, \rho_1]] \rho_k \rho_{k-1} \phi + [\Lambda, \rho_1] D^j \rho_k \rho_{k-1} \phi$$

$$\overset{\text{(Jacobi identity)}}{=} - [\rho_1, [\Lambda, D^j]] \rho_k \rho_{k-1} \phi - [D^j, [\rho_1, \Lambda]] \rho_k \rho_{k-1} \phi$$

$$+ [\Lambda, \rho_1] D^j \rho_k \rho_{k-1} \phi$$

$$\overset{(\Lambda, D^j \text{ commute})}{=} -0 - D^j [\rho_1, \Lambda] \rho_k \rho_{k-1} \phi$$

$$= D^j [\Lambda, \rho_1] \rho_k \phi.$$

As a result,

$$D^j \Lambda \rho_1 \rho_k \phi = D^j \rho_1 \Lambda \rho_k \phi + D^j [\Lambda, \rho_1] \rho_k \phi$$

$$= D^j \rho_1 \Lambda \rho_k \phi + [\Lambda, [D^j, \rho_1]] \rho_k \rho_{k-1} \phi + [\Lambda, \rho_1][D^j, \rho_k] \rho_{k-1} \phi$$

$$+ [\Lambda, \rho_1] \rho_k D^j \rho_{k-1} \phi.$$

Therefore, using (7.8.1.1),

$$\|\nabla \rho_k \phi\|^2_{(k-2)/2} = \|\nabla \Lambda \rho_1 \rho_k \phi\|^2_{-1/2}$$

$$\lesssim \|\nabla \rho_1 \Lambda \rho_k \phi\|^2_{-1/2} + \sum_j \| [\Lambda, [D^j, \rho_1]] \rho_k \rho_{k-1} \phi\|^2_{-1/2}$$

$$+ \sum_j \|[\Lambda, \rho_1][D^j, \rho_k] \rho_{k-1} \phi\|^2_{-1/2}$$

$$+ \sum_j \|[\Lambda, \rho_1] \rho_k D^j \rho_{k-1} \phi\|^2_{-1/2}$$

$$\lesssim \|\nabla \rho_1 \Lambda \rho_k \phi\|^2_{-1/2} + \|\rho_{k-1} \phi\|^2_{-1/2+(k-3)/2}$$

$$\|\rho_{k-1} \phi\|^2_{-1/2+(k-3)/2} + \|\nabla \rho_{k-1} \phi\|^2_{-1/2+(k-3)/2}$$

$$\lesssim \|\nabla \rho_1 \Lambda \rho_k \phi\|^2_{-1/2} + \|\nabla \rho_{k-1} \phi\|^2_{(k-3)/2}. \qquad (7.8.1.2)$$

Consider the term $\|\nabla \rho_1 \Lambda \rho_k \phi\|^2_{-1/2}$ on the right-hand side. Set

$$\rho_1 \Lambda \rho_k \equiv \rho_1 \Lambda_t^{(k-1)/2} \rho_k \equiv A.$$

Then

$$
\begin{aligned}
\|\nabla \rho_1 \Lambda \rho_k \phi\|^2_{-1/2} \quad &= \quad \|\nabla A \rho_{k-1} \phi\|^2_{-1/2} \\[4pt]
&\lesssim \quad Q(A\rho_{k-1}\phi, A\rho_{k-1}\phi) \\[4pt]
&= \quad \mathrm{Re}\, Q(\rho_{k-1}\phi, A'A\rho_{k-1}\phi) \\[4pt]
&\quad\; + \mathcal{O}\left(\|\nabla \rho_{k-1}\phi\|^2_{(k-3)/2}\right) \\[6pt]
&\overset{(\rho_{k-1}\underline{A'=A'})}{=} \mathrm{Re}\, Q(\phi, A'A\rho_{k-1}\phi) + \mathcal{O}\left(\|\nabla \rho_{k-1}\phi\|^2_{(k-3)/2}\right) \\[4pt]
&= \quad \mathrm{Re}\, \langle F\phi, A'A\rho_{k-1}\phi\rangle + \mathcal{O}(\|\nabla \rho_{k-1}\phi\|^2_{(k-3)/2} \\[4pt]
&= \quad \mathrm{Re}\, \langle AF\phi, A\rho_{k-1}\phi\rangle + \mathcal{O}\left(\|\nabla \rho_{k-1}\phi\|^2_{(k-3)/2}\right)
\end{aligned}
$$

$$\lesssim \quad \|A\rho_1 F\phi\|_{-1/2} \|A\phi\|_{1/2} + \mathcal{O}\big(\|\nabla\rho_{k-1}\phi\|^2_{(k-3)/2}\big)$$

$$= \quad \|\rho_1\Lambda_t^{(k-1)/2}\rho_k\rho_1 F\phi\|_{-1/2}\|\rho_1\Lambda_t^{(k-1)/2}\rho_k\phi\|_{1/2}$$
$$+ \mathcal{O}\big(\|\nabla\rho_{k-1}\phi\|^2_{(k-3)/2}\big)$$

$$\lesssim \quad \|\rho_1 F\phi\|_{(k-1)/2-1/2}\|\rho_k\phi\|_{(k-1)/2+1/2}$$
$$+ \mathcal{O}\big(\|\nabla\rho_{k-1}\phi\|^2_{(k-3)/2}\big)$$

$$\lesssim \quad \frac{1}{\epsilon}\|\rho_1 F\phi\|^2_{(k-2)/2} + \epsilon\|\nabla\rho_k\phi\|^2_{(k-2)/2}$$
$$+ \|\nabla\rho_{k-1}\phi\|^2_{(k-3)/2}.$$

Substituting this last inequality into (7.8.1.2) we obtain

$$\|\nabla\rho_k\phi\|^2_{(k-2)/2} \le \|\nabla\rho_1\Lambda\rho_k\phi\|^2_{-1/2} + \|\nabla\rho_{k-1}\phi\|^2_{(k-3)/2}$$

$$\lesssim \frac{1}{\epsilon}\|\rho_1 F\phi\|^2_{(k-2)/2} + \epsilon\|\nabla\rho_k\phi\|^2_{(k-2)/2} + \|\nabla\rho_{k-1}\phi\|_{(k-3)/2}.$$

Therefore

$$\|\nabla\rho_k\phi\|^2_{(k-2)/2} \quad \lesssim \quad \frac{1}{\epsilon}\|\rho_1 F\phi\|^2_{(k-2)/2} + \|\nabla\rho_{k-1}\phi\|^2_{(k-3)/2}$$

$$\overset{\text{(induction)}}{\lesssim} \frac{1}{\epsilon}\|\rho_1 F\phi\|^2_{(k-2)/2} + \|\rho_1 F\phi\|^2_{(k-3)/2} + \|F\phi\|^2_0$$

$$\lesssim \quad \|\rho_1 F\phi\|^2_{(k-2)/2} + \|F\phi\|^2_0.$$

This completes the proof. ∎

THEOREM 7.8.2
Assume that the basic estimate holds in $\mathcal{D}^{p,q}$. Let V be a special boundary neighborhood on which $\|\nabla\phi\|^2_{-1/2} \le E(\phi)^2$. Let $U \subset\subset V$ and choose a cutoff function $\rho_1 \in \bigwedge_c^{0,0}(V \cap \bar\Omega)$ such that $\rho_1 \equiv 1$ on U.
  Then for each $\rho \in \bigwedge_c^{0,0}(U \cap \bar\Omega)$ and each nonnegative integer s it holds that

$$\|\rho\phi\|^2_{s+1} \lesssim \|\rho_1 F\phi\|^2_s + \|F\phi\|^2_0$$

for all $\phi \in \operatorname{dom}(F) \cap \mathcal{D}^{p,q}$.

PROOF We proceed by induction on s. We apply the previous lemma with $k = 2, \rho_2 = \rho$, and $0 = \rho_3 = \rho_4 = \cdots$. It tells us that

$$\|\nabla \rho_2 \phi\|_0^2 \leq \|\rho_1 F \phi\|_0^2 + \|F \phi\|_0^2 \,,$$

that is,

$$\|\nabla \rho \phi\|_0^2 \leq \|\rho_1 F \phi\|_0^2 + \|F \phi\|_0^2 \,.$$

Therefore

$$\|\rho \phi\|_1^2 \approx \|\nabla \rho \phi\|_0^2 + \|\rho \phi\|_0^2$$

$$\lesssim \|\rho_1 F \phi\|_0^2 + \|F \phi\|_0^2 \,,$$

which is the statement that we wish to prove for $s = 0$.

Now suppose the statement to be true for s, some $s \geq 0$. Then

$$\|\rho \phi\|_{s+1}^2 \approx \sum_{|\alpha| \leq s+1} \|D^\alpha (\rho \phi)\|_0^2$$

$$= \sum_{|\alpha| = s+1} \|D^\alpha (\rho \phi)\|_0^2 + \|\rho \phi\|_s^2$$

$$\lesssim \sum_{|\alpha| = s+1} \|D^\alpha (\rho \phi)\|_0^2 + \|\rho_1 F \phi\|_{s-1}^2 + \|F \phi\|_0^2$$

$$\leq \sum_{|\alpha| = s+1} \|D^\alpha (\rho \phi)\|_0^2 + \|\rho_1 F \phi\|_s^2 + \|F \phi\|_0^2.$$

It remains to estimate the top order term. Pick a sequence of cutoff functions $\rho_1 \geq \rho_2 \geq \cdots \geq \rho_{2s+2} = \rho$ such that $\mathrm{supp}\, \rho_j \subset\subset \mathrm{supp}\, \rho_{j-1}, j = 2, \ldots, 2s + 2$ and set $\rho_j \equiv 0$ for $j > 2s + 2$. We apply 7.8.1 with $k = 2s + 2$ to obtain

$$\|\nabla \rho_{2s+2} \phi\|_s^2 \lesssim \|\rho_1 F \phi\|_s^2 + \|F \phi\|_0^2.$$

If $|\beta| = s + 1$ then

$$\|D_t^\beta \rho_{2s+2} \phi\|_0^2 \lesssim \|\nabla \rho_{2s+2} \phi\|_s^2$$

$$\lesssim \|\rho_1 F \phi\|_s^2 + \|F \phi\|_0^2$$

$$\lesssim \|\rho_1 F \phi\|_s^2 + \|F \phi\|_0^2.$$

For $|\beta| = s$ we have

$$\|D_t^\beta D_r \rho \phi\|_0^2 \lesssim \|\nabla \rho \phi\|_s^2$$

$$\lesssim \|\rho_1 F \phi\|_s^2 + \|F \phi\|_0^2.$$

Thus it remains to estimate $\|D_t^\beta D_r^m \rho \phi\|_0^2$ for $m \geq 2$ and $|\beta| + m = s + 1$.

We proceed by induction on m (this is a second induction within the first induction on s). We use the differential equation as follows: We know that $\square + I = F$. On the special boundary chart the function $F\phi$ can be expressed as

$$F\phi = A_{2n} D_r^2 + \left[\sum_{j=1}^{2n-1} A_j D_t^j D_r \phi + \sum_{j,k=1}^{2n-1} A_{j,k} D_t^j D_t^k \phi \right].$$

The operator F is strongly elliptic so that A_{2n} is an invertible matrix. Therefore we can express the second derivative $D_r^2 \phi$ in terms of the remaining expressions on the right side of the last equation (each of which involves at most *one* normal derivative) and in terms of $F\phi$ itself. This means that we may estimate expressions of the form $\|D_t^\beta D_r^2 \rho \phi\|_0^2$ in terms of $\|D_t^\beta D_r \rho \phi\|_0^2$ and $\|D_t^\beta \rho \phi\|_0^2$, both of which have already been estimated. Thus we have handled the case $m = 2$.

Inductively, we can handle any term of the form $\|D_t^\beta D_r^m \rho \phi\|_0^2$.

This concludes the induction on m, which in turn concludes the induction on s. The proof is therefore complete.   ∎

The final step of the last theorem is decisive. While the $\bar{\partial}$-Neumann boundary conditions are degenerate, the operator $F = \square + I$ is strongly elliptic—as nice an operator as you could want. One of the special features of such an operator L is that (as we learned in Chapter 4 by way of pseudodifferential operators) *any* second derivative of a function f can be controlled by Lf, modulo error terms.

Now recall the main estimate:

THEOREM 7.8.3   MAIN ESTIMATE
Let $\Omega \subseteq \mathbb{C}^n$ be a smoothly bounded domain. Assume that the basic estimate holds for elements of $\mathcal{D}^{p,q}$. For $\alpha \in H_0^{p,q}$, we let ϕ denote the unique solution to the equation $F\phi = \alpha$. Then we have:

1. *If W is a relatively open subset of $\bar{\Omega}$ and if $\alpha|_W \in \wedge^{p,q}(W)$, then $\phi|_W \in \wedge^{p,q}(W)$.*

2. *Let ρ, ρ_1 be smooth functions with supp $\rho \subseteq$ supp $\rho_1 \subseteq W$ and $\rho_1 \equiv 1$ on supp ρ. Then:*

(a)   If $W \cap \partial\Omega = \emptyset$ then $\forall s \geq 0$ there is a constant $c_s > 0$ (depending on ρ, ρ_1 but independent of α) such that

$$\|\rho\phi\|_{s+2}^2 \leq c_s \left(\|\rho_1\alpha\|_s^2 + \|\alpha\|_0^2 \right). \tag{7.8.3.2a}$$

(b)   If $W \cap \partial\Omega \neq \emptyset$ then $\forall s \geq 0$ there exists a constant $c_s \geq 0$ such that

$$\|\rho\phi\|_{s+1}^2 \leq c_s \left(\|\rho_1\alpha\|_s^2 + \|\alpha\|_0^2 \right). \tag{7.8.3.2b}$$

REMARK   In the theorem that we just proved, we established the *a priori* estimate

$$\|\rho\phi\|_{s+1}^2 \lesssim \|\rho_1 F\phi\|_s^2 + \|F\phi\|_0^2.$$

That is, we know that this inequality holds for a testing function ϕ. In the Main Estimate we are now addressing the problem of *existence*: given a $v \in H_0^{p,q}$ we want to know that there exists a ϕ which is smooth and satisfies (7.8.3.2a) and (7.8.3.2b). We *do* know that the ϕ given by Friedrichs is in L^2, but nothing further. It is in these arguments that the elliptic regularization technique comes to the rescue. In the original papers [KOH1] Kohn had to use delicate functional analysis techniques to address the existence issue. ∎

PROOF OF THE MAIN ESTIMATE   Let $\alpha \in H_0^{p,q}$. Let $\phi \in \tilde{\mathcal{D}}^{p,q}$ be the unique solution to the equation $F\phi = \alpha$. This will of course be the ϕ that we seek, but we must see that it is smooth. Observe that this assertion, namely part (1) of the theorem, follows from part (2).

Let U be a subregion of $\mathbb{C}^n$. Assume that

$$\alpha|_{U \cap \Omega} \in \bigwedge^{p,q} (U \cap \bar{\Omega}).$$

We need to see that

$$\phi|_{U \cap \Omega} \in \bigwedge^{p,q} (U \cap \bar{\Omega}).$$

The easy case is if $U \cap \partial\Omega = \emptyset$. For then $\phi|_U \in \bigwedge^{p,q}(U)$ by the interior elliptic regularity for $F = \Box + I$. This gives us the estimate (7.8.3.2a) as well. Refer to Chapter 4 for details.

More interesting is the case $U \cap \partial\Omega \neq \emptyset$. Then the last theorem tells us that

$$\|\rho\phi\|_{s+1}^2 \lesssim \|\rho\alpha\|_s^2 + \|\alpha\|_0^2$$

provided that we know in advance that ϕ is smooth on $U \cap \bar{\Omega}$. Again we emphasize the importance of this subtlety: we know that ϕ exists; we also know that *if ϕ is known to be smooth on $U \cap \bar{\Omega}$ then it satisfies the desired regularity estimates*. We need to pass from this *a priori* information to a general regularity result. To this task we now turn.

Let $0 < \delta \leq 1$ and let ϕ^δ be the solution of $F^\delta \phi = \alpha$. We know that F^δ is hypoelliptic up to the boundary (see 7.6.4). Recall that

$$Q^\delta(\phi, \psi) = Q(\phi, \psi) + \delta \sum_{j=1}^{2n} \langle D^j \phi, D^j \psi \rangle$$

and that

$$Q^\delta(\phi, \phi) \geq Q(\phi, \phi) \geq E(\phi)^2.$$

The proof of the last theorem then applies to F^δ and ϕ^δ; thus we have

$$\|\rho\phi^\delta\|_{s+1}^2 \lesssim \|\rho_1 \alpha\|_s^2 + \|\alpha\|_0^2.$$

The constant in the inequality is independent of δ since the estimate depends only on the majorization

$$|\nabla \rho\phi^\delta|_{-1/2}^2 \leq Q(\rho\phi^\delta, \rho\phi^\delta) \approx Q^\delta(\rho\phi^\delta, \rho\phi^\delta) - \delta\|\rho\phi^\delta\|_1^2.$$

(In fact, we shall provide the details of this important assertion in the appendix to this chapter.) Thus $\{\rho\phi^\delta\}_{0 \leq \delta \leq 1}$ is uniformly bounded in $\| \cdot \|_{s+1}$ for every s. Fix an s. By Rellich's lemma, there is a subsequence $\{\rho\phi^{\delta_n}\}$ that converges in $H_s^{p,q}$ as $n \to \infty$. By diagonalization, we may assume that $\{\rho\phi^{\delta_n}\}$ converges in $H_s^{p,q}$ for *every* s—to the same function $\rho\mu$. We wish to show that $\rho\mu = \rho\phi$, where ϕ is the function whose existence comes from the Fredholm theory. Then we will know, by the Sobolev theorem, that $\rho\phi$ is smooth and we will be done.

It suffices to show that $\phi^{\delta_n} \to \phi$ in the $H_0^{p,q}$ topology (for, of course, $\phi^{\delta_n} \to \mu$ in that topology). We know that the interior estimates hold uniformly in δ:

$$\|\rho\phi^\delta\|_2 \lesssim \|\alpha\|_0$$

for any ρ with compact support in Ω. By 7.6.3, $\phi^\delta \in H_1^{p,q}$. If α is globally smooth, we can apply a partitition of unity argument and the boundary estimate for $s = 0$ (see 7.8.2) to see that

$$\|\phi^\delta\|_1 \lesssim \|\alpha\|_0, \tag{7.8.3.3}$$

uniformly in δ as δ as $\delta \to 0$. By the density of smooth forms in the space $H_0^{p,q}$, we conclude that (7.8.3.3) holds for all elements $\alpha \in H_0^{p,q}$.

Next we calculate, for $\psi \in \mathcal{D}^{p,q}$, that

$$Q(\phi, \psi) = \langle \alpha, \psi \rangle = Q^\delta(\phi^\delta, \psi)$$

$$= Q(\phi^\delta, \psi) + \mathcal{O}(\delta\|\phi^\delta\|_1 \|\psi\|_1)$$

$$= Q(\phi^\delta, \psi) + \|\alpha\|_0 \|\psi\|_1 \cdot \mathcal{O}(\delta). \tag{7.8.3.4}$$

Give the equation

$$\langle \alpha, \psi \rangle = Q(\phi^\delta, \psi) + \|\alpha\|_0 \|\psi\|_1 \cdot \mathcal{O}(\delta)$$

the name $R(\delta)$. By subtracting $R(\delta')$ from $R(\delta)$ we find that

$$Q(\phi^\delta - \phi^{\delta'}, \psi) = \|\alpha\|_0 \|\psi\|_1 \cdot \mathcal{O}(\delta + \delta').$$

By 7.6.3 we can find a sequence $\{\psi_n\} \subseteq \mathcal{D}^{p,q}$ converging with respect to both the norms Q and $\| \cdot \|_1$ to $\phi^\delta - \phi^{\delta'}$. The result is that

$$Q(\phi^\delta - \phi^{\delta'}, \phi^\delta - \phi^{\delta'}) = \mathcal{O}(\delta + \delta')\|\alpha\|_0 \|\phi^\delta - \phi^{\delta'}\|_1$$

$$= \mathcal{O}(\delta + \delta') \cdot \|\alpha\|_0^2$$

$$\to 0$$

as $\delta, \delta' \to 0$.

We conclude that ϕ^δ converges in the topology of $\tilde{\mathcal{D}}^{p,q}$ as $\delta \to 0$ and (7.8.3.4) shows that the limit is ϕ. Consequently, $\phi^\delta \to \phi$ in $H_0^{p,q}$ and we are done. ∎

In the next section we shall develop the Main Estimate into some useful results about the original $\bar{\partial}$-Neumann problem.

7.9   The Solution of the $\bar{\partial}$-Neumann Problem

Our ultimate goal is to understand existence and regularity for the $\bar{\partial}$-Neumann problem. We begin with some remarks about the operator F. We assume throughout this section that the basic estimate holds on Ω.

PROPOSITION 7.9.1
Let $\alpha, \phi, U, \rho_1, \rho$ be as in the Main Estimate. Let $\alpha|_U \in H_s^{p,q}(U \cap \bar{\Omega})$. Then $\rho\phi \in H_{s+k}^{p,q}(U \cap \bar{\Omega})$, where k is either 1 or 2 according to whether $U \cap \partial\Omega \neq \emptyset$ or $U \cap \partial\Omega = \emptyset$. Also $\|\rho\phi\|_{s+k}^2 \lesssim \|\rho_1\alpha\|_s^2 + \|\alpha\|_0^2$.

PROOF Let ρ_0 be a smooth function with support in U such that $\rho_0 \equiv 1$ on $\text{supp}\,\rho_1$. See Figure 7.1. Let $\{\beta_n\}, \{\gamma_n\}$ be sequences of smooth forms such that

$$\text{supp}\,\beta_n \subseteq \text{supp}\,\rho_0,$$

$$\text{supp}\,\gamma_n \subseteq \text{supp}\,(1 - \rho_0),$$

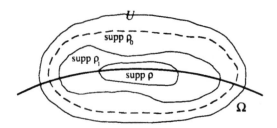

FIGURE 7.1

and

$$\beta_n \to \rho_0 \alpha \quad \text{in } H_s^{p,q},$$

$$\gamma_n \to (1 - \rho_0)\alpha \quad \text{in } H_0^{p,q}.$$

Then

$$\alpha_n = \beta_n + \gamma_n \to \alpha \quad \text{in } H_0^{p,q}$$

and

$$\rho_1 \alpha_n \to \rho_1 \alpha \quad \text{in } H_s^{p,q}.$$

Let $\phi_n \equiv F^{-1}\alpha_n$. Now F^{-1} is bounded in the Sobolev topology so that $\phi_n \to F^{-1}\alpha = \phi$ in $H_0^{p,q}$. Now we apply the Main Estimate to obtain

$$\|\rho(\phi_n - \phi_m)\|_{s+k} \lesssim \|\rho_1(\alpha_n - \alpha_m)\|_s + \|\alpha_n - \alpha_m\|_0.$$

Therefore

$$\lim_{n \to \infty} \rho\phi_n \in H_{s+k}^{p,q},$$

that is, $\rho\phi \in H_{s+k}^{p,q}$. The desired estimate therefore holds. ∎

PROPOSITION 7.9.2
If $F\phi = \alpha$ and $\alpha \in \bigwedge^{p,q}(\bar{\Omega})$ then $\phi \in \bigwedge^{p,q}(\bar{\Omega})$ and $\|\phi\|_{s+1}^2 \lesssim \|\alpha\|_s^2$ for every s.

PROOF This follows at once from the Main Estimate by taking $U \supseteq \bar{\Omega}$ and noticing that $\|\alpha\|_0 \le \|\alpha\|_s$. ∎

COROLLARY 7.9.3
If $F\phi = \alpha$ and $\alpha \in H_s^{p,q}$ then $\phi \in H_{s+1}^{p,q}$ and $\|\phi\|_{s+1} \lesssim \|\alpha\|_s$.

PROOF   Immediate.   ∎

COROLLARY 7.9.4
The operator F^{-1} is a compact operator on $H_s^{p,q}$.

PROOF   By the corollary we know that F^{-1} is bounded from $H_s^{p,q}$ to $H_{s+1}^{p,q}$ so we apply Rellich's lemma to obtain the result.   ∎

COROLLARY 7.9.5
The operator F has discrete spectrum with no finite limit point and each eigenvalue occurs with finite multiplicity.

PROOF   By the theory of compact operators (see, for instance, [WID]), we know that F^{-1} has countable, compact spectrum with 0 as its only possible limit point. Also each eigenvalue has finite multiplicity. Since λ is an eigenvalue for F if and only if λ^{-1} is an eigenvalue for F^{-1} we have proved the corollary.   ∎

PROPOSITION 7.9.6
Let U, ρ, ρ_1, α, and k be as in Proposition 7.9.1. Suppose that $\rho_1 \alpha \in H_s^{p,q}$ for some integer $s > 0$. If ϕ satisfies $(F - \lambda)\phi = \alpha$ for some constant λ then $\rho\phi \in H_{s+k}^{p,q}(\Omega)$.

PROOF   Consider the case $k = 1$. Set $\alpha' = \alpha + \lambda\phi$. Then ϕ satisfies the equation $F\phi = \alpha'$. Now $\alpha' \in H_0^{p,q}$ so that $\rho_1\phi \in H_1^{p,q}$ by Proposition 7.9.1. Let $\{\rho_k\}_{k=2}^s$ be a sequence of smooth functions with $\rho_s = \rho$ and $\rho_j \equiv 1$ on supp ρ_{j+1}. Inductively we may reason that

$$\rho_1\alpha' \in H_1^{p,q} \Rightarrow \rho_2\phi \in H_2^{p,q} \Rightarrow \rho_2\alpha' \in H_2^{p,q} \ldots.$$

The result follows.
    The case $k = 2$ is similar.   ∎

COROLLARY 7.9.7
The operator $F - \lambda I$ is hypoelliptic.

PROOF   Immediate from the proposition and Sobolev's theorem.   ∎

COROLLARY 7.9.8
The eigenforms of F are all smooth.

PROOF Obvious.   ▮

PROPOSITION 7.9.9
The space $H_0^{p,q}$ has a complete orthonormal basis of eigenforms for the operator $\Box_F$ that are smooth up to the boundary of Ω. The eigenvalues are nonnegative, with no finite accumulation point, and occur with finite multiplicity. Moreover, for each s,

$$\|\phi\|_{s+1}^2 \lesssim \|\Box\phi\|_s^2 + \|\phi\|$$

for all $\phi \in dom\,(F) \cap \bigwedge^{p,q}(\bar{\Omega})$.

PROOF Recall that $\Box_F = F - I$ is the restriction of $\Box$ to the domain of F. We know that $H_0^{p,q}$ has a complete orthonormal basis of eigenforms for F^{-1} (just because it is a compact operator on a Hilbert space). Then the same holds for F and thus for $F - I$. We also have that the eigenvalues are nonnegative, with no finite accumulation point and with finite multiplicity.

The desired estimates follow by induction on s and by the global regularity statement for F:

$$\|\phi\|_1^2 \lesssim \|F\phi\|_0^2$$

$$\lesssim \|\Box\phi\|_0^2 + \|\phi\|_0^2;$$

$$\|\phi\|_2^2 \lesssim \|F\phi\|_1^2$$

$$\lesssim \|\Box\phi\|_1^2 + \|\phi\|_1^2$$

$$\lesssim \|\Box\phi\|_1^2 + \|\Box\phi\|_0^2 + \|\phi\|_0^2$$

$$\lesssim \|\Box\phi\|_1^2 + \|\phi\|_0^2;$$

and so forth.   ▮

PROPOSITION 7.9.10
The space $H_0^{p,q}$ admits the strong orthogonal decomposition

$$H_0^{p,q} = range\,(\Box_F) \oplus kernel\,(\Box_F)$$

$$= \bar{\partial}\vartheta dom\,(F) \oplus \vartheta\bar{\partial} dom\,(F) \oplus kernel\,(\Box_F).$$

PROOF First of all we need to show that range $(\Box_F)$ is closed. Set $\mathcal{H}^{p,q} \equiv$ kernel $(\Box_F)$. Then $\mathcal{H}^{p,q}$ is the eigenspace corresponding to the eigenvalue 0. The orthogonal complement of $\mathcal{H}^{p,q}$ is $\bigoplus_{|\lambda|>0} V_\lambda$, where V_λ is the eigenspace corresponding to the eigenvalue λ. Then $\Box_F$ is bounded away from 0 on $(\mathcal{H}^{p,q})^\perp$ and it is one-to-one on this space. Thus $\Box_F$ restricted to the closure of the range of $\Box_F$ has a continuous inverse which we call L.

Let $\Box_F x_n \to y$. Then $L\Box_F x_n \to Ly$, that is, $x_n \to Ly$ and $\Box_F(Ly) = y$. Thus $y \in$ range $\Box_F$ and range $\Box_F$ is closed. Since range $\Box_F = (\mathcal{H}^{p,q})^\perp$, the first equality follows.

For the second equality, notice that $\bar\partial^2 = 0$ hence range $\bar\partial \perp$ range $\bar\partial^*$. Also $\bar\partial^* = \vartheta|_{\text{dom } \bar\partial}$. and the second equality follows as well.   ∎

COROLLARY 7.9.11
The range of $\bar\partial$ on dom $(\bar\partial) \cap H_0^{p,q-1}$ is closed.

PROOF Since range $\bar\partial \perp$ kernel $(\bar\partial^*)$ and $\bar\partial^*(\vartheta\bar\partial \text{ dom } (F) \oplus \mathcal{H}^{p,q}) = 0$, we may conclude that range $\bar\partial = \bar\partial\vartheta \text{ dom } (F)$.   ∎

We are engaged in setting up a Hodge theory for the $\bar\partial$ operator. For analogous material in the classical setting of the exterior differential operator d we refer the reader to [CON].

Now we define the harmonic projector H to be the orthogonal projection from $H_0^{p,q}$ onto $\mathcal{H}^{p,q}$. We use that operator in turn to define the $\bar\partial$-Neumann operator.

DEFINITION 7.9.12 *The Neumann operator $N : H_0^{p,q} \to \text{dom}\,(F)$ is defined by*

$$N\alpha = 0 \qquad \text{if } \alpha \in \mathcal{H}^{p,q}$$

$$N\alpha = \phi \qquad \text{if } \alpha \in \text{range } \Box_F \text{ and } \phi \text{ is the unique}$$
$$\text{solution of } \Box_F\phi = \alpha \text{ with } \phi \perp \mathcal{H}^{p,q}.$$

Then we extend N to all of $H_0^{p,q}$ by linearity.

Notice that $N\alpha$ is the unique solution ϕ to the equations

$$H\phi = 0$$
$$\Box_F\phi = \alpha - H\alpha.$$

Finally, we obtain the solution to the $\bar\partial$-Neumann problem:

THEOREM 7.9.13

1. *The operator N is compact.*
2. *For any $\alpha \in H_0^{p,q}$ we have $\alpha = \bar{\partial}\vartheta N\alpha + \vartheta\bar{\partial}N\alpha + H\alpha$.*
3. *$NH = HN = 0, N\square = \square N = I - H$ on $\operatorname{dom} F$, $N\bar{\partial} = \bar{\partial}N$ on $\operatorname{dom}\bar{\partial}$, and $N\vartheta = \vartheta N$ on $\operatorname{dom}\bar{\partial}^{*}$.*
4. *$N(\bigwedge^{p,q}(\bar{\Omega})) \subseteq \bigwedge^{p,q}(\bar{\Omega})$ and for each s the inequality*

$$\|N\alpha\|_{s+1} \lesssim \|\alpha\|_s$$

holds for all $\alpha \in \bigwedge^{p,q}(\bar{\Omega})$.

PROOF Statement (2) is part of Proposition 7.9.10. It is also immediate from the definitions that $NH = HN = 0$. Next, $N\square = \square N = I - H$ follows from part (2).

If $\alpha \in \operatorname{dom}\bar{\partial}$ then, since $\bar{\partial}^2 = 0$ and $\bar{\partial}H = 0$, we have

$$
\begin{aligned}
N\bar{\partial}\alpha &= N\bar{\partial}(\vartheta\bar{\partial}N\alpha + \bar{\partial}\vartheta N\alpha + H\alpha) \\
&= N\bar{\partial}(\vartheta\bar{\partial}N\alpha) \\
&= N(\bar{\partial}\vartheta + \vartheta\bar{\partial})\bar{\partial}N\alpha \\
&= N\square\bar{\partial}N\alpha \\
&= (I - H)\bar{\partial}N\alpha \\
&= \bar{\partial}N\alpha.
\end{aligned}
$$

The same reasoning applies to see that $\vartheta N = N\vartheta$.

The first statement of part (4) follows because $H\alpha$ is smooth whenever α is (by part (2)) and $\phi = N(\alpha - H\alpha)$ satisfies $\square_F\phi = \alpha - N\alpha$. Hence this ϕ is smooth. Furthermore, 7.9.9 implies that

$$
\begin{aligned}
\|N\alpha\|_{s+1}^2 &\lesssim \|\square N\alpha\|_s^2 + \|N\alpha\|_0^2 \\
&\leq \|\alpha\|_s^2 + \|H\alpha\|_s^2 + \|N\alpha\|_0^2 \\
&\lesssim \|\alpha\|_s^2 + \|H\alpha\|_0^2 + \|N\alpha\|_0^2 \\
&\lesssim \|\alpha\|_s^2.
\end{aligned}
$$

(We use here the fact that all norms on the finite-dimensional space $\mathcal{H}^{p,q}$ are equivalent.) That proves the second statement of part (4).

Finally, (1) follows from (4) and Rellich's lemma. ∎

Next we want to solve the inhomogeneous Cauchy–Riemann equation $\bar{\partial}\phi = \alpha$. Notice that there is no hope to solve this equation unless $\alpha \perp$ kernel $(\bar{\partial}^*)$ or equivalently $\bar{\partial}\alpha = 0$ and $H\alpha = 0$.

THEOREM 7.9.14
Suppose that $q \geq 1$, $\alpha \in H_0^{p,q}$, $\bar{\partial}\alpha = 0$, and $H\alpha = 0$. Then there exists a unique $\phi \in H_0^{p,q-1}$ such that $\phi \perp$ kernel $(\bar{\partial})$ and $\bar{\partial}\phi = \alpha$. If $\alpha \in \bigwedge^{p,q}(\bar{\Omega})$ then $\phi \in \bigwedge^{p,q-1}(\bar{\Omega})$ and

$$\|\phi\|_s \lesssim \|\alpha\|_s.$$

PROOF By the conditions on α we have that $\alpha = \bar{\partial}\vartheta N\alpha$. Thus we take $\phi = \vartheta N\alpha$ and $\phi \perp$ kernel $(\bar{\partial})$ implies uniqueness. By part (4) of the last theorem, we know that $N\alpha \in \bigwedge^{p,q}$ if $\alpha \in \bigwedge^{p,q}$. Hence $\phi \in \bigwedge^{p,q-1}$ and

$$\|\phi\|_s \equiv \|\vartheta N\alpha\|_s$$
$$\lesssim \|N\alpha\|_{s+1}$$
$$\leq \|\alpha\|_s. \qquad \blacksquare$$

It is in fact the case that, on a domain in Euclidean space, the harmonic space $\mathcal{H}^{p,q}$ is zero dimensional. Thus the condition $H\alpha = 0$ is vacuous. There is no known elementary way to see this assertion. It follows from the Kodaira vanishing theorem (see [WEL]), or from solving the $\bar{\partial}$-Neumann problem with certain weights. A third way to see the assertion appears in [FOK]. We shall say no more about it here.

In fact, it is possible to prove a stronger result than 7.9.14: if α has H^s coefficients, then ϕ has $H^{s+1/2}$ coefficients. This gain of order $1/2$ is sharp for Ω strongly pseudoconvex. We refer the reader to [FOK, p. 53] for details.

Appendix to Section 7.8: Uniform estimates for F^δ and ϕ^δ

Refer to Section 7.8—especially the proof of the Main Estimate—for terminology. The purpose of this appendix is to prove that the estimate

$$\|\rho\phi^\delta\|_{s+1} \lesssim \|\rho_1 F^\delta \phi^\delta\|_s + \|F^\delta \phi^\delta\|_0 \qquad (7.A.1)$$

holds with a constant that is independent of δ. We begin as follows:

PROPOSITION 7.A.2 *We have*

$$Q^\delta(A\phi, A\phi) - \mathrm{Re}\, Q^\delta(\phi, A'A\phi) = \mathcal{O}(\|\nabla\phi\|_{k-1}^2),$$

where the constant in $\mathcal{O}$ is independent of δ.

PROOF The proof of 7.7.2 goes through, with $\bar{\partial}$ replaced by any first-order differential operator $\tilde{D}$ with constant coefficients, without any change. Thus

$$\langle \tilde{D} A\phi, \tilde{D} A\phi \rangle - \mathrm{Re}\,\langle \tilde{D}\phi, \tilde{D} A' A\phi \rangle = \mathcal{O}(\|\nabla\phi\|_{k-1}^2).$$

Then

$$Q^\delta(A\phi, A\phi) - \mathrm{Re}\,Q^\delta(\phi, A'A\phi) = Q(A\phi, A\phi) + \delta\sum_j \langle D^j A\phi, D^j A\phi \rangle$$

$$- \mathrm{Re}\left[Q(\phi, A'A\phi) + \delta\sum_j \langle D^j \phi, D^j A' A\phi \rangle \right]$$

$$= Q(A\phi, A\phi) - \mathrm{Re}\,Q(\phi, A'A\phi)$$

$$+ \delta\sum_j \left[\langle D^j A\phi, D^j A\phi \rangle - \langle D^j \phi, D^j A' A\phi \rangle \right]$$

$$= (1+\delta)\mathcal{O}\left(\|\nabla\phi\|_{k-1}^2\right)$$

$$= \mathcal{O}\left(\|\nabla\phi\|_{k-1}^2\right),$$

where the constant in $\mathcal{O}$ is independent of δ. ∎

We need one more preliminary result:

PROPOSITION 7.A.3 *Let hypotheses be as in 7.8.1 with $\phi \in \tilde{\mathcal{D}}_\delta^{p,q}$. Then*

$$\|\nabla\rho_k \phi^\delta\|_{(k-2)/2} \lesssim \|\rho_1 F^\delta \phi^\delta\|_{(k-2)/2}^2 + \|F^\delta \phi^\delta\|_0^2, \qquad (7.A.3.1)$$

where the constants are independent of δ.

PROOF We follow the proof of 7.8.1 closely, checking that all constants that arise are independent of δ. We induct on k. First let $k=1$. Then

$$\|\nabla\rho_1 \phi^\delta\|_{-1/2}^2 \overset{(7.7.4)}{\lesssim} E(\rho_1 \phi^\delta)$$

$$\overset{\text{(basic estimate)}}{\leq} Q(\rho_1 \phi^\delta, \rho_1 \phi^\delta)$$

$$\leq Q^\delta(\rho_1 \phi^\delta, \rho_1 \phi^\delta)$$

$$\overset{(7.A.2)}{=} \mathrm{Re}\,Q^\delta(\phi^\delta, \rho_1^2 \phi^\delta) + \mathcal{O}\left(\|\phi^\delta\|_0^2\right)$$

$$\overset{\text{(Friedrichs)}}{=} \mathrm{Re}\,\left(F^\delta \phi^\delta, \rho_1^2 \phi^\delta\right) + \mathcal{O}\left(\|\phi^\delta\|_0^2\right)$$

$$\leq \|F^\delta \phi^\delta\|_0 \|\phi^\delta\|_0 + \mathcal{O}\left(\|\phi^\delta\|_0^2\right).$$

Notice that $T^\delta \equiv (F^\delta)^{-1}$ is bounded on L^2 uniformly in δ. Indeed, by the

Friedrichs theorem we have

$$\|T^\delta \alpha\|_0^2 \le Q^\delta(T^\delta \alpha, T^\delta \alpha)$$
$$= \langle \alpha, T^\delta \alpha \rangle$$
$$\le \|\alpha\|_0 \|T^\delta \alpha\|_0.$$

It follows that $\|T^\delta\|_{\text{op}} \le 1$. Therefore

$$\|F^\delta \phi^\delta\|_0 \|\phi^\delta\|_0 + \mathcal{O}\left(\|\phi^\delta\|_0^2\right) \le \|F^\delta \phi^\delta\|_0^2 + \mathcal{O}\left(\|F^\delta \phi^\delta\|_0^2\right)$$
$$\lesssim \|F^\delta \phi^\delta\|_0^2$$
$$\lesssim \|\rho_1 F^\delta \phi^\delta\|_{-1/2}^2 + \|F^\delta \phi^\delta\|_0^2.$$

This proves the case $k = 1$.

Next, as in the proof of 7.8.1, we have for the tangential Bessel potential $\Lambda \equiv \Lambda_t^{(k-1)/2}$ that

$$\|\nabla \rho_k \phi^\delta\|_{(k-2)/2} \le \|\nabla \rho_1 \Lambda \rho_k \phi^\delta\|_{-1/2}^2 + \|\nabla \rho_{k-1} \phi^\delta\|_{(k-3)/2}^2. \qquad (7.A.3.2)$$

Setting $\rho_1 \Lambda \rho_k \equiv A$ we now calculate that

$$\|\nabla A \rho_{k-1} \phi^\delta\|_{-1/2}^2 \le Q^\delta(A\rho_{k-1}\phi^\delta, A\rho_{k-1}\phi^\delta)$$
$$= \text{Re}\, Q^\delta(\rho_{k-1}\phi^\delta, A'A\rho_{k-1}\phi^\delta) + \mathcal{O}\left(\|\nabla \rho_{k-1}\phi^\delta\|_{(k-3)/2}^2\right)$$
$$= \text{Re}\, Q^\delta(\phi^\delta, A'A\rho_{k-1}\phi^\delta) + \mathcal{O}\left(\|\nabla \rho_{k-1}\phi^\delta\|_{(k-3)/2}^2\right)$$
$$= \text{Re}\, \langle F^\delta \phi^\delta, A'A\rho_{k-1}\phi^\delta \rangle + \mathcal{O}\left(\|\nabla \rho_{k-1}\phi^\delta\|_{(k-3)/2}^2\right)$$
$$= \text{Re}\, \langle AF^\delta \phi^\delta, A\rho_{k-1}\phi^\delta \rangle + \mathcal{O}\left(\|\nabla \rho_{k-1}\phi^\delta\|_{(k-3)/2}^2\right)$$
$$\le \|AF^\delta \phi^\delta\|_{-1/2}\|A\rho_{k-1}\phi^\delta\|_{1/2} + \mathcal{O}\left(\|\nabla \rho_{k-1}\phi^\delta\|_{(k-3)/2}^2\right)$$
$$\le \|\rho_1 F^\delta \phi^\delta\|_{(k-1)/2-1/2}\|\rho_k \phi^\delta\|_{(k-1)/2+1/2}$$
$$\quad + \mathcal{O}\left(\|\nabla \rho_{k-1}\phi^\delta\|_{(k-3)/2}^2\right)$$
$$\le \frac{1}{\epsilon}\|\rho_1 F^\delta \phi^\delta\|_{(k-2)/2} + \epsilon\|\nabla \rho_k \phi^\delta\|_{(k-2)/2}^2$$
$$\quad + \mathcal{O}\left(\|\nabla \rho_{k-1}\phi^\delta\|_{(k-3)/2}^2\right).$$

Substituting into (7.A.3.2) and using induction we get

$$\|\nabla \rho_k \phi^\delta\|^2_{(k-2)/2} \lesssim \|\rho_1 F^\delta \phi^\delta\|^2_{(k-2)/2} + \|F^\delta \phi^\delta\|^2_0,$$

where the constants are independent of δ. That completes the proof of (7.A.3.1). ∎

Now proving the inequality (7.A.1) is straightforward, for we imitate the proof of 7.8.1 with obvious changes.

8

Applications of the $\bar{\partial}$-Neumann Problem

8.1 An Application to the Bergman Projection

In recent years the Bergman projection $P : L^2(\Omega) \to A^2(\Omega)$ has been an object of intense study. The reason for this interest is primarily that Bell and Ligocka [BEL], [BE1], [BE2] have demonstrated that the boundary behavior of biholomorphic mappings of domains may be studied by means of the regularity theory of this projection mapping. Of central importance in these considerations is the following:

DEFINITION 8.1.1 CONDITION R *Let $\Omega \subseteq \mathbb{C}^n$ be a smoothly bounded domain. We say that Ω satisfies Condition R if P maps $C^\infty(\bar{\Omega})$ to $C^\infty(\bar{\Omega})$.*

A representative theorem in the subject is the following:

THEOREM 8.1.2 BELL
Let Ω_1, Ω_2 be smooth, pseudoconvex domains in $\mathbb{C}^n$. Let $\Phi : \Omega_1 \to \Omega_2$ be a biholomorphic mapping. If at least one of the two domains satisfies Condition R then Φ extends to a C^∞ diffeomorphism of $\bar{\Omega}_1$ to $\bar{\Omega}_2$.

There are roughly two known methods to establish Condition R for a domain. One is to use symmetries, as in [BAR] and [BEB]. The more powerful method is to exploit the $\bar{\partial}$-Neumann problem. That is the technique we treat here. Let us begin with some general discussion.

Let $\Omega \subset\subset \mathbb{C}^n$ be a fixed domain on which the equation $\bar{\partial}u = \alpha$ is always solvable when α is a $\bar{\partial}$-closed $(0,1)$-form (e.g., a domain of holomorphy—in other words, a pseudoconvex domain). Let $P : L^2(\Omega) \to A^2(\Omega)$ be the Bergman projection. If u is any solution to $\bar{\partial}u = \alpha$ then $w = w_\alpha = u - Pu$ is the unique solution that is orthogonal to holomorphic functions. Thus w is well defined, independent of the choice of u. Define the mapping

$$T : \alpha \mapsto w_\alpha.$$

Then, for $f \in L^2(\Omega)$ it holds that

$$Pf = f - T(\bar{\partial}f). \tag{8.1.3}$$

To see this, first notice that $\bar{\partial}[f - T(\bar{\partial}f)] = \bar{\partial}f - \bar{\partial}f = 0$, where all derivatives are interpreted in the weak sense. Thus $f - T(\bar{\partial}f)$ is holomorphic. Also $f - [f - T(\bar{\partial}f)]$ is orthogonal to holomorphic functions by design. This establishes the identity (8.1.3). But we have a more useful way of expressing T: namely $T = \bar{\partial}^* N$. Thus we have derived the following important result:

$$P = I - \bar{\partial}^* N \bar{\partial}. \tag{8.1.4}$$

Now suppose that our domain is strongly pseudoconvex. Then we know that N maps H^s to H^{s+1} for every s. Recall that $\bar{\partial}$ and $\bar{\partial}^*$ are first-order differential operators. Then a trivial calculation with (8.1.4) shows that

$$P : H^s \to H^{s-1}$$

for every s. By the Sobolev imbedding theorem, a strongly pseudoconvex domain therefore satisfies Condition R. Thus, thanks to the program of Bell and Ligocka (see [BEL], [KR1]), we know that biholomorphic mappings of strongly pseudoconvex domains extend to be diffeomorphisms of their closures.

It is often convenient, and certainly aesthetically more pleasing, to be able to prove that $P : H^s \to H^s$. This is known to be true on strongly pseudoconvex domains. We now describe the proof, due to J. J. Kohn [KOH3], of this assertion.

THEOREM 8.1.5
Let Ω be a smoothly bounded strongly pseudoconvex domain in $\mathbb{C}^n$. Then for each $s \in \mathbb{R}$ there is a constant $C = C(s)$ such that

$$\|Pf\|_{H^s} \le C \cdot \|f\|_{H^s}. \tag{8.1.5.1}$$

REMARK   In fact, the specific property of a strongly pseudoconvex domain that will be used is the following: For every $\epsilon > 0$ there is a $C(\epsilon) > 0$ so that the inequality

$$\|\phi\|^2 \le \epsilon Q(\phi, \phi) + C(\epsilon)\|\phi\|^2_{-1} \tag{8.1.5.2}$$

for all $\phi \in \mathcal{D} \equiv \bigwedge^{0,1} \cap \operatorname{dom} \bar{\partial} \cap \operatorname{dom} \bar{\partial}^*$. We leave it as an exercise for the reader to check that property (8.1.5.2) is equivalent to the norm Q being compact in the following sense: if $\{\phi_j\}$ is bounded in the Q norm then it has a convergent subsequence in the L^2 norm.

The theorem that we are about to prove is in fact true on any smoothly bounded domain with the property (8.1.5.2). Property (8.1.5.2) is known to hold for a large class of domains, including domains of finite type (see [CAT1], [CAT2], [DAN1], [DAN2], [DAN3], [KR1]) and, in particular, domains with real analytic boundary ([DF]). ∎

PROOF OF THE THEOREM   We have already observed that the Bergman projection of a strongly pseudoconvex domain maps functions in $C^\infty(\bar{\Omega})$ to functions in $C^\infty(\bar{\Omega})$. Thus it suffices to prove our estimate (8.1.5.1) for $f \in C^\infty(\bar{\Omega})$.

Let r be a smooth defining function for Ω. Let $\zeta \in \partial\Omega$ and let $U \subseteq \mathbb{C}^n$ be a neighborhood of ζ. We may select a smooth function w on U such that $\omega^n \equiv w \cdot \partial r$ satisfies $|\omega^n| \equiv 1$ on U. We select $\omega^1, \ldots, \omega^{n-1}$ on U such that $\omega^1, \ldots, \omega^n$ forms an orthonormal basis of the $(1,0)$-forms on U. Thus any $\phi \in \mathcal{D}^{0,1}$ can be expressed, on $\bar{\Omega} \cap U$, as a linear combination

$$\phi = \sum_j \phi_j \bar{\omega}^j.$$

Of course, $\phi \in \mathcal{D}^{0,1}$ if and only if $\phi_n = 0$ on $\partial\Omega$.

Let Λ_t^s be the tangential Bessel potential of order s, as defined in Section 7.5. If η is any real-valued cutoff function supported in U then, whenever $\phi \in \mathcal{D}^{0,1}$ we have $\eta\Lambda_t^s(\eta\phi) \in \mathcal{D}^{0,1}$ as well. The identity $Q(N\alpha, \psi) = \langle \alpha, \psi \rangle$, with $\alpha = \bar{\partial}f$ and $\psi = \eta^3 \Lambda^{2s} \eta N \bar{\partial}f$, yields that

$$Q(N\bar{\partial}f, \eta^3\Lambda^{2s}\eta N\bar{\partial}f) = \langle \bar{\partial}f, \eta^3\Lambda^{2s}\eta N\bar{\partial}f \rangle. \tag{8.1.5.3}$$

Now we apply the compactness inequality (8.1.5.2) with $\phi = \eta\Lambda_t^s(\eta N\bar{\partial}f)$ to obtain

$$\|\eta\Lambda_t^s(\eta N\bar{\partial}f)\|^2 \leq \epsilon Q(\eta\Lambda_t^s(\eta N\bar{\partial}f), \eta\Lambda_t^s(\eta N\bar{\partial}f)) + C(\epsilon)\|\eta\Lambda_t^s(\eta N\bar{\partial}f)\|_{-1}^2$$

$$\leq \epsilon Q(N\bar{\partial}f, \eta^3\Lambda_t^{2s}\eta N\bar{\partial}f) + \epsilon C\|N\bar{\partial}f\|_s^2 + C'(\epsilon)\|N\bar{\partial}f\|_{s-1}^2.$$

Of course in the last estimate we have done two things: First, we have moved η and Λ_t^s across the inner product Q at the expense of creating certain acceptable error terms (which are controlled by the term $\epsilon C\|N\bar{\partial}f\|_s^2$). Second, we have used the fact that $\|\Lambda_t^s g\|_0^2 \leq \|g\|_s^2$ by definition. Now, using (8.1.5.3), we see that

$$\|\eta\Lambda_t^s(\eta N\bar{\partial}f)\|^2 \leq \epsilon\langle f, \bar{\partial}^*\eta^3\Lambda_t^{2s}\eta N\bar{\partial}f\rangle + \epsilon C\|N\bar{\partial}f\|_s^2$$

$$+C'(\epsilon)\|N\bar{\partial}f\|_{s-1}^2. \tag{8.1.5.4}$$

Now we may cover $\bar{\Omega}$ with boundary neighborhoods U as above plus an interior patch on which our problem is strongly elliptic. We obtain an estimate like (8.1.5.4) on each of these patches. We may sum the estimates, using (as we did in the solution of the $\bar{\partial}$-Neumann problem) the fact that $\partial\Omega$ is noncharacteristic for Q, to obtain

$$\|N\bar{\partial}f\|_s^2 \le \epsilon C\|\bar{\partial}^* N\bar{\partial}f\|_s^2 + C'(\epsilon)(\|f\|_s^2 + \|N\bar{\partial}f\|_{s-1}^2).$$

Applying this inequality, with s replaced by $s-1$, to the last term on the right, and then repeating, we may finally derive that

$$\|N\bar{\partial}f\|_s^2 \le \epsilon C\|\bar{\partial}^* N\bar{\partial}f\|_s^2 + C'(\epsilon)(\|f\|_s^2 + \|N\bar{\partial}f\|_0^2). \tag{8.1.5.5}$$

We know that $\bar{\partial}\bar{\partial}^* N\bar{\partial}f = \bar{\partial}f$. As a result,

$$
\begin{aligned}
\|\eta\Lambda_t^s\eta\bar{\partial}^* N\bar{\partial}f\|^2 &= \langle N\bar{\partial}f, \eta^3\Lambda_t^{2s}\eta\bar{\partial}\bar{\partial}^* N\bar{\partial}f\rangle + \mathcal{O}\left(\|N\bar{\partial}f\|_s\|\eta\Lambda_t^s\eta\bar{\partial}^* N\bar{\partial}f\|\right) \\
&= \langle N\bar{\partial}f, \eta^3\Lambda_t^{2s}\eta\bar{\partial}f\rangle + \mathcal{O}\left(\|N\bar{\partial}f\|_s\|\eta\Lambda_t^s\eta\bar{\partial}^* N\bar{\partial}f\|\right) \\
&= \mathcal{O}\left(\left(\|N\bar{\partial}f\|_s + \|f\|_s\right)\|\eta\Lambda_t^s\eta\bar{\partial}^* N\bar{\partial}f\|\right).
\end{aligned}
$$

Summing as before, we obtain the estimate

$$\|\bar{\partial}^* N\bar{\partial}f\|_s \le C(\|N\bar{\partial}f\|_s + \|f\|_s).$$

Putting (8.1.5.5) into this last estimate gives

$$\|\bar{\partial}^* N\bar{\partial}f\|_s \le \epsilon C\|\bar{\partial}^* N\bar{\partial}f\|_s^2 + C'(\epsilon)(\|f\|_s^2 + \|N\bar{\partial}f\|_0^2).$$

If we choose $\epsilon > 0$ small enough, then we may absorb the first term on the right into the left-hand side and obtain

$$\|\bar{\partial}^* N\bar{\partial}f\|_s \le C \cdot \left(\|f\|_s + \|N\bar{\partial}f\|_0\right). \tag{8.1.5.6}$$

But the operator $\bar{\partial}^*$ is closed since the adjoint of a densely defined operator is always closed. It follows from the open mapping principle that

$$\|N\bar{\partial}f\| \le C\|\bar{\partial}^* N\bar{\partial}f\|.$$

On the other hand, $\bar{\partial}^* N\bar{\partial}$ is projection onto the orthogonal complement of $A^2(\Omega)$. Thus it is bounded in L^2 and we see that

$$\|N\bar{\partial}f\| \le C\|f\|_0.$$

Putting this information into (8.1.5.6) gives

$$\|\bar{\partial}^* N\bar{\partial}f\|_s \le C\|f\|_s.$$

If we recall that $P = I - \bar{\partial}^* N \bar{\partial}$ then we may finally conclude that

$$\|Pf\|_s \leq C\|f\|_s.$$

That concludes the proof. $\blacksquare$

8.2   Smoothness to the Boundary of Biholomorphic Mappings

In this section we shall use the fact that Condition R holds on any strongly pseudoconvex domain (Theorem 8.1.5) to prove the following theorem of Fefferman (this generalizes the one-variable result from Section 1.5):

THEOREM 8.2.1
Let $\Omega \subseteq \mathbb{C}^n$ be a strongly pseudoconvex domain with smooth boundary and $\Phi : \Omega \to \Omega$ a biholomorphic mapping. Then Φ extends uniquely to a diffeomorphism of $\bar{\Omega}$ to $\bar{\Omega}$.

As already indicated, the proof given here is that of Bell [BE2] and Bell and Ligocka [BEL]. It will be particularly convenient for us to use the following form of Condition R that we proved in the last section: If P is the Bergman projection on the strongly pseudoconvex domain Ω and $s \in \mathbb{R}$ then $P : H^s \to H^s$.

Observe that the "uniqueness" portion of Theorem 8.2.1 is virtually a tautology and we leave its consideration to the reader. We build now a sequence of lemmas leading to the more interesting "smoothness" assertion of Fefferman's theorem. We begin with some notation.

If $\Omega \subset\subset \mathbb{C}^n$ is any smoothly bounded domain and if $j \in \mathbf{N}$, we let

$$\mathcal{H}^j(\Omega) = H^j(\Omega) \cap \{\text{holomorphic functions on } \Omega\},$$

$$\mathcal{H}^\infty(\Omega) = \bigcap_{j=1}^\infty \mathcal{H}^j(\Omega) = C^\infty(\bar{\Omega}) \cap \{\text{holomorphic functions on } \Omega\}.$$

Here H^j is the standard Sobolev space on a domain. Let $H_0^j(\Omega)$ be the H^j closure of $C_c^\infty(\Omega)$. (Exercise: if j is sufficiently large, then the Sobolev imbedding theorem implies trivially that $H_0^j(\Omega)$ is a proper subset of $H^j(\Omega)$.)

Let us say that $u, v \in C^\infty(\bar{\Omega})$ agree up to order k on $\partial\Omega$ if

$$\left(\frac{\partial}{\partial z}\right)^\alpha \left(\frac{\partial}{\partial \bar{z}}\right)^\beta (u - v)\Bigg|_{\partial\Omega} = 0 \quad \forall \alpha, \beta \quad \text{with} \quad |\alpha| + |\beta| \leq k.$$

LEMMA 8.2.2
Let $\Omega \subset\subset \mathbb{C}^n$ be smoothly bounded and strongly pseudoconvex. Let $w \in \Omega$ be fixed. Let K denote the Bergman kernel. Then there is a constant $C_w > 0$ such that

$$\|K(w, \cdot)\|_{\sup} \leq C_w.$$

PROOF Observe that the function $K(z, \cdot)$ is harmonic. Let $\phi : \Omega \to \mathbb{R}$ be a radial, C_c^∞ function centered at w (that is, $\phi(\zeta^1) = \phi(\zeta^2)$ whenever $|\zeta^1 - w| = |\zeta^2 - w|$). Assume that $\phi \geq 0$ and $\int \phi(\zeta)\, dV(\zeta) = 1$. Then the mean value property (use polar coordinates) implies that

$$K(z, w) = \int_\Omega K(z, \zeta)\phi(\zeta)\, dV(\zeta).$$

But the last expression equals $P\phi(z)$. Therefore

$$\|K(w, \cdot)\|_{\sup} = \sup_{z \in \Omega} |K(w, z)|$$

$$= \sup_{z \in \Omega} |K(z, w)|$$

$$= \sup_{z \in \Omega} |P\phi(z)|.$$

By Sobolev's theorem, this is

$$\leq C(\Omega) \cdot \|P\phi\|_{\mathcal{H}^{2n+1}}.$$

By Condition R, this is

$$\leq C(\Omega) \cdot \|\phi\|_{H^{2n+1}} \equiv C_w. \qquad \blacksquare$$

LEMMA 8.2.3
Let $u \in C^\infty(\bar{\Omega})$ be arbitrary. Let $s \in \{0, 1, 2, \ldots\}$. Then there is a $v \in C^\infty(\bar{\Omega})$ such that $Pv = 0$ and the functions u and v agree to order s on $\partial\Omega$.

PROOF After a partition of unity, it suffices to prove the assertion in a small neighborhood U of $z_0 \in \partial\Omega$. After a rotation, we may suppose that $\partial\rho/\partial z_1 \neq 0$ on $U \cap \bar{\Omega}$, where ρ is a defining function for Ω. Define the differential operator

$$\nu = \frac{\text{Re}\left\{\sum_{j=1}^n \frac{\partial\rho}{\partial z_j} \frac{\partial}{\partial \bar{z}_j}\right\}}{\sum_{j=1}^n \left|\frac{\partial\rho}{\partial z_j}\right|^2}.$$

Notice that $\nu\rho = 1$. Now we define v by induction on s.
For the case $s = 0$, let

$$w_0 = \frac{\rho u}{\partial\rho/\partial \zeta_1}.$$

Define

$$v_0 = \frac{\partial}{\partial \zeta_1} w_0$$
$$= u + O(\rho).$$

Then u and v_0 agree to order 0 on $\partial\Omega$. Also

$$Pv_0(z) = \int K(z,\zeta) \frac{\partial}{\partial \zeta_1} w_0(\zeta) \, dV(\zeta).$$

This equals, by integration by parts,

$$- \int \frac{\partial}{\partial \zeta_1} K(z,\zeta) w_0(\zeta) \, dV(\zeta).$$

Notice that the integration by parts is valid by Lemma 8.2.2 and because $w_0|_{\partial\Omega} = 0$. Also, the integrand in this last line is zero because $K(z,\cdot)$ is conjugate holomorphic. Thus $Pv_0 \equiv 0$ as desired.

Suppose inductively that $w_{s-1} = w_{s-2} + \theta_{s-1}\rho^s$ and $v_{s-1} = (\partial/\partial z_1)(w_{s-1})$ have been constructed. We show that there is a w_s of the form

$$w_s = w_{s-1} + \theta_s \cdot \rho^{s+1}$$

such that $v_s = (\partial/\partial z_1)(w_s)$ agrees to order s with u on $\partial\Omega$. By the inductive hypothesis,

$$v_s = \frac{\partial}{\partial z_1} w_s$$

$$= \frac{\partial w_{s-1}}{\partial z_1} + \frac{\partial}{\partial z_1} \left[\theta_s \cdot \rho^{s+1} \right]$$

$$= v_{s-1} + \rho^s \left[(s+1)\theta_s \frac{\partial \rho}{\partial z_1} + \rho \cdot \frac{\partial \theta_s}{\partial z_1} \right]$$

agrees to order $s-1$ with u on $\partial\Omega$ so long as θ_s is smooth. So we need to examine $D(u - v_s)$, where D is an s-order differential operator. But if D involves a tangential derivative D_0, then write $D = D_0 \cdot D_1$. It follows that $D(u - v_s) = D_0(\alpha)$, where α vanishes on $\partial\Omega$ so that $D_0\alpha = 0$ on $\partial\Omega$. So we need only check $D = \nu^s$.

We have seen that θ_s must be chosen so that

$$\nu^s(u - v_s) = 0 \qquad \text{on} \quad \partial\Omega.$$

Equivalently,

$$\nu^s(u - v_{s-1}) - \nu^s \left(\frac{\partial}{\partial z_1} \right) (\theta_s \rho^{s+1}) = 0 \quad \text{on} \quad \partial\Omega$$

or

$$\nu^s(u - v_{s-1}) - \theta_s \left(\nu^s \frac{\partial}{\partial z_1} \rho^{s+1} \right) = 0 \quad \text{on} \quad \partial\Omega$$

or

$$\nu^s(u - v_{s-1}) - \theta_s \cdot (s+1)! \frac{\partial\rho}{\partial z_1} = 0 \quad \text{on} \quad \partial\Omega.$$

It follows that we must choose

$$\theta_s = \frac{\nu^s(u - v_{s-1})}{(s+1)! \frac{\partial\rho}{\partial z_1}},$$

which is indeed smooth on U. As in the case $s = 0$, it holds that $Pv_s = 0$. This completes the induction and the proof. ∎

REMARK   In this proof we have in fact constructed v by subtracting from u a Taylor type expansion in powers of ρ. ∎

LEMMA 8.2.4
For each $s \in \mathbb{N}$ we have $\mathcal{H}^\infty(\Omega) \subseteq P(H_0^s(\Omega))$.

PROOF   Let $u \in C^\infty(\bar{\Omega})$. Choose v according to Lemma 8.2.3. Then $u - v \in H_0^s$ and $Pu = P(u - v)$. Therefore

$$P(H_0^s) \supseteq P(C^\infty(\bar{\Omega})) \supseteq P(\mathcal{H}^\infty(\Omega)) = \mathcal{H}^\infty(\Omega). \quad ∎$$

Henceforth, let Ω_1, Ω_2 be fixed C^∞ strongly pseudoconvex domains in $\mathbb{C}^n$, with K_1, K_2 their Bergman kernels and P_1, P_2 the corresponding Bergman projections. Let $\Phi : \Omega_1 \to \Omega_2$ be a biholomorphic mapping, and let $u = \det \mathrm{Jac}_{\mathbb{C}} \Phi$. For $j = 1, 2$, let $\delta_j(z) = \delta_{\Omega_j}(z) = \mathrm{dist}(z, {}^c\Omega_j)$.

LEMMA 8.2.5
For any $g \in L^2(\Omega_2)$ we have

$$P_1(u \cdot (g \circ \Phi)) = u \cdot ((P_2(g)) \circ \Phi).$$

PROOF   Notice that $u \cdot (g \circ \Phi) \in L^2(\Omega_1)$ by change of variables (see Lemma 6.2.9). Therefore, by 6.2.8,

$$P_1(u \cdot (g \circ \Phi))(z) = \int_{\Omega_1} K_1(z, \zeta) u(\zeta) g(\Phi(\zeta)) \, dV(\zeta)$$

$$= \int_{\Omega_1} u(z) K_2(\Phi(z), \Phi(\zeta)) \overline{u(\zeta)} u(\zeta) g(\Phi(\zeta)) \, dV(\zeta).$$

Change of variable now yields

$$P_1(u \cdot (g \circ \Phi))(z) = u(z) \int_{\Omega_2} K_2(\Phi(z), \xi) g(\xi) \, dV(\xi)$$

$$= u(z) \cdot \left[(P_2(g)) \circ \Phi \right](z). \qquad \blacksquare$$

Exercise: Let $\Omega \subset\subset \mathbb{C}^n$ be a smoothly bounded domain. Let $j \in \mathbb{N}$. There is an $N = N(j)$ so large that $g \in H_0^N$ implies that g vanishes to order j on $\partial\Omega$.

LEMMA 8.2.6
Let $\psi : \Omega_1 \to \Omega_2$ be a C^j diffeomorphism that satisfies

$$\left| \frac{\partial^\alpha \psi}{\partial z^\alpha}(z) \right| \leq C \cdot (\delta_1(z))^{-|\alpha|}, \qquad (8.2.6.1)$$

for all multiindices α with $|\alpha| \leq j \in \mathbb{N}$ and

$$|\nabla \psi^{-1}(w)| \leq C(\delta_2(w))^{-1}. \qquad (8.2.6.2)$$

Suppose also that

$$\delta_2(\psi(z)) \leq C\delta_1(z). \qquad (8.2.6.3)$$

Then there is a number $J = J(j)$ such that, whenever $g \in H_0^{j+J}(\Omega_2)$, then $g \circ \psi \in H_0^j(\Omega_1)$.

PROOF   The subscript 0 causes no trouble by the definition of H_0^j. Therefore it suffices to prove an estimate of the form

$$\|g \circ \psi\|_{H_0^j} \leq C \|g\|_{H_0^{j+J}}, \quad \text{all } g \in C_c^\infty(\Omega).$$

By the chain rule and Leibniz's rule, if α is a multiindex of modulus not exceeding j, then

$$\left(\frac{\partial}{\partial z} \right)^\alpha (g \circ \psi) = \sum \left[(D^\beta g) \circ \psi \right] \cdot D^{\gamma_1} \psi \cdots D^{\gamma_c} \psi,$$

where $|\beta| \leq |\alpha|, \sum |\gamma_i| \leq |\alpha|$, and the number of terms in the sum depends only on α (a classical formula of Faà de Bruno—see [ROM]—actually gives this sum quite explicitly, but we do not require such detail). Note here that $D^{\gamma_l}\psi$ is used to denote a derivative of *some component* of ψ. By hypothesis, it follows that

$$\left|\left(\frac{\partial}{\partial z}\right)^\alpha (g \circ \psi)\right| \le C \sum |(D^\beta g) \circ \psi| \cdot (\delta_1(z))^{-j}.$$

Therefore

$$\int_{\Omega_1} \left|\left(\frac{\partial}{\partial z}\right)^\alpha (g \circ \psi)(z)\right|^2 dV(z) \le C \sum \int_{\Omega_1} |(D^\beta g) \circ \psi(z)|^2 (\delta_1(z))^{-2j} dV(z)$$

$$= C \sum \int_{\Omega_2} |D^\beta g(w)|^2 \delta_1 \left(\psi^{-1}(w)\right)^{-2j}$$

$$\times |\det J_{\mathbf{C}} \psi^{-1}|^2 dV(w).$$

But (8.2.6.2) and (8.2.6.3) imply that the last line is majorized by

$$C \sum \int_{\Omega_2} |D^\beta g(w)|^2 \delta_2(w)^{-2j} \delta_2(w)^{-2n} dV(w). \tag{8.2.6.4}$$

Now if J is large enough, depending on the Sobolev imbedding theorem, then

$$|D^\beta g(w)| \le C \|g\|_{H_0^{j+J}} \cdot \delta_2(w)^{2n+2j}.$$

(Remember that g is supported in Ω_2 and vanishing at the boundary.) Hence (8.2.6.4) is majorized by $C\|g\|_{H_0^{j+J}}$. ∎

LEMMA 8.2.7
For each $j \in \mathbf{N}$, there is an integer J so large that if $g \in H_0^{j+J}(\Omega_2)$, then $g \circ \Phi \in H_0^j(\Omega_1)$.

PROOF   The Cauchy estimates give (since Φ is bounded) that

$$\left|\frac{\partial^\alpha \Phi_\ell}{\partial z^\alpha}(z)\right| \le C \cdot (\delta_1(z))^{-|\alpha|}, \qquad \ell = 1, \ldots, n \tag{8.2.7.1}$$

and

$$|\nabla \Phi^{-1}(w)| \le C(\delta_2(w))^{-1}, \tag{8.2.7.2}$$

where $\Phi = (\Phi_1, \ldots, \Phi_n)$. We will prove that

$$C \cdot \delta_1(z) \ge \delta_2(\Phi(z)). \tag{8.2.7.3}$$

Then Lemma 8.2.6 gives the result.

To prove (8.2.6.3), let ρ be a smooth, strictly plurisubharmonic defining function for Ω_1. Then $\rho \circ \Phi^{-1}$ is a smooth plurisubharmonic function on Ω_2. Since ρ vanishes on $\partial\Omega_1$ and since Φ^{-1} is proper, we conclude that $\rho \circ \Phi^{-1}$ extends continuously to $\bar{\Omega}_2$. If $P \in \partial\Omega_2$ and ν_P is the unit outward normal to $\partial\Omega_2$ at P, then Hopf's lemma (see our treatment in Section 1.5 or, for a different point of view, consult [KRA1, Chapter 1]) implies that the (lower) one-sided derivative $(\partial/\partial\nu_P)(\rho \circ \Phi^{-1})$ satisfies

$$\frac{\partial}{\partial\nu_P}(\rho \circ \Phi^{-1}(P)) \geq C.$$

So, for $w = P - \epsilon\nu_P, \epsilon > 0$ small, it holds that

$$-\rho \circ \Phi^{-1}(w) \geq C \cdot \delta_2(w).$$

These estimates are uniform in $P \in \partial\Omega_2$. Using the comparability of $|\rho|$ and δ_1 yields

$$C\delta_1(\Phi^{-1}(w)) \geq \delta_2(w).$$

Setting $z = \Phi^{-1}(w)$ now gives

$$C'\delta_1(z) \geq \delta_2(\Phi(z)),$$

which is (8.2.6.3).   ∎

LEMMA 8.2.8
The function u is in $C^\infty(\bar{\Omega}_1)$.

PROOF   It suffices to show that $u \in H^j(\Omega_1)$, every j. So fix j. According to (8.2.7.1), $|u(z)| \leq C\delta_1(z)^{-2n}$. Then, by Lemma 8.2.7 and the exercise for the reader preceding 8.2.6, there is a J so large that $g \in H_0^{j+J}(\Omega_2)$ implies $u \cdot (g \circ \Phi) \in H^j(\Omega_1)$. Choose, by Lemma 8.2.4, a $g \in H_0^{m+J}(\Omega_2)$ such that $P_2 g \equiv 1$. Then Lemma 8.2.5 yields

$$P_1(u \cdot (g \circ \Phi)) = u.$$

By Condition R, it follows that $u \in H^j(\Omega_1)$.   ∎

LEMMA 8.2.9
The function u is bounded from 0 on $\bar{\Omega}_1$.

PROOF   By symmetry, we may apply Lemma 8.2.8 to Φ^{-1} and $\det J_{\mathbb{C}}(\Phi^{-1}) = 1/u$. We conclude that $1/u \in C^\infty(\bar{\Omega}_2)$. Thus u is nonvanishing on $\bar{\Omega}$.   ∎

PROOF OF FEFFERMAN'S THEOREM (THEOREM 8.2.1) Use the notation of the proof of Lemma 8.2.8. Choose $g_1, \ldots, g_n \in H_0^{j+J}(\Omega_2)$ such that $P_2 g_i(w) = w_i$ (here w_i is the ith coordinate function). Then Lemmas 8.2.5 and 8.2.7 yield that $u \cdot \Phi_i \in H^j(\Omega_1), i = 1, \ldots, n$. By Lemma 8.2.9, $\Phi_i \in H^j(\Omega_1), i = 1, \ldots, n$. By symmetry, $\Phi^{-1} \in H^j(\Omega_2)$. Since j is arbitrary, the Sobolev imbedding theorem finishes the proof. ∎

8.3 Other Applications of $\bar{\partial}$ Techniques

The unifying theme of this book has been the theory of holomorphic mappings. We have seen that regularity for the classical Dirichlet problem for the Laplacian provided the key to understanding boundary regularity of conformal mappings in the complex plane. Likewise, boundary regularity for the $\bar{\partial}$-Neumann problem has provided the key for determining the boundary behavior of biholomorphic mappings of several complex variables.

Since we have gone to so much trouble to derive estimates for the $\bar{\partial}$-Neumann problem, it seems appropriate at this time to depart from our principal theme and discuss some other applications of our results on the $\bar{\partial}$ problem.

A point that hampered the theory of the Bergman kernel for years was that of boundary regularity. On the ball, for example, one sees that the Bergman kernel blows up if $z, \zeta \in B$ approach the same boundary point. But on $\bar{B} \times \bar{B} \setminus$ (diagonal) the Bergman kernel is smooth. One might hope that a similar result is true on, say, strongly pseudoconvex domains (the question is open for general smoothly bounded domains). Heartening partial results on this problem were made by Diederich in [DIE]. However, Kerzman [KER2] realized that the result follows easily from $\bar{\partial}$-Neumann considerations. Here is a part of what Kerzman proved (this result is implicit in one of the lemmas from the last section):

PROPOSITION 8.3.1
Let $\Omega \subseteq \mathbb{C}^n$ be smoothly bounded and strongly pseudoconvex. Fix $z \in \Omega$. Then $K(z, \cdot)$ is in $C^\infty(\bar{\Omega})$.

PROOF Let $\phi \in C_c^\infty(\mathbb{C}^n)$ be a nonnegative radial function of total mass one and with support in the unit ball. Define $\phi_\epsilon(\zeta) = \epsilon^{-2n}\phi(\zeta/\epsilon)$. Choose ϵ to be a positive number less than the Euclidean distance of z to the boundary of Ω.

Because K is harmonic in the first variable we have, by the mean value property, that

$$K(w, z) = \int K(w, \xi)\phi_\epsilon(z - \xi)\, dV(\xi).$$

But this last (a function of w) also equals the Bergman projection of $\phi_\epsilon(z - \cdot)$. Since $\phi_\epsilon(z - \cdot) \in C^\infty(\bar{\Omega})$ and Ω satisfies condition R, we may conclude that

$K(\cdot, z) \in C^\infty(\bar{\Omega})$. But the Bergman kernel is conjugate symmetric in its variables; that finishes the proof. ∎

A fundamental problem in the complex function theory of both one and several complex variables is to take a local construction of a holomorphic function and turn it into a global construction. In one complex variable we have the theorems of Weierstrass and Mittag–Leffler, Cauchy formulas, conformal mapping, and many other devices for achieving this end. In several complex variables, the $\bar{\partial}$-Neumann problem is certainly one of the most important tools (along with sheaf theory and integral formulas) for this type of problem.

To illustrate this circle of ideas, let Ω be a smoothly bounded, strongly pseudoconvex and smooth. Then for any defining function ρ and $P \in \partial\Omega$ we know that the Levi form

$$\sum_{j,k=1}^{n} \frac{\partial^2}{\partial z_j \partial \bar{z}_k} \rho(P) w_j \bar{w}_k$$

is positive definite on all $w = (w_1, \ldots, w_n)$ satisfying

$$\sum_{j=1}^{n} \frac{\partial \rho}{\partial z_j}(P) w_j = 0. \tag{8.3.2}$$

This is the *definition* of strong pseudoconvexity. However it is an elementary exercise in calculus to see that if $A > 0$ is large enough then the new defining function

$$\rho_A(z) = \frac{\exp(A\rho(z)) - 1}{A}$$

satisfies

$$\sum_{j,k=1}^{n} \frac{\partial^2}{\partial z_j \partial \bar{z}_k} \rho_A(P) w_j \bar{w}_k \geq C|w|^2 \tag{8.3.3}$$

for *all* $w \in \mathbb{C}^n$ and some $C > 0$. Details of this construction may be found in [KRA1, Chapter 3].

Now with such a function ρ_A in hand we define the *Levi polynomial*

$$L_P(z) = \sum_{j=1}^{n} \frac{\partial \rho_A}{\partial z_j}(P)(z_j - P_j) + \frac{1}{2} \sum_{j,k=1}^{n} \frac{\partial^2 \rho_A}{\partial z_j \partial z_k}(P)(z_j - P_j)(z_k - P_k).$$

The key technical result is this:

PROPOSITION 8.3.4
With Ω, P, L_P as above there is a $\delta > 0$ such that if $|z - P| < \delta, z \in \bar{\Omega}$, and $L_P(z) = 0$ then $z = P$. See Figure 8.1.

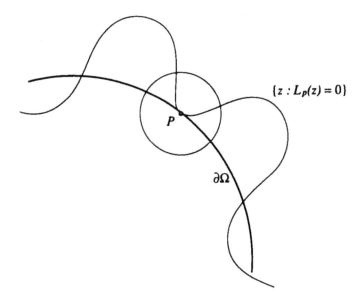

$\{z : L_p(z) = 0\}$

P

$\partial\Omega$

FIGURE 8.1

PROOF We write the Taylor expansion for ρ_A about the point P in complex notation:

$$\rho(z) = \rho(P) + \sum_{j=1}^{n} \frac{\partial \rho_A}{\partial z_j}(P)(z_j - P_j) + \sum_{j=1}^{n} \frac{\partial \rho_A}{\partial \bar{z}_j}(P)(\overline{z_j - P_j})$$

$$+ \frac{1}{2} \sum_{j,k=1}^{n} \frac{\partial^2 \rho_A}{\partial z_j \partial z_k}(P)(z_j - P_j)(z_k - P_k)$$

$$+ \frac{1}{2} \sum_{j,k=1}^{n} \frac{\partial^2 \rho_A}{\partial \bar{z}_j \partial \bar{z}_k}(P)(\overline{z_j - P_j})(\overline{z_k - P_k})$$

$$+ \sum_{j,k=1}^{n} \frac{\partial^2 \rho_A}{\partial z_j \partial \bar{z}_k}(P)(z_j - P_j)(\overline{z_k - P_k}) + \mathcal{O}(|z - P|^3)$$

$$= 2\mathrm{Re}\, L_P(z) + \sum_{j,k=1}^{n} \frac{\partial^2 \rho_A}{\partial z_j \partial \bar{z}_k}(P)(z_j - P_j)(\overline{z_k - P_k}) + \mathcal{O}(|z - P|^3).$$

Now if z is such that $L_P(z) = 0$ then we find that

$$\rho(z) = \sum_{j,k=1}^{n} \frac{\partial^2 \rho_A}{\partial z_j \partial \bar{z}_k}(P)(z_j - P_j)(\overline{z_k - P_k}) + \mathcal{O}(|z - P|^3).$$

Of course the sum is nothing other than that on the left-hand side of (8.3.3) with $w = z - P$. So in fact we have

$$\rho(z) \geq C \cdot |z - P|^2 + \mathcal{O}(|z - P|^3).$$

If $|z - P|$ is sufficiently small then this last quantity is greater than or equal to $(C/2)|z - P|^2$. That is what we wished to prove. ∎

Now the upshot of the technical construction that we have just achieved is the following:

THEOREM 8.3.5
Let Ω be smoothly bounded and strongly pseudoconvex. Let $P \in \partial\Omega$. Then there is a function that is holomorphic on Ω and that cannot be analytically continued to any neighborhood of P.

PROOF With δ as in the preceding lemma, let $\phi \in C_c^\infty(\mathbb{C}^n)$ satisfy $\phi \equiv 1$ near P and ϕ is supported in the ball of center P and radius δ. Let

$$g(z) = \frac{\phi(z)}{L_P(z)}.$$

Notice that g is holomorphic on the intersection of a small neighborhood of P with Ω, but it is certainly not holomorphic on all of Ω. Observe also that, by the choice of the support of ϕ, the function g is well defined because we have not divided by zero. Finally, g blows up at P because $L_P(P) = 0$.

Consider the $\bar\partial$ problem

$$\bar\partial u = -\frac{\bar\partial \phi(z)}{L_P(z)}.$$

The right-hand side has coefficients that are smooth on $\bar\Omega$ because $\bar\partial\phi$ vanishes in a neighborhood of P. And, by inspection, the right-hand side of this equation is $\bar\partial$-closed. By our theory of the $\bar\partial$ problem on strongly pseudoconvex domains, there is a function u that satisfies this equation and is C^∞ on $\bar\Omega$.

Now the function

$$G(z) \equiv \frac{\phi(z)}{L_P(z)} + u(z)$$

has the property that it is holomorphic (since $\bar\partial G \equiv 0$ by design). Moreover, G blows up (because ϕ/L_P does and u does not) at P. We conclude that the holomorphic function G blows up at P hence cannot be analytically continued past P. That completes the proof. ∎

The Levi problem consists in showing that all pseudoconvex domains are domains of holomorphy. This can be reduced, by relatively elementary means, to proving the result for strongly pseudoconvex domains (see [KR1] for the whole story). That in turn can be reduced to showing that each boundary point P of a strongly pseudoconvex domain is essential: that is, there is a *globally defined holomorphic function G* on Ω that cannot be analytically continued past P. In fact, that is what we have just proved.

It is not difficult to see that, on the ball with center P and radius δ, the real part of L_P is of one sign. Thus we may take a fractional root of this holomorphic function. As a result, it may be arranged that $g \in L^2(\Omega)$. We leave it as an exercise for the reader to provide details of the following assertion:

PROPOSITION 8.3.6
Let Ω be a smoothly bounded, strongly pseudoconvex domain. Let $P \in \partial\Omega$. Then there is an L^2 holomorphic function on Ω (an element of $A^2(\Omega)$) that cannot be analytically continued past P.

It is in fact possible to construct a function that is C^∞ on $\bar{\Omega}$ and holomorphic on Ω that cannot be continued past *any* boundary point. This requires additional techniques. See [HAS], [CAT3] for details of this result. Both the result of the last proposition, and the result of this paragraph, are true on any smoothly bounded (weakly) pseudoconvex domain.

We conclude this section with another illustration of $\bar{\partial}$ techniques in an application to the extension of holomorphic functions:

THEOREM 8.3.7
Let $\Omega \subseteq \mathbb{C}^n$ be smoothly bounded and strongly pseudoconvex. Let $M = \{z \in \mathbb{C}^n : z_n = 0\}$. Let $\omega = \Omega \cap M$. If f is a holomorphic function on ω then there is a holomorphic function F on Ω such that $F\big|_\omega = f$.

PROOF Let $\pi : \mathbb{C}^n \to M$ be given by $\pi(z_1, \ldots, z_n) = (z_1, \ldots, z_{n-1}, 0)$. Define $\omega = \Omega \cap M$ and $B = \{z \in \Omega : \pi(z) \notin \omega\}$. Then B and ω are disjoint and relatively closed in Ω. Refer to Figure 8.2.

By the C^∞ Urysohn lemma ([HIR]) there is a C^∞ function ϕ on Ω such that $\phi = 1$ in a neighborhood of ω and $\phi = 0$ in a neighborhood of B. Now define

$$H(z) = \phi(z) \cdot f(\pi(z)).$$

Since the support of ϕ lies in the complement of B, the function H is well-defined. And it extends f. But of course it is not holomorphic.

We endeavor to make H holomorphic by adding on a correction term: set

$$F(z) = H(z) + z_n \cdot w(z). \tag{8.3.7.1}$$

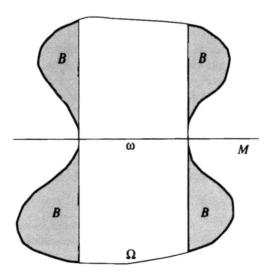

FIGURE 8.2

We seek $w(z)$ such that $\bar{\partial}F = 0$ on Ω. This leads to the $\bar{\partial}$ equation

$$\bar{\partial}w = -\frac{\bar{\partial}H}{z_n} \, . \qquad\qquad (8.3.7.2)$$

The right-hand side is smooth and well defined because H is holomorphic on a neighborhood of the set $\{z \in \Omega : z_n = 0\}$. Moreover, the right-hand side is $\bar{\partial}$-closed by inspection. Finally, it is an exercise in calculus to see that there is an $s \in \mathbb{R}$ such that the coefficients of the right-hand side lie in the Sobolev space H^s.

By our theory of the $\bar{\partial}$ problem on a strongly pseudoconvex domain, there exists a $w \in H^s$ that solves the equation (8.3.7.2). Since the $\bar{\partial}$ problem is in fact elliptic on the interior, we see that w is in fact a classical smooth function on Ω. Thus the equation (8.3.7.1) defines a holomorphic function on Ω that plainly has the property that $F\big|_\omega = f$. The proof is complete. ∎

9

The Local Solvability Issue and a Look Back

9.1 Some Remarks about Local Solvability

In the nineteenth century and the first half of the twentieth, it was generally believed that any partial differential equation with smooth coefficients and smooth data would—at least locally—have smooth solutions. This belief was fueled, at least in part, by the Cauchy–Kowalewski theorem, which says that this assertion, with "smooth" replaced by "real analytic," is always true.

The Cauchy–Kowalewski theorem is the only general existence and regularity theorem in the entire theory of partial differential equations (see [KRP] for a treatment of this result). While fifty years ago it was thought to be typical, we now realize that it represents the exception. Jacobowitz and Treves [JAT] have shown that, in a reasonable sense, nonlocally solvable equations are generic.

H. Lewy presented the first nonlocally solvable partial differential equation in [LEW2]. Lewy's equation is astonishingly simple. If the coordinates on $\mathbb{R}^3$ are given by $(x, y, t) \approx (x + iy, t) \approx (z, t)$, then the equation is

$$Lu = \frac{\partial u}{\partial z} + i\bar{z}\frac{\partial u}{\partial t} = f.$$

Although it is not made explicit in that paper, Lewy's discovery grew naturally out of analytic continuation considerations for holomorphic functions of two complex variables. This assertion becomes clearer when the accompanying paper [LEW1] is consulted.

The local solvability issue is also intimately connected with integrability for systems of vector fields—specifically the vector fields arising as the real and imaginary parts of holomorphic tangent vector fields to a strongly pseudoconvex hypersurface. In particular, the Lewy equation is, in local coordinates, just the tangential Cauchy–Riemann equations on a spherical cap; the issue of local solvability for these equations is essentially equivalent (because the Levi form is nondegenerate) to the issue of analytically continuing CR functions to the (pseudo)convex side of the spherical cap.

The local solvability question has received a great deal of attention since Lewy's work. Hörmander gave necessary conditions for the local solvability of a linear partial differential operator (see [HOR1, Section 6.1]. Nirenberg and Treves [NTR] gave necessary conditions for a partial differential operator of principal type with smooth coefficients to be locally solvable. They also gave sufficient conditions when the coefficients are real analytic. Beals and Fefferman [BEF1] showed that the Nirenberg–Treves condition is sufficient in general.

Our intention in this brief chapter is to exposit the basic ideas concerning local solvability, with special emphasis on the connections with complex analysis. Although much of this material can be presented in an entirely elementary fashion (see Section 9.4), we position the subject last in the book so that we may draw on the earlier chapters both for concepts and for motivation. We refer the reader to [KOH2] and [NIR] for more on these matters; some of the material presented here is drawn from those references.

9.2   The Szegö Projection and Local Solvability

In order to be as explicit as possible, we shall work with the Siegel upper half-space

$$\mathcal{U} = \{z \in \mathbb{C}^2 : \operatorname{Im} z_2 > |z_1|^2\}$$

with boundary

$$M = \{z \in \mathbb{C}^2 : \operatorname{Im} z_2 = |z_1|^2\}.$$

It is worth noting explicitly that $\mathcal{U}$ is biholomorphic to the ball B: the mapping

$$\Phi_1(z) = \frac{z_2}{1 + z_1}$$

$$\Phi_2(z) = i \cdot \frac{1 - z_1}{1 + z_1}$$

provides an explicit mapping of B onto $\mathcal{U}$. In fact, $\mathcal{U}$ is an example of a Siegel domain of type two; all such domains have bounded realizations (see [KAN]).

We shall identify M with $\mathbb{R}^3$ as follows: Let the coordinates on $\mathbb{R}^3$ be given by $(x, y, t) \approx (x + iy, t) \approx (z, t)$. Let

$$\psi : \mathbb{R}^3 \to M$$

$$(x, y, t) \mapsto (z, t + i|z|^2).$$

The Jacobian of this mapping transforms the Lewy operator L (see Section 1) to the operator

$$L' = \frac{\partial}{\partial z_1} + 2i\bar{z}_1 \frac{\partial}{\partial z_2}.$$

More explicitly,

$$\frac{\partial}{\partial z} = \frac{\partial z_1}{\partial z}\frac{\partial}{\partial z_1} + \frac{\partial z_2}{\partial z}\frac{\partial}{\partial z_2} + \frac{\partial \bar{z}_1}{\partial z}\frac{\partial}{\partial \bar{z}_1} + \frac{\partial \bar{z}_2}{\partial z}\frac{\partial}{\partial \bar{z}_2}$$

$$= \frac{\partial}{\partial z_1} + i\bar{z}\frac{\partial}{\partial z_2} + (-i)\bar{z}\frac{\partial}{\partial \bar{z}_2}$$

and

$$\frac{\partial}{\partial t} = \frac{\partial z_1}{\partial t}\frac{\partial}{\partial z_1} + \frac{\partial z_2}{\partial t}\frac{\partial}{\partial z_2} + \frac{\partial \bar{z}_1}{\partial t}\frac{\partial}{\partial \bar{z}_1} + \frac{\partial \bar{z}_2}{\partial t}\frac{\partial}{\partial \bar{z}_2}$$

$$= \frac{\partial}{\partial z_2} + \frac{\partial}{\partial \bar{z}_2} \,.$$

The formula for L' follows.

We shall pass back and forth freely between statements on $\mathbb{R}^3$ and statements on M. In particular, we shall need to take statements about holomorphic functions on the ambient space in which M lives and interpret them in the coordinates on $\mathbb{R}^3$.

We should certainly note at this stage that if $\rho(z) = \text{Im}\, z_2 - |z_1|^2$ is a defining function for $\mathcal{U}$, then L' annihilates ρ at points of M. This says that L' is a *tangential holomorphic vector field*. By linear algebra, any other tangential holomorphic vector field is a scalar multiple of L'. The operator $\bar{L}'$ is frequently termed the *tangential Cauchy–Riemann operator*. We shall develop that usage as the chapter progresses.

Recall the space H^2 from Chapter 6. We recast it in our present language. Equip M with the area measure $dx\,dy\,dt$ inherited from $\mathbb{R}^3$. Define $L^2(M)$ with respect to this measure. Let $H^2(M)$ denote the subspace of $L^2(M)$ consisting of boundary values of holomorphic functions on $\mathcal{U}$. Equivalently, $H^2(M)$ consists of the $L^2(M)$ closure of the boundary functions of those functions that are smooth on $\bar{\mathcal{U}}$, holomorphic on $\mathcal{U}$, and decay fast enough at ∞. The equivalence, and naturality, of these various definitions is treated in [STBV] and [KR1]. Let $P : L^2(M) \to H^2(M)$ be orthogonal projection. It is convenient also to have a separate notation for the mapping that assigns to each $f \in L^2(M)$ the *holomorphic extension* of Pf to $\mathcal{U}$. Call this mapping $\tilde{P}$.

Then, as indicated in Chapter 6, $\tilde{P}$ is just the Szegö integral:

$$\tilde{P}f(z_1, z_2) = \int_M S(z_1, z_2, x, y, t) f(x, y, t)\, dx\, dy\, dt, \qquad (9.2.1)$$

where

$$S(z_1, z_2, x, y, t) = \frac{1}{\pi^2}\frac{1}{(i(\bar{w}_2 - z_2) - 2\bar{w}_1 z_1)^2} \,.$$

In fact, one may derive this formula by pulling back the Szegö formula on the ball that we derived in Chapter 6. It is worth noting that the identification

$(z, t) \leftrightarrow (z, t + i|z|^2)$ is in fact the canonical one coming from the simple transitive action of $\mathbb{R}^3$ acting as the Heisenberg group on M. This action is explained in detail in [KR1, Ch. 2] and [CHK]. In particular, the Haar measure on the Heisenberg group is just the standard Lebesgue measue on this realization of $\mathbb{R}^3$. While this point of view is not essential to an understanding of the present chapter, it helps to explain formula (9.2.1). For more on the Heisenberg group, see [CHK]. See also [KOV].

PROPOSITION 9.2.2
Let $\zeta \in M$ and $f \in L^2(M)$. Then $\tilde{P}f$ is analytically continuable past ζ (into the complement of $\overline{U}$) if and only if Pf is real analytic on M near ζ.

PROOF This is essentially a tautology. ∎

Now the main result of this section is the following:

THEOREM 9.2.3
Let $f \in L^2(M)$ and let $\zeta \in M$. Then the equation

$$Lu = f$$

has a C^1 solution in a neighborhood of $\zeta \in M$ if and only if Pf is real analytic in a neighborhood of ζ.

COROLLARY 9.2.4
Fix $\zeta \in M$. There exist functions $f \in L^2(M) \cap C(M)$ such that the equation

$$Lu = f$$

has no C^1 solution in a neighborhood of ζ. Indeed, such f are generic.

PROOF OF THE COROLLARY Suppose without loss of generality that ζ is the image under the biholomorphism Φ of the point $(1, 0) \in \partial B$. Let $F(z) = \mu \circ \Phi^{-1}(z)$, where $\mu(w) = \sqrt{1 - w_1}$. Then F is plainly continuous near ζ and not real analytic in any neighborhood of ζ. Set $f = F\big|_M$. Then of course $Pf = f$ and $\tilde{P}f = F$. This f does the job.

The fact that such functions f are generic follows from elementary considerations: Once we have peaking functions (actually $1 - \mu$ is a peaking function) as in the preceding paragraph then the density of f for which the partial differential equation is not solvable near P follows from Stone–Weierstrass considerations. For further details on the genericity of f for which the differential equation is not solvable, we refer the reader to [JAT]. ∎

The remainder of this section, and also the next section, will be devoted to a proof of the theorem. For some parts of the proof we shall have to refer the reader to the literature, but we can certainly describe the key ideas.

We will make use of the following property of the Szegö kernel. Because we have an explicit formula for S on $\mathcal{U}$, this property follows immediately by inspection. However, we wish to isolate it because our arguments will apply on any domain on which the property holds.

Condition A: Let $W \subseteq M$ be a relatively open set. Then the Szegö kernel extends to be a continuous function on $W \times (\overline{\mathcal{U}} \setminus \overline{W})$. Furthermore, if $K \subseteq W$ is compact then there is an open set $V \subseteq \mathbb{C}^n$ such that $K \subseteq V \cap M \subseteq W$ and, for each fixed $w \in \overline{\mathcal{U}} \setminus \overline{W}$, the function $S(\,\cdot\,, w)$ continuous analytically to V.

This condition is slightly at variance with an analogous condition enunciated in [KOH2, p. 216]. However, it is best suited to our purposes. Condition A is known to hold, for instance, on smoothly bounded strongly pseudoconvex domains. See [FEF], [BDS], [TAR], and [TRE1] for details.

Now consider the partial differential operator

$$\overline{L'} = \frac{\partial}{\partial \bar{z}_1} - 2iz_1 \frac{\partial}{\partial \bar{z}_2} \; .$$

If $\rho(z)$ is the defining function for $\mathcal{U}$ as above, then $\overline{L'}\rho = 0$ on $\partial\mathcal{U}$. Of course $\overline{L'}$ is nothing other than the tangential Cauchy–Riemann operator. It makes sense to restrict its action to functions on M. In particular, $\overline{L'}$ annihilates any $H^2(M)$ function.

We may consider its formal adjoint $\overline{L'}^*$, which is given by

$$\langle \overline{L'}^* u, v \rangle = \langle u, \overline{L'}v \rangle$$

for all $u, v \in C_c^\infty(M)$. Notice that $\overline{L'}^*$ is a first order homogeneous partial differential operator because L' is.

PROPOSITION 9.2.5
Let $\zeta \in M$. Let $f \in L^2(M)$. If there is an L^2 function u on a neighborhood $U \cap M$ of ζ satisfying

$$\overline{L'}^* u = f$$

on U then $\tilde{P}f$ has a holomorphic extension past ζ; that is, there is an open set $V \subseteq \mathbb{C}^2$ with $\zeta \in V$ such that $\tilde{P}f$ continues analytically to V.

PROOF If g is an L^2 function on M that is 0 in a neighborhood W of $\zeta \in M$ then it is immediate from Condition A that $\tilde{P}g$ extends analytically past ζ.

Now let $\eta \in C_c^\infty(M)$ satisfy $\eta \equiv 1$ in a neighborhood of ζ. Then, in that same neighborhood,

$$f - \overline{L'}^*(\eta u) = f - [\overline{L'}^*(\eta)]u - \eta[\overline{L'}^* u]$$
$$= f - 0 \cdot u - 1 \cdot f = 0.$$

By the preceding paragraph, $\tilde{P}(f - \overline{L'}^*(\eta u))$ continues analytically past ζ. But $\overline{L'}^*(\eta u) \perp H^2(M)$ hence $\tilde{P}(f - \overline{L'}^*(\eta u)) = \tilde{P}f$. That completes the proof. ∎

Now notice that $\overline{L'}^* = L'$. Thus we have

COROLLARY 9.2.6
Fix $\zeta \in M$ and let $f \in L^2(M)$. If $\tilde{P}f$ is not real analytic in a neighborhood of ζ, then the equation

$$L'u = f$$

is not solvable in a neighborhood of ζ.

PROOF Since $\tilde{P}f$ is not real analytic in a neighborhood of ζ then (as has been noted above), $\tilde{P}f$ does not continue analytically past ζ. The proposition then gives the result. ∎

Of course the proposition is just the forward direction of Theorem 9.2.3. In order to prove the converse direction, we shall need the basic Hodge theory for the tangential Cauchy–Riemann operator. That will be developed in the next section.

9.3 The Hodge Theory for the Tangential Cauchy-Riemann Complex

What we are about to do here closely parallels the Hodge theory that we developed for the $\bar{\partial}$-Neumann complex in Chapter 7. Therefore we shall be brief. A thorough treatment of CR manifolds and the tangential Cauchy–Riemann operator may be found in [BOG].

Intuitively what we want to do is to look at the restriction of the $\bar{\partial}$ complex to M. This means that we want to restrict attention to only certain types of forms. The convenient language for formulating these ideas precisely is to pass to a quotient. Let $p, q \in \{0, 1, 2, \ldots\}$. Following the notation of Kohn ([FOK], [KOH2]), we shift gears in this section and let $\mathcal{A}^{p,q}$ denote the (p, q)-forms with C^∞ coefficients on $\overline{\mathcal{U}}$. Now set

$$\mathcal{C}^{p,q} = \{f \in \mathcal{A}^{p,q} : \bar{\partial}\rho \wedge f = 0 \text{ on } M\}.$$

Here $\rho(z) = \operatorname{Im} z_2 - |z_1|^2$ is the usual defining function for $\mathcal{U}$. Clearly,

$$\bar{\partial} : C^{p,q} \to C^{p,q+1}. \tag{*}$$

Set

$$\mathcal{B}^{p,q} = \mathcal{A}^{p,q}/C^{p,q}.$$

By $(*)$ we see that $\bar{\partial}$ induces a well-defined operator on $\mathcal{B}^{p,q}$ which we denote by $\bar{\partial}_b$. We call

$$\cdots \mathcal{B}^{p,q} \xrightarrow{\bar{\partial}_b} \mathcal{B}^{p,q+1} \xrightarrow{\bar{\partial}_b} \mathcal{B}^{p,q+2} \cdots$$

the *boundary complex* and $\bar{\partial}_b$ the *tangential Cauchy–Riemann operator*. (Exercise: What does this tangential Cauchy–Riemann operator have to do with the operator from Section 9.2?)

Let $T^{1,0}(M) = T(M) \cap T^{1,0}$. Here T is the complexified tangent space described earlier in this book. Thus $\alpha \in T_\zeta^{1,0}(M)$ if and only if

$$\alpha = \sum a_j \frac{\partial}{\partial z_j} \qquad \text{and} \qquad \sum a_j(P) \frac{\partial \rho}{\partial z_j}(\zeta) = 0.$$

Likewise $T^{0,1}(M) = T(M) \cap T^{0,1}$. Of course $\mathcal{B}^{1,0}$ and $\mathcal{B}^{0,1}$ are, respectively, dual to $T^{1,0}$ and $T^{0,1}$. The standard hermitian inner product on $T(M)$ induces inner products on $T_\zeta^{1,0}$ and $T_\zeta^{0,1}$. It follows that an inner product is induced on each $\mathcal{B}_\zeta^{p,q}$; using the L^2 structure on M, we then obtain an inner product

$$\langle \alpha, \beta \rangle = \int_M \langle \alpha_\zeta, \beta_\zeta \rangle_\zeta \, dx \, dy \, dt$$

on the forms in $\mathcal{B}^{p,q}$. We invite the reader to verify that $\mathcal{B}^{p,q}$ is nothing other than the space of forms restricted to M that are pointwise orthogonal to the ideal generated by $\bar{\partial}\rho$.

We define the *formal adjoint* ϑ_b of $\bar{\partial}_b$ by the relation

$$\langle \vartheta_b \alpha, \beta \rangle = \langle \alpha, \bar{\partial}_b \beta \rangle$$

for forms α, β with C_c^∞ coefficients on M. Thus $\vartheta_b : \mathcal{B}^{p,q} \to \mathcal{B}^{p,q-1}$ when $q \geq 1$. In analogy with the development in Chapter 7, we define the $\bar{\partial}_b$ Laplacian by

$$\Box_b = \bar{\partial}_b \vartheta_b + \vartheta_b \bar{\partial}_b.$$

If we let $L_2^{p,q}$ denote the competion (in the L^2 topology) of $\mathcal{B}^{p,q}$ then we may define

$$\operatorname{Dom}(\Box_b^{p,q}) = \{\alpha \in L_2^{p,q} : \alpha \text{ in the domain of } \Box_b\}.$$

In analogy with our work in Chapter 7, it can be shown that

$$\operatorname{Dom}(\Box_b) = \operatorname{Dom}(\bar{\partial}_b) \cap \operatorname{Dom}(\vartheta_b) \cap \{\alpha : \bar{\partial}_b \alpha \in \operatorname{Dom}(\vartheta_b), \vartheta_b \alpha \in \operatorname{Dom}(\bar{\partial}_b)\}.$$

Set

$$\mathcal{H}_b^{p,q} = \{\alpha \in \mathbf{Dom}(\Box_b^{p,q}) : \Box_b^{p,q}\alpha = 0\}.$$

This is the harmonic space. Now let ϑ_b denote the closure of the operator ϑ_b that we defined above. It follows (again see Chapter 7) that

$$(\bar{\partial}_b)^* = \vartheta_b,$$

where $(\bar{\partial}_b)^*$ is the L^2 adjoint of $\bar{\partial}_b$. Notice that

$$\langle \Box_b^{p,q}\alpha, \alpha \rangle = \|\bar{\partial}_b\phi\|_{L^2}^2 + \|\vartheta_b\phi\|_{L^2}^2.$$

Then it follows that

PROPOSITION 9.3.1
We have that

$$\mathcal{H}_b^{p,q} = \{\alpha \in Dom(\bar{\partial}_b) \cap Dom(\vartheta_b) \cap L_2^{p,q}(M) : \bar{\partial}_b\alpha = \vartheta_b\alpha = 0\}.$$

Of course the operator ϑ_b on functions is just the zero operator. It follows that $\mathcal{H}_b^{0,0}$ is nothing other than $H^2(M)$. By the same token, on functions,

$$\Box_b\alpha = \vartheta_b\bar{\partial}_b\alpha.$$

Now we have a local solvability statement for $\Box_b$:

THEOREM 9.3.2
Let $f \in L^2(M)$. Let $\zeta \in M$ and let U be a boundary neighborhood of ζ. If there is a function $u \in L^2(U \cap M)$ such that

$$\Box_b u = f,$$

then $\tilde{P}f$ has an analytic continuation past ζ.

PROOF   This is the same as the forward direction of the proof of Theorem 9.2.3. ∎

But, using the additional machinery we have developed, there is the following strictly stronger result:

THEOREM 9.3.3
Assume that the range of the L^2 closure of $\Box_b$ on functions is closed. Fix $\zeta \in M$. Then the equation

$$\Box_b u = f$$

is solvable in a neighborhood of ζ if and only if $\tilde{P}f$ is analytically continuable past ζ.

PROOF If B is a subspace of the Hilbert space A, then we let $A \ominus B$ denote the orthogonal complement of B in A.

The hypothesis that the closed operator (still denoted by $\Box_b$) has closed range means that there is a bounded self-adjoint operator

$$N_b : L^2(M) \ominus H^2(M) \to \mathrm{Dom}(\Box^{0,0}) \ominus H^2(M)$$

such that

$$\Box_b N_b \beta = \beta.$$

Thus N_b is a right inverse for $\Box_b$. (This is the Neumann operator for $\Box_b$; again refer to our work in Chapter 7 for details.)

Of course we may extend N_b to all of L^2 by setting $N_b = 0$ on H^2 and extending by linearity. Then we have the orthogonal decomposition

$$f = \Box_b(N_b f) + Pf.$$

If Pf is analytic in a neighborhood of ζ then the Cauchy–Kowalewski theorem guarantees (because $\Box_b$ has analytic coefficients) that there is a function v on a neighborhood of ζ such that $\Box_b v = Pf$. But then

$$f = \Box_b(N_b f + v),$$

proving the local solvability of our partial differential equation.

Of course the converse direction is contained in the preceding theorem. ∎

Now one may check from the definitions that, on functions,

$$\Box_b = L\bar{L}$$

(think about the canonical dual object in the cotangent space at each point to the vector field L and check that it is pointwise orthogonal on M to the ideal generate by $\bar{\partial}\rho$). Thus if f on M has the property that Pf is analytic near ζ and if v is a solution to $\Box_b v = f$ as provided by the last theorem, then $\bar{L}v$ is a solution to $Lw = f$. That proves the converse direction of Theorem 9.2.3 and completes our discussion of local solvability for the operator L.

9.4 Commutators, Integrability, and Local Solvability

In this section we shall examine the local solvability question from a more elementary point of view. In the end, however, we shall tie it into the preceding discussions.

Consider a partial differential operator

$$P = P_1 + iP_2,$$

where P_1, P_2 are linear, real partial differential operators. Let us assume that P_1, P_2 are linearly independent at each point. An example of such an operator is

$$P = \frac{1}{2}\frac{\partial}{\partial x} + \frac{i}{2}\frac{\partial}{\partial y} \ .$$

This is nothing other than the operator $\partial/\partial \bar{z}$. If f is a given C_c^1 function then the formula

$$u(z) = -\frac{1}{\pi} \iint \frac{f(\zeta)}{\zeta - z}\, d\xi\, d\eta$$

gives a solution to the equation $Pu = f$. See [KR1] for details.

It turns out that for arbitrary P satisfying the above hypotheses, it is possible to find a local change of coordinates so that, in these coordinates, the operator P becomes (a nonvanishing multiple of) the Cauchy–Riemann operator as above. Indeed, if $z = x + iy$ satisfies $Pz = 0$ and if $\mathrm{Re}\,\nabla z, \mathrm{Im}\,\nabla z$ are linearly independent, then we may write

$$P = \rho\frac{\partial}{\partial z} + \omega\frac{\partial}{\partial \bar{z}} \ .$$

Applying both sides of this equation to the function z yields

$$0 = \rho$$

so that

$$P = \omega\frac{\partial}{\partial \bar{z}} \ .$$

Here ω is a nonvanishing function. The existence of such a function z, given an operator P, is not elementary. The necessary variational techniques may be found in [COH, Ch. 4, Section 8].

In any event, the study of operators P of the given form in dimension 2 is straightforward and reduces to consideration of the well-understood Cauchy–Riemann operator of classical complex analysis.

Now suppose that the dimension is at least three. One might hope, in analogy with the classical method of integral curves (see [ZAU]), to find integral surfaces of the vector fields P_1 and P_2. The operator P would then act on each of these surfaces much as P acted on $\mathbb{R}^2$ in the two-dimensional case. Then we could analyze the three-dimensional case by studying the problem on each two-dimensional integral surface and amalgamating the results.

However, the Frobenius integrability theorem [KOM] provides conditions that are both necessary and sufficient for the preceding program to be feasible. Indeed, it is required that the commutator $[P_1, P_2] = P_1 P_2 - P_2 P_1$ lie, at each point, in the span over $\mathbb{R}$ of P_1, P_2. And in fact this integrability condition goes to the heart of the matter. Hörmander (see [HOR1]) showed that the integrability condition is necessary for local solvability of the partial differential operator P.

Given Hörmander's theorem, we see that the operator

$$L = \frac{\partial}{\partial z} + i\bar{z}\frac{\partial}{\partial t}$$

$$= \frac{1}{2}\left(\frac{\partial}{\partial x} - i\frac{\partial}{\partial y}\right) + i(x - iy)\left(\frac{\partial}{\partial t}\right)$$

$$= \left[\frac{1}{2}\frac{\partial}{\partial x} + y\frac{\partial}{\partial t}\right] + i \cdot \left[-\frac{1}{2}\frac{\partial}{\partial y} + x\frac{\partial}{\partial t}\right]$$

satisfies

$$L_1 = \left[\frac{1}{2}\frac{\partial}{\partial x} + y\frac{\partial}{\partial t}\right]$$

$$L_2 = \left[-\frac{1}{2}\frac{\partial}{\partial y} + x\frac{\partial}{\partial t}\right].$$

Thus

$$[L_1, L_2] = \frac{1}{2}\frac{\partial}{\partial t} + \frac{1}{2}\frac{\partial}{\partial t} = \frac{\partial}{\partial t},$$

and we see that $[L_1, L_2]$ is not in the span of L_1, L_2. By Hörmander's theorem, L is not locally solvable.

A slightly more elementary operator that is amenable to Hörmander's analysis is the Grushin operator [GRU], which is closely related to an example of Garabedian [GAB2]. The operator, on $\mathbb{R}^2$, is given by

$$M = \frac{\partial}{\partial x} + ix\frac{\partial}{\partial y}.$$

The reader may check that the integrability condition is not satisfied at the origin.

Let us conclude this chaper with an elementary proof that the equation $Mu = f$ is not always locally solvable in a neighborhood of the origin. The argument presented here is taken from [NIR]. It is of particular interest in that it proceeds by attempting to carry out the program of changing the operator M into the Cauchy–Riemann operator and then dealing with the resulting singularity. We shall choose a particular *non–real analytic* f to facilitate the argument.

To wit, let (x, y) be the coordinates on $\mathbb{R}^2$. Let $D_n \subseteq \mathbb{R}^2$ be the closed disc $\bar{D}(1/n, 4^{-n}), n = 1, 2, \ldots$ Let f be a C^∞ function on $\mathbb{R}^2$ that is compactly supported, even in the x-variable, and vanishes outside of the discs D_n. Assume that f is positive on the interior of each D_n. It is plain that f is not real analytic in any neighborhood of the origin.

PROPOSITION 9.4.1
With the operator M and function f as above, the equation

$$Mu = f$$

has no C^1 solution in any neighborhood of the origin.

PROOF   Seeking a contradiction, suppose that u is a C^1 solution of the equation in a neighborhood U of the origin. Write $u = e + o$, where e is even in the x-variable and o is odd in the x-variable. Then the even (in the x-variable) part of the equation

$$Mu = f$$

is

$$\frac{\partial}{\partial x}o + ix\frac{\partial}{\partial y}o = f(x, y). \tag{9.4.1.1}$$

In particular, M annihilates e and we may take $u = o$.

Let us restrict attention to the region $\{(x, y) : x \geq 0\}$. Notice that o vanishes on the boundary of this region. Introduce the new variable $s = x^2/2$. Hence

$$\frac{\partial}{\partial s} = \frac{1}{x}\frac{\partial}{\partial x} \;.$$

Dividing the equation (9.4.1.1) by x and making suitable substitutions yields

$$\frac{\partial u}{\partial s} + i\frac{\partial u}{\partial y} = \frac{1}{\sqrt{2s}} \cdot f(\sqrt{2s}, y). \tag{9.4.1.2}$$

Moreover, $u = 0$ when $s = 0$.

Observe that the left-hand side of (9.4.1.2) is just two times the Cauchy–Riemann operator applied to u in the coordinates (s, y). Thus we see that u is a holomorphic function of the complex variable $s+iy$ on $W = \mathbb{R}^2 \setminus [\cup_n (D_n \cup \tilde{D}_n)]$, where $\tilde{D}_n$ is the reflection of D_n in the y-axis. But we also know that u vanishes on the y-axis. By analytic continuation, it follows that $u \equiv 0$ on W. In particular, $u = 0$ on the boundary of each D_n.

Return now to the (x, y) coordinates. Then Stokes's theorem yields a contradiction:

$$0 \neq \iint_{D_n} f(x, y)\, dx\, dy = \iint_{D_n} (u_x + ixu_y)\, dx\, dy = \oint_{\partial D_n} [u\, dy - ixu\, dx] = 0.$$

This contradicton completes the proof.   ∎

Let us conclude by noting the connection between the point of view of the present section with that of the last section. We already know that the operator L from Section 2 is essentially the tangential Cauchy–Riemann operator. Saying that its real and imaginary parts satisfy the Frobenius integrability condition (in the sense that the real tangent space is in the span of $\operatorname{Re} L, \operatorname{Im} L$ and their commutators) is, in the language of several complex variables, saying precisely that the boundary of $\mathcal{U}$ is of finite commutator type in the sense of Kohn (see [KR1] for details and background of these ideas). And this is implied, for instance, by the hypothesis that the boundary is strongly pseudoconvex. Indeed, the fact that the commutator has a component in the complex normal direction is just the same as saying that the Levi form is definite, as is exhibited explicitly by the formula

$$\mathcal{L}(Z, W) = \langle [Z, \overline{W}], \partial\rho \rangle.$$

Here $\mathcal{L}$ denotes the Levi form and ρ is the defining function for $\mathcal{U}$.

Thus we see that as soon as a hypersurface in $\mathbb{C}^2$ exhibits strong complex "convexity" then the complex analysis gives rise in a natural fashion to an unsolvable partial differential operator. Such a phenomenon could not arise in the context of one complex variable, since the boundary of a domain in $\mathbb{C}^1$ has no complex structure.

Table of Notation

Notation	Meaning	Page Number
$\| \ \|_{H^2}$	Hardy space norm	138
H^2	Hardy space	139
$S(z,\zeta)$	Szegö kernel	139
$\mathcal{P}(z,\zeta)$	Poisson–Szegö kernel	140
$B(z,r)$	Isotropic ball	141
$\beta(z,r)$	Non-isotropic ball	142, 150
$\ell_g(\gamma)$	Length of the curve γ	143
$\triangle_B$	Bergman Laplacian	147
$\mathcal{P}_k$	Homogeneous polynomials of degree k	154
d_k	Dimension of $\mathcal{P}_k$	154
$\langle P, Q \rangle$	Inner product on $\mathcal{P}_k$	154
Σ_{N-1}	Unit sphere in $\mathbb{R}^N$	156
$\mathcal{H}_k$	Surface spherical harmonics of degree k	157
$\mathcal{A}_k$	Solid spherical harmonics	157
$Y^{(k)}$	Spherical harmonic	157
$Z_{x'}^{(k)}$	Zonal harmonic	161
$P_k^\lambda(t)$	Gegenbauer polynomial	169
$\mathcal{H}^{p,q}$	Bigraded spherical harmonics	173
$D(p,q;n)$	Dimension of $\mathcal{H}^{p,q}$	173
$\pi_{p,q}$	Projection onto $\mathcal{H}^{p,q}$	173
$F(a,b,c;x)$	Hypergeometric function	175
$S_n^{p,q}(r)$	Radial solution of the Bergman Laplacian	175, 178
$\mathbb{C}T_P(\mathbb{C}^n)$	Complexified tangent space	185
$\bigwedge^{p,q}$	Bigraded differential forms	186, 188
$\bigwedge^r$	Differential forms of degree r	186
∂	Holomorphic exterior derivative	186
$\bar{\partial}$	Antiholomorphic exterior derivative	186
dV	Volume element	187
$\bigwedge_c^{p,q}$	Forms with coefficients in C_c^∞	188
$H_s^{p,q}$	Forms with coefficients in H^s	188
γ	Volume form	188
$*$	Hodge star operator	188
$\Pi_{0,1}$	Projection into $\bigwedge^{0,1}$	191
ϑ	Formal adjoint of $\bar{\partial}$	191
$\mathcal{D}^{p,q}$	$\bigwedge^{p,q}(\bar{\Omega}) \cap \mathrm{dom}\,\bar{\partial}^*$	195
$Q(\phi,\phi)$	Quadratic form	196, 197
$\mathcal{D}_c^{p,q}$	$\bigwedge_c^{p,q} \cap \mathrm{dom}\,\bar{\partial}^*$	198
$\tilde{\mathcal{D}}^{p,q}$	Closure of $\mathcal{D}^{p,q}$ in the Q-topology	198
$\Box$	$\bar{\partial}$-Neumann Laplacian	199

Notation	Meaning	Page Number
F	Friedrichs operator	196, 199
L_p	Levi form	205
$E(\phi)$	Special Sobolev norm	206, 228
L_j	Vector fields	209
$\tilde{L}_j$	Vector fields	210
$\tilde{\tilde{L}}_j$	Vector fields	210
Λ^r	Bessel potential	212
Λ_t^s	Tangential Bessel potential	217
$\| \ \|_s$	Tangential Sobolev norm	217
$\tilde{\Delta}_H^\beta$	Tangential difference operator	217
Q^δ	Regularized quadratic form	218
F^δ	Regularized Friedrichs operator	219
∇	Gradient	225
ρ_k	Cutoff functions	234
$\mathcal{H}^{p,q}$	Harmonic space	246
H	Harmonic projection	246
N	Neumann operator	246
P	Bergman projection	252
Condition R	Bergman regularity	252
$\mathcal{H}^j(\Omega), \mathcal{H}^\infty(\Omega)$	Sobolev holomorphic functions	256
δ_j	Distance to boundary of Ω_j	259
L_P	Levi polynomial	264
$\mathcal{U}$	Siegel upper half-space	270
M	$\partial\mathcal{U}$	270
Φ	Cayley map	270
$\mathcal{A}^{p,q}$	(p,q) forms with $C^\infty(\mathcal{U})$ coefficients	274
$\mathcal{B}^{p,q}$	Domain of $\bar{\partial}_b$	275
$\bar{\partial}_b$	Tangential Cauchy–Riemann operator	275
T	Complexified tangent space	275
$\mathcal{H}_b^{p,q}$	Tangential harmonic space	276
$\Box_b$	Tangential Neumann Laplacian	276
N_b	Tangential Neumann operator	277
$\mathcal{L}$	Levi form	281

Bibliography

[ARO]   N. Aronszajn, Theory of reproducing kernels, *Trans. Am. Math. Soc.* 68(1950), 337–404.

[BAC]   G. Bachman, *Elements of Abstract Harmonic Analysis*, Academic Press, 1964.

[BAR]   D. Barrett, Regularity of the Bergman projection on domains with transverse symmetries, *Math. Annalen* 258(1982), 441–446.

[BEA1] R. Beals, A general calculus of pseudo-differential operators, *Duke Math. J.* 42(1975), 1–42.

[BEA2] R. Beals, Characterization of pseudodifferential operators and applications, *Duke J.* 44(1977), 45–57; correction, ibid. 46(1979), 215.

[BEA3] R. Beals, Weighted distribution spaces and pseudodifferential operators, *J. d'Analyse Mathematique* 39(1981), 131–187.

[BEF1] R. Beals and C. Fefferman, On local solvability of linear partial differential equations, *Ann. Math.* 97(1973), 482–498.

[BEF2] R. Beals and C. Fefferman, Spatially inhomogeneous pseudo-differential operators I, *Comm. Pure Appl. Math.* 27(1974), 1–24.

[BEA]   F. Beatrous, L^p estimates for extensions of holomorphic functions, *Mich. J. Math.* 32(1985), 361–380.

[BEC]   W. Beckner, Inequalities in Fourier analysis, *Ann. Math.* 102(1975), 159–182.

[BD]   E. Bedford, Proper holomorphic maps, *Bull. A.M.S.* 10(1984), 157–175.

[BE1]   S. Bell, Analytic hypoellipticity of the $\bar{\partial}$-Neumann problem and extendability of holomorphic mappings, *Acta Math.* 147(1981), 109–116.

[BE2]   S. Bell, Biholomorphic mappings and the $\bar{\partial}$ problem, *Ann. Math.* 114(1981), 103–113.

[BEB]   S. Bell and H. Boas, Regularity of the Bergman projection in weakly pseudo-convex domains, *Math. Annalen* 257(1981), 23–30.

[BKR]   S. Bell and S. G. Krantz, Smoothness to the boundary of conformal maps, *Rocky Mt. J. Math.* 17(1987), 23–40.

[BEL]   S. Bell and E. Ligocka, A simplification and extension of Fefferman's theorem on biholomorphic mappings, *Invent. Math.* 57(1980), 283–289.

[BGL]   J. Bergh and J. Löfström, *Interpolation Spaces: An Introduction*, Springer-Verlag, Berlin, 1976.

[BO1]   H. Boas, Sobolev space projections in weakly pseudoconvex domains, *Trans. A.M.S.* 288(1985), 227–240.

[BO2]   H. Boas, Regularity of the Szegö projection in weakly pseudoconvex domains, *Indiana Univ. Math. J.* 34(1985), 217–223.

[BOK]   J. Bokobza and A. Unterberger, Sur une généralization des opérateurs de Calderó–Zygmund et des espaces H^s, *C. R. Acad. Sci. Paris* 260(1965), 3265–3267.

[BOG]   A. Boggess, *Tangential Cauchy–Riemann Equations and CR Manifolds on CR Manifolds*, CRC Press, Boca Raton, 1991.

[BDS]   L. Boutet de Monvel and J. Sjöstrand, Sur la singularité des noyaux de Bergman et Szegö, *Soc. Mat. de France Asterisque* 34–35(1976), 123–164.

[CAL]   A. P. Calderón, unpublished lecture notes.

[CZ1]   A. P. Calderón and A. Zygmund, On the existence of certain singular integrals, *Acta Math.* 88(1952), 85–139.

[CZ2]   A. P. Calderón and A. Zygmund, Singular integral operators and differential equations, *Am. J. Math.* 79(1957), 901–921.

[CAR]   C. Carathéodory, *Gesammelte Mathematische Schriften*, 1955, Beck, Munich.

[CCP]   G. Carrier, M. Crook, and C. Pearson, *Functions of a Complex Variable*, McGraw-Hill, New York, 1966.

[CAT1] D. Catlin, Necessary conditions for subellipticity of the $\bar{\partial}$-Neumann problem, *Ann. Math.* 117(1983), 147–172.

[CAT2] D. Catlin, Subelliptic estimates for the $\bar{\partial}$ Neumann problem, *Ann. Math.* 126(1987), 131–192.

[CAT3] D. Catlin, Boundary behavior of holomorphic functions on pseudoconvex domains, *J. Diff. Geom.* 15(1980), 605–625.

[CHK]   D. C. Chang and S. G. Krantz, *Analysis on the Heisenberg Group and Applications*, preprint.

[COL]   E. Coddington and N. Levinson, *Theory of Ordinary Differential Equations*, McGraw-Hill, New York, 1955.

[CMB]   J. F. Colombeau, *New Generalized Functions and Multiplication of Distributions*, North Holland, Amsterdam and New York, 1984.

[CON]   P. E. Conner, *The Neumann's Problem for Differential Forms on Riemannian Manifolds*, Mem. Am. Math. Soc. #20, 1956.

[COR] H. Cordes, On pseudodifferential operators and smoothness of special Lie-group representations, *Manuscripta Math.* 28(1979) 51–69.

[COW] H. Cordes and D. Williams, An algebra of pseudodifferential operators with nonsmooth symbol, *Pac. J. Math.* 78(1978), 279–290.

[COH] R. Courant and D. Hilbert, *Methods of Mathematical Physics*, Interscience, New York, 1962.

[DAN1] J. P. D'Angelo, Real hypersurfaces, orders of contact, and applications, *Ann. Math.* 115(1982), 615–637.

[DAN2] J. P. D'Angelo, Intersection theory and the $\bar{\partial}$-Neumann problem, *Proc. Symp. Pure Math.* 41(1984), 51–58.

[DAN3] J. P. D'Angelo, Finite type conditions for real hypersurfaces in $\mathbb{C}^n$. In *Complex Analysis Seminar*, Springer Lecture Notes Vol. 1268, Springer-Verlag, 1987, 83–102.

[DER] G. de Rham, *Varieties Differentiables*, 3rd ed., Hermann, Paris, 1973.

[DIE] K. Diederich, Über die 1. and 2. Ableitungen der Bergmanschen Kernfunktion und ihr Randverhalten, *Math. Annalen* 203(1973), 129–170.

[DF] K. Diederich and J. E. Fornæss, Pseudoconvex domains with real-analytic boundary, *Ann. Math.* 107(1978), 371–384.

[ERD] A. Erdelyi, et al., *Higher Transcendental Functions*, McGraw-Hill, New York, 1953.

[FED] H. Federer, *Geometric Measure Theory*, Springer-Verlag, Berlin, 1969.

[FEF1] C. Fefferman, The Bergman kernel and biholomorphic mappings of pseudoconvex domains, *Invent. Math.* 26(1974), 1–65.

[FEG] C. Fefferman, The uncertainty principle, *Bull. A.M.S.* 9(1983), 129–206.

[FKP] R. Fefferman, C. Kenig, and J. Pipher, preprint.

[FOL] G. B. Folland, Spherical harmonic expansion of the Poisson–Szegö kernel for the ball, *Proc. Am. Math. Soc.* 47(1975), 401–408.

[FOK] G. B. Folland and J. J. Kohn, *The Neumann Problem for the Cauchy–Riemann Complex*, Princeton University Press, Princeton, NJ, 1972.

[GAB1] P. Garabedian, *Partial Differential Equations*, Wiley, New York, 1964.

[GAB2] P. Garabedian, An unsolvable equation, *Proc. A.M.S.* 25(1970), 207–208.

[GIL] P. Gilkey, *The Index Theorem and the Heat Equation*, Publish or Perish Press, Boston, 1974.

[RGR1] C. Robin Graham, The Dirichlet problem for the Bergman Laplacian I, *Communications in Partial Differential Equations* 8(1983), 433–476; part II, ibid. 8(1983), 563–641.

[GRL] C. Robin Graham and J. M. Lee, Smooth solutions of degenerate Laplacians on strictly pseudoconvex domains, *Duke J. Math.* 57(1988), 697–720.

[GRE] P. Greiner, Subelliptic estimates for the $\bar{\partial}$-Neumann problem in $\mathbb{C}^2$, *J. Diff. Geom.* 9(1974), 239–250.

[GRS] P. Greiner and E. M. Stein, *Estimates for the $\bar{\partial}$-Neumann Problem*, Princeton University Press, Princeton, 1977.

[GRU] V. Grushin, A certain example of a differential equation without solutions, *Mat. Zametki* 10(1971), 125-8; transl. *Math. Notes* 10(1971), 499–501.

[GUP] M. de Guzman and I. Peral, *Fourier Analysis*, Proceedings of the El Escorial Conference, Asociacion Matematica Española, 1979.

[HAD] J. Hadamard, Le problème de Cauchy et les équations aux dérivées partielles linéaires hyperboliques, Paris, 1932.

[HAS] M. Hakim and N. Sibony, Spectre de $A(\bar{\Omega})$ pour des domaines bornés faiblement pseudoconvexes réguliers, *J. Funct. Anal.* 37(1980), 127–135.

[HEI] M. Heins, *Complex Function Theory*, Academic Press, New York, 1968.

[HEL] S. Helgason, *Differential Geometry and Symmetric Spaces*, Academic Press, New York, 1962.

[HIR] M. Hirsch, *Differential Topology*, Springer-Verlag, Berlin, 1976.

[HOR1] L. Hörmander, *Linear Partial Differential Equations*, Springer-Verlag, Berlin, 1969.

[HOR2] L. Hörmander, Pseudo-differential operators, *Comm. Pure Appl. Math.* 18(1965), 501–517.

[HOR3] L. Hörmander, Pseudo-differential operators and non-elliptic boundary problems, *Ann. Math.* 83((1966), 129–209.

[HOR4] L. Hörmander, *The Analysis of Linear Partial Differential Operators*, in four volumes, Springer-Verlag, Berlin, 1983–1985.

[HOR5] L. Hörmander, The Weyl calculus of pseudo-differential operators, *Comm. Pure Appl. Math.* 32(1979), 359–443.

[INCE] E. L. Ince, *Ordinary Differential Equations*, Longmans, Green, and Co., London and New York, 1927.

[JAT] H. Jacobowitz and F. Treves, Nowhere solvable homogeneous partial differential equations, *Bull. A.M.S.* 8(1983), 467–469.

[KAN] S. Kaneyuki, *Homogeneous Domains and Siegel Domains*, Springer Lecture Notes no. 241, Springer-Verlag, Berlin, 1971.

[KAT]   Y. Katznelson, *An Introduction to Harmonic Analysis*, Wiley, New York, 1968.

[KEL1] O. D. Kellogg, *Foundations of Potential Theory*, Springer-Verlag, Berlin, 1929.

[KEL2] O. D. Kellogg, Harmonic functions and Green's integral, *Trans. Am. Math. Soc.* 13(1912), 109–132.

[KER1] N. Kerzman, A Monge-Ampère equation in complex analysis, *Proc. Symp. Pure Math.* , Vol. 30, Part 1, American Mathematical Society, Providence, RI, 1977, 161–168.

[KER2] N. Kerzman, The Bergman kernel function. Differentiability at the boundary, *Math. Annalen* 195(1972), 149–158.

[KOM]   S. Kobayashi and K. Nomizu, *Foundations of Differential Geometry*, Vols. I and II, Interscience, New York, 1963, 1969.

[KOH1] J. J. Kohn, Harmonic integrals on strongly pseudoconvex manifolds I, *Ann. Math.* 78(1963), 112–148; II, ibid. 79(1964), 450–472.

[KOH2] J. J. Kohn, Methods of partial differential equations in complex analysis, *Proc. Symp. Pure Math.*, Vol. 30, American Mathematical Society, Providence, RI, 1977, 215–237.

[KOH3] J. J. Kohn, A survey of the $\bar{\partial}$-Neumann problem, *Proceedings of Symposia in Pure Math.* 41(1984), 137–145.

[KON1] J. J. Kohn and L. Nirenberg, On the algebra of pseudo-differential operators, *Comm. Pure Appl. Math.* 18(1965), 269–305.

[KON2] J. J. Kohn and L. Nirenberg, Non-coercive boundary value problems, *Comm. Pure and Appl. Math.* 18(1965), 443–492.

[KOV]   A. Koranyi and S. Vagi, Singular integrals in homogeneous spaces and some problems of classical analysis, *Ann. Scuola Norm. Sup. Pisa* 25(1971), 575–648.

[KR1]   S. G. Krantz, *Function Theory of Several Complex Variables*, 2nd Ed., Wadsworth Publishing, Belmont, 1992.

[KR2]   S. G. Krantz, Lipschitz spaces, smoothness of functions, and approximation theory, *Expositiones Math.* 3(1983), 193–260.

[KR3]   S. G. Krantz, Geometric Lipschitz spaces and applications to complex function theory and nilpotent groups, *J. Funct. Anal.* 34(1979), 456–471.

[KR4]   S. G. Krantz, Characterization of various domains of holomorphy via $\bar{\partial}$ estimates and applications to a problem of Kohn, *Ill. J. Math.* 23(1979), 267–286.

[KRP]   S. G. Krantz and H. R. Parks, *A Primer of Real Analytic Functions*, Birkhäuser, Basel, 1992, to appear.

[LEW1] H. Lewy, On the local character of the solutions of an atypical linear differential equation in three variables and a related theorem for regular functions of two complex variables, *Ann. Math.* 64(1956), 514–522.

[LEW2] H. Lewy, An example of a smooth linear partial differential equation without solution, *Ann. Math.* 66(1957), 155–158.

[LUM] Lumer, Expaces de Hardy en plusieurs variables complexes, *C. R. Acad. Sci. Paris Sér. A–B* 273(1971), A151–A154.

[MEY] Y. Meyer, Estimations L^2 pour les opérateurs pseudo-differentiels, *Séminaire d'Analyse Harmonique (1977/1978)*, 47–53, *Publ. Math. Orsay* 78, 12, Univ. Paris XI, Orsay, 1978.

[MIK1] S. G. Mikhlin, On the multipliers of Fourier integrals, *Dokl. Akad. Nauk SSSR* 109(1956), 701–703.

[MIK2] S. G. Mikhlin, *Multidimensional Singular Integral Equations*, Pergammon Press, New York, 1965.

[MIS] B. Mityagin and E. M. Semenov, The space C^k is not an interpolation space between C and $C^n, 0 < k < n$, *Sov. Math. Dokl.* 17(1976), 778–782.

[MOR] C. B. Morrey, *Multiple Integrals in the Calculus of Variations*, Springer-Verlag, Berlin, 1966.

[MOS] J. Moser, On Harnack's theorem for elliptic differential equations, *Comm. Pure Appl. Math.* 14(1961), 577–591.

[NIR] L. Nirenberg, *Lectures on Linear Partial Differential Equations*, American Mathematical Society, Providence, RI, 1973.

[NTR] L. Nirenberg and F. Treves, On local solvability of linear partial differential equations I. Necessary conditions; II. Sufficient Conditions, *Comm. Pure and Appl. Math.* 23(1970), 1–38; 459–510; Correction, ibid. 24(1971), 279–288.

[OSG] W. F. Osgood, *Lehrbuch der Funktionentheorie*, vols. 1 and 2, B. G. Teubner, Leipzig, 1912.

[PAI] P. Painlevé, Sur les lignes singulières des functions analytiques, *Thèse*, Gauthier-Villars, Paris, 1887.

[PEE] J. Peetre, Réctifications à l'article "Une caractérisation abstraite des opérateurs différentiels', *Math. Scand.* 8(1960), 116–120.

[RAN] R. M. Range, *Holomorphic Functions and Integral Representations in Several Complex Variables*, Springer-Verlag, Berlin, 1986.

[RIE] M. Riesz, L'intégrale de Riemann-Liouville et le problème de Cauchy, *Acta Math.* 81(1949), 1–223.

[ROM] S. Roman, The formula of Faà di Bruno, *Am. Math. Monthly* 87(1980), 805–809.

[RUD1] W. Rudin, *Principles of Mathematical Analysis*, 3rd Ed., McGraw-Hill, New York, 1976.

[RUD2] W. Rudin, *Function Theory in the Unit Ball of* C^n, Springer-Verlag, Berlin, 1980.

[RUD3] W. Rudin, *Real and Complex Analysis*, McGraw-Hill, New York, 1966.

[SCH] L. Schwartz, *Théorie des Distributions*, I, II, Hermann, Paris, 1950–51.

[SPE] D. C. Spencer, Overdetermined systems of linear partial differential equations, *Bull. Am. Math. Soc.* 75(1969), 179–239.

[STSI] E. M. Stein, *Singular Integrals and Differentiability Properties of Functions*, Princeton University Press, 1970.

[STBV] E. M. Stein, *The Boundary Behavior of Holomorphic Functions of Several Complex Variables*, Princeton University Press, Princeton, 1972.

[STW] E. M. Stein and G. Weiss, *Introduction to Fourier Analysis on Euclidean Spaces*, Princeton University Press, Princeton, 1971.

[STR] K. Stromberg, An Introduction to Classical Real Analysis, Wadsworth, Belmont, 1981.

[TAR] D. Tartakoff, The local real analyticity of solutions to $\Box_b$ and the $\bar{\partial}$-Neumann problem, *Acta Math.* 145(1980), 177–204.

[TAY] M. Taylor, *Pseudodifferential Operators*, Princeton University Press, Princeton, 1981.

[TRE1] F. Treves, Analytic hypo-ellipticity of a class of pseudodifferential operators with double characteristics and applications to the $\bar{\partial}$- Neumann problem, *Comm. Partial Diff. Eqs.* 3:6–7(1978), 475–642.

[TRE2] F. Treves, *Introduction to Pseudodifferential and Fourier Integral Operators*, 2 vols., Plenum, New York, 1980.

[TRI] H. Triebel, *Theory of Function Spaces*, Birkhauser, Basel and Boston, 1983.

[UNT] A. Unterberger, Résolution d'équations aux derivées partielles dans les espaces de distributions d'ordre de regularité variable, *Ann. Inst Fourier* 21(1971) 85–128.

[WEL] R. O. Wells, *Differential Analysis on Complex Manifolds*, 2nd ed., Springer-Verlag, 1979.

[WHW] Whittaker and Watson, *A Course of Modern Analysis*, 4th ed., Cambridge Univ. Press, London, 1935.

[WID] H. Widom, *Lectures on Integral Equations*, Van Nostrand Reinhold, New York, 1969.

[ZAU] E. Zauderer, *Partial Differential Equations of Applied Mathematics*, John Wiley and Sons, New York, 1

[ZYG] A. Zygmund, *Trigonometric Series*, Cambridge University Press, Cambridge, UK, 1968.

Index